Coated and Laminated Textiles for Aerostats and Airships

Coated and Laminated Textiles for Aerostats and Airships

Material Challenges and Technology

Edited by
Mangala Joshi

CRC Press is an imprint of the
Taylor & Francis Group, an **informa** business

First edition published 2022
by CRC Press
6000 Broken Sound Parkway NW, Suite 300, Boca Raton, FL 33487-2742

and by CRC Press
2 Park Square, Milton Park, Abingdon, Oxon, OX14 4RN

CRC Press is an imprint of Taylor & Francis Group, LLC

ISBN: 978-1-138-36069-3 (hbk)
ISBN: 978-1-032-19658-9 (pbk)
ISBN: 978-0-429-43299-6 (ebk)

DOI: 10.1201/9780429432996

Typeset in Palatino
by SPi Technologies India Pvt Ltd (Straive)

Contents

Editor

Mangala Joshi is Professor in the Department of Fibre and Textile Engineering at IIT Delhi. She specializes in polymer/fibre science and technology and is an alumnus of IIT Delhi, having obtained her MTech and PhD degrees in polymer science and technology in 1986 and 1992 respectively. Her current research interests are nanotechnology applications in textiles, polymer nanocomposites: fibres and coatings, nano-biomaterials, nanocoatings and nanofibres, bioactive and functional textiles, and environmentally friendly technologies. She has been involved in investigating several sponsored research projects funded by DST (Nanomission), Govt. of India, Aerial Delivery Research and Development Establishment (ADRDE, Agra) DRDO, Department of Biotechnology (DBT – Govt. of India) and Industry both national and international. She is the recipient of the 3rd National Award (2012–13) for 'Technology Innovation' under the category of R&D in the field of 'Polymer Science and Technology' awarded by the Ministry of Chemicals and Fertilizers, Govt. of India.

She is the coordinator of research vertical on 'Smart and Intelligent Textiles' under the Joint Advanced Technology Centre (JATC) set up at IIT Delhi by the Defense Research and Development Organization (DRDO) under an MOU. She is the principal investigator of the research project on 'Development of Materials for Aerostat and Airship Envelope' under JATC. She was also a member of Nano Research Group at IIT Delhi working on 'Bioactive Nanocomposites' under a sponsored project by US multinational Lockheed Martin wherein she holds a US patent, and also worked on a sponsored project from GAIL India on 'Multifunctional Hybrid Polymer Nanocomposites'. She has guided many PhD, MTech, and BTech projects at IIT Delhi related to these research areas. She has more than 95 refereed publications in international and national journals and over 125 papers in conferences, and also authored several book chapters. She has traveled widely abroad and presented her work at prestigious international as well as national conferences as an invited speaker. She served as Core member for the Expert Committee on 'Materials, Mining and Metallurgy' and Co-opted member for expert committee on 'Engineering Sciences' of 'Science and Engineering Research Board' SERB, DST, Govt. of India for the years 2015–18. She is a member of Research Advisory Council of ADRDE, DRDO, Agra, DMSRDE, Kanpur, Northern India Textile Research Association, NITRA, Ghaziabad and PSG Institute of Advanced Research, Coimbatore. She is an executive member of professional bodies such as Polymer Processing Academy (PPA) and Fiber Forum India (FFI), Textile Association of India.

Contributors

Bapan Adak
Kusumgar Corporates Pvt Ltd
Gujarat, India

S. Wazed Ali
Department of Textile and Fibre Engineering
Indian Institute of Technology
New Delhi, India

B. S. Butola
Department of Textile and Fibre Engineering,
Indian Institute of Technology
New Delhi, India

Upashana Chatterjee
Department of Textile and Fibre Engineering
Indian Institute of Technology
New Delhi, India

Shikha Chouhan
Department of Textile and Fibre Engineering
Indian Institute of Technology
New Delhi, India

Biswa R. Das
Defence Materials & Stores Research & Development Establishment
Defence Research and Development Organisation
Kanpur, India

Mangala Joshi
Department of Textile and Fibre Engineering
Indian Institute of Technology
New Delhi, India

Abhijit Majumdar
Department of Textile and Fibre Engineering
Indian Institute of Technology
New Delhi, India

Subhash Mandal
Defence Materials & Stores Research & Development Establishment
Defence Research and Development Organisation
Kanpur, India

Neeraj Mandlekar
Department of Textile and Fibre Engineering
Indian Institute of Technology
New Delhi, India

Unsanhame Mawkhlieng
Department of Textile and Fibre Engineering
Indian Institute of Technology
New Delhi, India

Kingsuk Mukhopadhyay
Defence Materials & Stores Research & Development Establishment
Defence Research and Development Organisation
Kanpur, India

S. Parasuram
Department of Textile and Fibre Engineering
Indian Institute of Technology
New Delhi, India

N. Eswara Prasad
Defence Materials & Stores Research & Development Establishment
Defence Research and Development Organisation
Kanpur, India

Debmalya Roy
Defence Materials & Stores Research & Development Establishment
Defence Research and Development Organisation
Kanpur, India

Dipak K. Setua
Defence Materials & Stores Research & Development Establishment
Defence Research and Development Organisation
Kanpur, India

Gaurav Singh
Aerial Delivery Research and Development Establishment
Defence Research and Development Organisation
Agra, India

Rishabh Tiwari
Department of Textile and Fibre Engineering
Indian Institute of Technology
New Delhi, India

Siddhanth Varshney
Department of Textile and Fibre Engineering
Indian Institute of Technology
New Delhi, India

1

Introduction to LTA Systems: Aerostats and Airships

Mangala Joshi
Indian Institute of Technology, New Delhi, India
Bapan Adak
Kusumgar Corporates Pvt Ltd, Gujarat, India

CONTENTS

DOI: 10.1201/9780429432996-1

1.1 Introduction

Lighter-than-air (LTA) aircrafts such as balloons, aerostat, blimps, airship and dirigibles are designed for various applications. Aerostat and airship are basically used in defence fields such as military surveillance, detection of aerial threats, etc., and also may be used for many other purposes such as network monitoring, weather forecasting, broadcasting, etc. [1, 2].

The LTA systems are filled with LTA gases such as hydrogen or helium, and they are continuously exposed under UV radiation coming from sunlight. Additionally, airships are also exposed under intense ozone as they work at much higher altitude compared to aerostat [3]. Moreover, the effects of temperature variation, pressure variation, rain and humidity also cause deterioration of envelope materials of LTA systems. Therefore, material development of LTA systems faces significant challenges, especially with regard to stratospheric airships, which require fulfilment of necessary criteria such as high strength, light weight, flexibility at low-temperature and capability of containing helium or non-flammable hydrogen gas for a long time. More importantly, it has to be weather resistant, which means it has to sustain under the exposure of harmful radiations such as UV and ozone, for providing a long service life. Here, multi-layered coated and laminated fabrics are used to fulfil all the requirements for this particular application, where a specific layer is used for a specific function [1, 4–6].

This chapter is all about the different types of LTA systems, the basic structure of the multi-layered envelope for an airship or aerostat, the requirements for each layer and different potential materials (polymers, fibres and fabrics) for different layers and challenges.

1.2 Different Types of LTA Systems

LTA systems can be classified in different ways. On the basis of working altitude, it may be two types – low-altitude aerostat and high-altitude airship (HAA). An aerostat is a tethered system; its working altitude generally varies between 2 and 5 km, and it operates from a fixed location by a mooring system. It is lifted solely by hydrogen or helium gas filled inside the envelope. No power is required to drive tethered LTA systems for station keeping or for controlling altitude, and it can be recovered easily for payload maintenance [1]. On the contrary, 'Airship' or HAA, is an untethered system and works at a higher altitude (about 17–55 km) from sea level. Airships are power driven and free-flight aircraft systems where lift is provided by a combined effect of lifting gas (such as hydrogen or helium) and aerodynamics. Recently, HAA systems operated above an altitude of 20 km from the earth surface are receiving more attention for multiple purposes. The working altitude of an LTA system is chosen based on many factors, such as severe weather conditions, jet stream or wind speed, and the presence of the "Federal Aviation Administration" (FAA) air-traffic layer [1, 7].

On the basis of the hull configuration or rigidity of the structure, an LTA system can be classified as: (i) rigid, (ii) semi-rigid and (iii) flexible. Non-rigid airship systems are also called blimps. Figure 1.1 shows different types of airships and their different components.

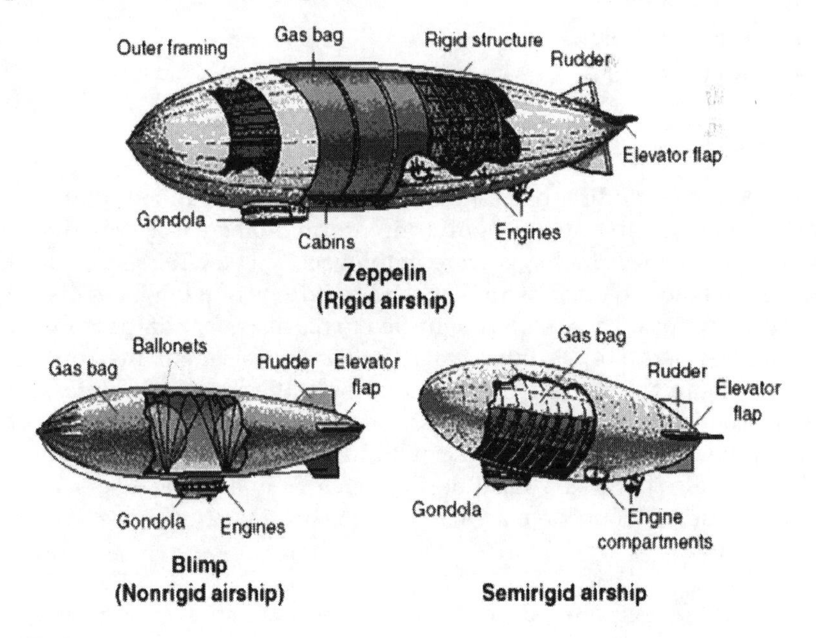

FIGURE 1.1
Different types of airships and their components.

Source: Taken from https://ruor.uottawa.ca/handle/10393/36594 [8].

Conventional airships can be further classified on the basis of (i) payload capacity (heavy-lift and medium-lift) and (ii) the way of producing vertical force (heavier-than-air, lighter-than-air, and hybrid). Because of their light weight, low cost, simple structure, easy fabrication and less maintenance requirement, non-rigid, flexible-type LTA systems are preferred over others [9–11].

1.3 Applications of Aerial Platforms

Recent advancements in aerostat technology have achieved great success by making the application versatile. With the flexibility in volume, size and shape, the advanced aerostats have attracted tremendous applications in different domains. These structures are economical, handy and stable and can fly steadily over land as well as over sea in almost all weather conditions. These characteristics of aerostats brought profoundness in real-life solutions. The application domains have been broadly classifieds into five categories [1, 4, 10, 12, 13] and are discussed in their respective sections.

1. Atmospheric estimations/studies
2. Surveillance/reconnaissance
3. Telecommunications
4. Scientific research
5. Energy harvesting and power generation

1.3.1 Atmospheric Estimations/Studies

The weather/sounding-balloon is a high-altitude balloon that carries instruments aloft to transfer significant information on atmospheric conditions. In 1896, these balloons were employed by Léon Teisserenc de Bort, the French meteorologist, who launched hundreds of them from his observatory in Trappes. These experiments led to his discovery of the tropopause and stratosphere [14]. Military and civilian government meteorological agencies share all the data produced through these balloons, internationally. The balloons are usually made of highly flexible latex material such as Chloroprene and are usually filled with helium or hydrogen gas. The control over the gas pressure facilitates the change in altitude as per requirements. These balloons have a variety of applications; some of them are as follows:

a) These balloons are launched around the world to diagnose climatic conditions, viz. atmospheric pressure, temperature, humidity and wind speed, through a measuring device called a radiosonde using RADAR/navigation systems. The data are eventually processed by human forecasters and computer models for weather forecasting. About 800 locations around the globe do routine releases, usually at 0000 UTC and 1200 UTC.

b) There also exist specialized applications for aviation interests, pollution monitoring, photography/videography and research.

c) Moreover, field research programs often use mobile launchers from land vehicles as well as ships and aircraft. In recent years, they have also been used for scattering human ashes at high altitude as well as to create the fictional entity "Rover" in the TV series *The Prisoner*, filmed in Portmeirion [15].

d) Transosonde balloons (that stay at a constant altitude for a long period) helped in diagnosing radioactive debris from atomic fallout through experimentation in 1958.

Further, extending the potential of ballooning over the atmosphere of Venus, the VEGA balloon mission was launched by the Soviet Union in 1985. The hostile atmosphere on Venus had challenged the aerostat technology in creating exploration opportunities differently. In the mission, two aerostats were successfully deployed on Venus with 3.5 m super-pressure balloons at an altitude of about 55 km with the total payload being 6.9 kg; each was tracked from Earth for about two earth days.

In this period there were also several US and European proposals to fly such missions with technological advancements for meticulous explorations. Such missions had a further objective in developing an aerostat with larger payload capability than VEGA. The Jet Propulsion Laboratory (JPL) has been developing super-pressure balloons tolerant of both the sulfuric acid environment on Venus and capable of accommodating the diurnal stresses induced on the balloon. A 5.5 m balloon with a payload capability of 45 kg is at TRL 5, whereas a 7.0 m balloon with a payload of 110 kg is now under development [16]. In the past, several proposals have been made to apply this technology to the VALOR (Venus Aerostatic-Lift Observatories for in-situ Research) [17] and the European Venus Explorer (EVE).

Other, more ambitious concepts involved the deployment of sondes from the aerostat such that the aerostat could serve as both a platform for precise deployment of the sondes and as a communications relay. This enables greater data return. With this idea, in 2011, the Planetary Science Decadal Survey recommended a Venus Climate Mission (VCM) as a small flagship mission, comprising an aerostat, a deep probe, and two sondes. The deep probe would be released during initial descent and provide atmospheric and chemical data into the deep atmosphere.

Recent work also suggests that valuable geoscience studies can be performed from the aerostat itself. Infrasound signatures of earthquakes can be detected in the atmosphere [18], and natural-source electromagnetic sounding can probe the upper mantle [19]. Together, these techniques can constrain the geodynamics of Venus without ever touching the surface.

1.3.2 Surveillance/Reconnaissance

The various projects and applications of aerostat in the domain of surveillance are discussed here.

1.3.2.1 To Overcome Military Surveillance Challenges

Aerostat systems find major applications for military/army surveillance challenges. Aerostats have made it advantageous to conduct surveillance beyond the horizon. This leads to large applications in the military domain. For CCTV surveillance of base force protection (US Army, Afghanistan), a small tactical multi-purpose aerostat system (STMPAS) (with 75 m^3 volume) was successfully deployed. It was capable of carrying a video camera to detect and target the positions of insurgents from long distances over a long duration in all weathering conditions [20]. Another major application is by the Australian Defense Force (ADF) for long-distance digital radio-relay for proper maintenance of military/emergency services. Though ADF's first attempt at using large aerostats to lift "Micro-Light" radios to 1000 ft proved to be uneconomical and non-viable, during the second attempt the use of small-sized aerostats (named Desert Star Aerostats) became successful, less time consuming and handy in this regard. This resulted in lifting Microlight (a unique and robust technique for precise positioning) with ease, achieving 42 miles of range with high bandwidth at 1000 ft elevation [21]. In another application, the Bowman Aerostat Beyond Line-of-Sight (BABLOS) has proved to be successful. Within a very short duration (a few minutes) of inflation and deployment, the small, stealthy, BABLOS system could potentially lift various accessories such as helium cylinders, Aerostat, launch systems, flying lines, coax cables and antennae with ease of mobility to an elevation of 200 ft. This has helped in achieving a range of 17 miles giving 908 square miles of coverage being the important factor for BABLOS [22]. The Royal Navy could deploy Desert Star Aerostats (15 m^3 capacity) over 900 ft altitude with necessary accessories, providing over-the-horizon radio communications between the shore and the unmanned boat. This allowed them to go more than 20 nautical miles over the horizon by providing accurate and precise data and increasing the safety of ships and crews [23]. Before the first Gulf War, the Kuwaiti government turned to the TCOM system, which is the leader in persistent airborne surveillance in solving the border surveillance issues using an aerostat system, making its borders safer [24].

1.3.2.2 Towards Non-Military Applications

Despite having major military and defence applications, tiny aerostats with a capacity of only a few m^3 have been involved in lots of other surveillance applications. One such application, handled by Allsopp Aerostats Ltd, is in the process of developing a maritime airborne aerostat camera system (an Ocean-Eye) which is suitable for aiding oil spill detection and clean-up. As oil spills are hard to spot from sea level, these systems are of great help in determining the extent and thickness of the oil over the sea. The system provides real-time streaming aerial video of the oil slick as the clean-up boom is positioned to pick it up; this allows for accurate positioning of the boom and helps in yielding more oil in a shorter time. This, in turn, reduces damage to the marine ecology and also clean-up costs [25].

Another profound application is seen in the aerial photographic survey of ancient Amarna in Egypt, proposed by the Archaeology Department at the University of Cambridge. Despite hot weather conditions, the aerostat performed perfectly, producing better quality of photographs at cheaper costs [26]. Dome Petroleum (Canada) made an application of a STARS system which is equipped with marine radar to search for large ice floes which were a threat to Dome oil drilling operations. That helped them to detect, track and avoid large ice floes and safely continue at-sea oil drilling operations [24]. Another challenging application of aerostat is seen in lifting multiple Tethersondes for meteorological research carried at the University of Millersville, USA. Tethersondes (attached to the aerostats) is used to measure wind speed, humidity and temperature at high altitude simultaneously. This is essential for the understanding of the boundary layer of air situated just above the land, studying the weather conditions and estimating air pollution.

1.3.2.3 Stratobus as the Sentinel of the Stratosphere

A renowned multi-mission program, Stratobus was launched by Thales Alenia Space, a Franco-Italian aerospace manufacturer and its partners in April 2016 [27]. This program shows potential in serving as a missing link between drones and satellites. It offers a geostationary platform for floating/flying at high altitude (around 20 km), widely known as High Altitude Pseudo-Satellite (HAPS), that allows a vision over 500 km aiming simply at complementing the global satellite coverage. It is an autonomous stratospheric platform concept that usually provides locations to payloads typically for space experiments or instruments. It would ideally be operated between the two tropics with minimum gust that allows it to be effective as well as highly stable.

With the dimensions of 140 m by 32 m and 5 tons of self-weight (Figure 1.2), it can accommodate payloads of 250 kg with a power of 5 kW for five years of its estimated life [28]. Powered with solar energy and having clean energies

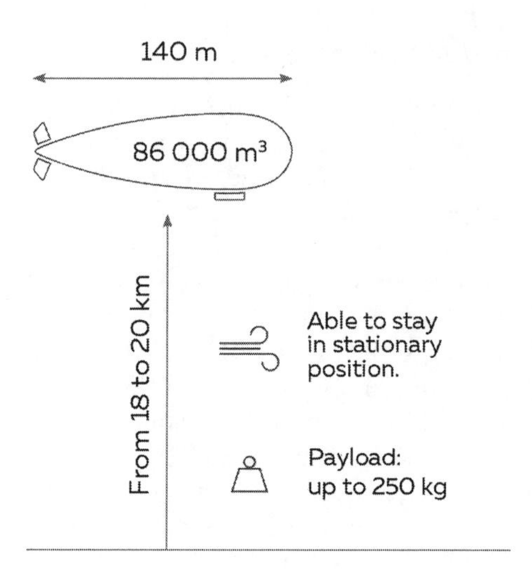

FIGURE 1.2
Typical diagram of high-altitude pseudo satellite.

Source: Taken from, https://cnim-air-space.com/en/technologies-for-stratobus/ [28].

on board, Stratobus will have a low carbon footprint. Detached from its three cables, it can take off from a platform the size of a football field, be piloted from a mobile station and, finally, be controlled from a fixed command post without heavy installations, unlike drones and satellites. Without any launcher for this helium-inflated balloon, it moves with four electric motors powered by photovoltaic cells. Moving vertically, it reaches its altitude in four hours, and it descends gradually in six hours towards its point of the launch where it must be recovered and moored. The descent is piloted but, in the future, it could be made by a drone that retrieves it and brings it back to the launching point. It is required to evolve autonomously and permanently in the lower layer of the stratosphere (12 to 50 km in altitude). Its evolution zone is particularly prone to the surrounding weather conditions with cold temperatures, very present ozone, gusts of wind, significant sunlight and heating components and aggressive ultraviolet rays. But if the external conditions are hard, this zone also has advantages of no air traffic and non-existence of aeronautical regulations.

The program finds its major applications in the field of observations. There, it mainly focuses on the surveillance of sensible industrial sites such as the oil industry, as it could identify a threat and its location with up to 10 cm resolution using its visible and infra-red sensors as well as its embedded RADAR system in almost all-weather conditions. Further, it could be employed in continuous monitoring of borders, the anticipation of maritime piracy, and managing the environment by detecting marine pollution, measuring weather conditions and managing maritime traffic.

The platform can also be strongly employed in telecommunications via 4G and 5G networks in regards to terrestrial networks by covering a larger area with less interference in a shorter time of deployment. Hence, this could appeal to civilian and military observation and surveillance applications as the Stratobus can stay above a point for a year. Moreover, it could serve to reduce the digital divide in geographical areas where the internet is not yet accessible, strengthen the GSM (Global System for Mobile Communications) in case of major events such as the Olympic Games, and restore internet and telephone connections in a nominal way for better management of humanitarian action during natural disasters such as earthquakes and floods. Finally, in terms of navigation, it could offer the possibility of increasing GPS coverage in areas of heavy traffic. Compared to satellites, these platforms have lower latency (transmission delay) and the possibility of return for maintenance or payload reconfiguration. With better persistence and resolution than satellites, these projects show potential applications for remote sensing as well.

However, these platforms face major technological challenges, such as insufficiency in a lightweight structure, inability to generate and store energy, poor thermal management and a lesser reliability factor. Another important drawback is the poor coordination with the aerospace control agency, especially during the launch and recovery phases. Also, the aspects of international law related to the overflight of other countries need to be analysed.

1.3.2.4 CNIM Air Space

CNIM Air Space, with its 40 years of experience, is the pioneer in designing and manufacturing fully equipped envelope for Stratobus using the most powerful, lightweight and resistant fabric in the world. The organization is an expert in the assembly of films and hermetic fabrics. Most of its production was used by the National Centre for Space Studies (CNES), making it a huge success [27]. Following are some of its products introduced in brief:

A. *Stratospheric balloons*: These balloons are mostly used by the scientific community to study the atmosphere, its chemistry and dynamics. They are also valuable tools for astronomers and biologists as well as for demonstrating technologies. Since 1971, the organization has acquired a strong experience in textile design and manufacturing of these balloons. With its capacity to lift anything from a few kilograms to several tons can operate at an altitude from a few hundred meters to 40 kilometres.

B. *Open stratospheric balloons*: These balloons, filled with helium, have one or several openings which enable the balance between atmospheric pressure and the gas inside the balloon. The duration of flight ranges from a few hours to several days. Also, with their capacity to lift to 1700 kg, they can operate up to 45 km.

C. *Pressurized stratospheric balloons*: These balloons' envelopes benefit from a very high level of tightness and strong reliability. Thus, long-term flights above inhabited continents are quite possible with this system. The duration of flight ranges from a few weeks to several months. Also, with their capacity to lift to 50 kg, they can operate up to 20 km.

The airship project has begun its development phase and is planning to fly on a small scale. It has been certified by both the Pégase and the Techtera competitiveness clusters. The first qualifying flight of a full-size model would occur in 2022. Since 2013, Airbus Defence and Space, Thales Alenia Space, Google and Facebook have invested in these projects, aiming to supply internet in areas with poor telecommunications infrastructure, and thus bringing new hope to achieve the establishment of a HAPS industry. Its future will be driven mainly by the evolution of technologies of potential competitors, such as microsatellite constellations, and the availability of financial resources to overcome HAPS's technological challenges.

1.3.2.5 SkyStar 180

SkyStar 180, a small-sized mobile aerostat (Figure 1.3), is manufactured by the Israeli-based aerostat company RT LTA. It is specially designed for tactical mid-range surveillance and public safety, as well as for police and military applications. The system uses a stabilized day/night electro-optical payload suspended from a helium-filled aerostat and tethered to a ground system [29].

SkyStar operates continuously at wind velocity of up to 40 knots and can lift a payload of up to 20 kg, providing surveillance coverage from an altitude of up to 1000 ft. for up to 72 hours, after which it is brought down for a 20-minute helium refill. Only two people are required to fully maintain the system. It is ideal for police and intelligence applications, HLS, defence, border control, strategic infrastructure protection and public safety missions.

The SkyStar systems have already accumulated more than 1,500,000 million operational hours worldwide, offering availability of over 85% in any given area. Together with the SkyStar 180 aerostat system, RT will also provide a small tactical unmanned aerial vehicle (UAV), in cooperation with the UAV manufacturer.

A high-altitude balloon is pertinent to multiple disciplines. This is a worldwide topic, with major balloon programs and/or facilities located on every continent. A high-altitude (>20 km) balloon platform is nearly ideal for carrying out scientific observations in a space-like environment, flight qualifying novel instrumentation, and transporting humans to the edge of space. The astrophysics, heliophysics, planetary and earth science disciplines have been instrumental in fostering a renaissance in stratospheric ballooning. The

FIGURE 1.3
Deployment of SkyStar 180.

Source: Taken from https://www.rt.co.il/skystar-180.

increasing interest in a robust platform to carry out a large variety of stratospheric research has driven the community to develop improved capabilities for payloads to fly at high altitudes for longer durations (>100 days), competing with some space missions. To accomplish this, improvements needed to be made in the design and materials used to make the balloon.

For a more detailed history of high-altitude ballooning, one can refer to SBPfotzer (1972) and Ubertini (2008). The Montgolfier balloon was made of sackcloth (linen lined with paper) and held together with buttons and a large (heavy) net of cords. This balloon flew to an altitude of ~2 km, descending within minutes after launch. In December of that same year, J. Charles and the Robert brothers switched out the sackcloth with a rubberized silk cloth, impermeable to hydrogen. The balloon travelled over 40 km and rose to an altitude of around 2 km. On this flight, the first scientific measurement (air temperature as a function of altitude) was made. The next ~160 years showed little improvement in balloon technologies, resulting in fairly low altitudes and short-duration flights. Even at these lower altitudes, there is a huge advantage over Earth-based astronomical observations due to the reduced atmosphere.

Some scientific discoveries and measurements using balloon technology include the following. Sivel and CroceSpinelli (1874) made the first astronomical measurement from a balloon. They used a spectroscope to determine the origin of the water vapour observed during ground-based spectroscopic observations of the Sun. Their measurements confirmed that the origin of the water vapour was Earth's atmosphere, rather than water vapour in the Sun's atmosphere [30]. In addition to larger balloons being flown, scientists frequently used small rubber balloons (sondes), which became available in the 1890s [31]. These super-pressure balloons (SPB) facilitated experiments in atmospheric physics and were used to study the Sun and cosmic rays. However, these balloons could not carry large payloads and were short-lived, as they were meant to burst when they reached a certain altitude. As such, open-basket manned flights were still the method of choice for carrying out scientific investigations. In 1912, Austrian physicist Victor Hess made one of the most significant discoveries in high-energy particle physics – the discovery of cosmic rays [32]. Even at a relatively low altitude of 4.8 km, Hess's electroscopes were able to measure an increase in the radiation as a function of altitude. The desire to ascend even to higher altitudes would require an improved balloon design and a pressurized gondola.

Standard NASA scientific balloons are constructed of polyethene film having a thickness of only 20 microns, which is same as an ordinary sandwich wrap. The film is cut into banana-peel-shaped sections called gores and heat-sealed together to form the balloon. Up to 180 gores are used to make NASA's largest balloons. These standard, zero-pressure balloons are open to the atmosphere at the bottom to equalize the internal pressure. Balloons can carry payloads of up to ~3600 kg or altitudes up to ~49 km and/or fly for several weeks, but they cannot do all at the same time [33].

1.3.3 Telecommunications

In the domain of telecommunication, different unique and innovative applications of aerostat are discussed in this section.

1.3.3.1 Loon LLC

Aiming at creating a strong and diverse network of internet, Loon LLC has not only succeeded in bringing internet access to rural areas but also significantly improved communication especially during natural disasters in affected regions [34, 35]. Initially, the company had begun as a research and development project by X (formerly Google X) but later spun out into a separate company in July 2018 [36]. Rich DeVaul, the chief technical architect, is a key person involved in the project. The whole infrastructure is based on LTE and the eNodeB component has carried in the balloon.

The Loon LLC is a super-pressure balloon filled with helium gas with dimensions of 15 m by 12 m, in its inflated form. This enables control over altitude (18–25 km) and creates an aerial wireless network (4G LTE) and ultimately transmits the signals to ground stations [37, 38]. These balloons, designed and manufactured by Raven Aerostar, are composed of metalized Mylar, BoPET, or a highly flexible latex material of about 0.076 mm thickness, with a typical service life of about 100 days [39, 40].

Typically, the system consists of a small box (weighing 10 kg) containing electronics and communication boards for navigation and control and to communicate with other balloons. Along with this, it also reserves a space for batteries to store solar energy enabling it to operate during the night. A parachute, called Raven Aerostar Payload Recovery Parachute, is also attached to the top which allows controlled and descent landing and payload recovery when a balloon is taken out of service [39, 41]. In the case of an unexpected failure, it deploys automatically [42].

The project technology could also allow countries to cut down the costs involved by avoiding the use of expensive fibre cable installed underground to allow users to connect to the Internet. Alphabet feels this will greatly increase Internet usage in developing countries in regions such as Africa and Southeast Asia that can't afford to lay underground fibre-optic cable [43].

1.3.3.2 Aerostat for Wireless Communication in Remote Areas

This application of aerostat serves as a platform for cost-effective wireless communication as an alternative to fixed tower networks. This helps in filling the digital devoid of the communication in sharing the knowledge and information from urban to rural areas in India. The aim is to develop a platform to hold wireless communication equipment at an altitude of 100–250 m AGL, which helps in establishing connectivity over a radius of 10 km. Along with the aerostat, the basic components of the system are wireless bridges for connecting network interfaces, omnidirectional antennas to maintain the wireless link, cables for connectivity and power supply and amplifiers.

Based on the preliminary work [44], a methodology for initial sizing and conceptual design of an aerostat system is developed to arrive at the required geometrical parameters and the detailed mass breakup of an aerostat system, given the values of some operation, configuration and performance-related parameters. This methodology implements spreadsheet form of Microsoft Excel and named PADS. PADS accepts all the input parameters, constant parameters and some geometrical and operation-related options such as envelope profile selection, gas pressure management by ballonets or symmetrically expandable elastic strip, and type of LTA gas used. The objective behind providing this facility for selection of optional parameters was to make the methodology more flexible and adaptive for any future modification in the aerostat system and also to make sensitivity analyses much more comprehensive. In an aerostat, the geometry of the envelope has a profound effect on its aerodynamic characteristics, and hence on the stability and payload carrying ability. Some standard shapes of the aerostat envelopes exist, and their profiles were incorporated in the input part of the PADS.

1.3.3.3 Aerostats in Australia

Aerostats All Australia (AAA) aims to extend the mobile coverage of Australia's landmass and surrounding sea by deploying high-altitude aerostats in reduced cost. The mission proposes a multiyear plan to reach 70% of Australia, comprising the huge potential of the country. This includes various aerial platforms, listed as follows:

A. Platforms with propulsion
B. Free-floating balloons
C. Tethered aerostats

For the comparison of aerial platforms, the European Union conducted extensive research [45]. It says that the main drawback of airborne platforms with propulsion and free-floating balloons in comparison to tethered aerostats is the high cost, low capacity and potentially long latency backhaul which would have to be provided by microwave or satellite being inferior to fibre-optic terrestrial backhaul.

Moreover, aerial platforms study CORDIS [45] also provides an overview of airborne solutions including aerostats. The US Homeland Security System Assessment and Validation for Emergency Responders (SAVER) report on Tethered Aerostat Systems [46] gives an overview of the aerostat industry. The complete system of the tethered Helikite aerostat consists of the aerostat, mooring station, winch and tether cable. In locations with no electrical power supply, the aerostat may carry its solar cells and battery. The increased aerostat weight is offset by a lighter tether that does not have to carry electricity. This allows higher altitude and therefore greater 4G range. The trade-off depends on the length of the tether to be carried. Whilst the aim is to maximize equipment on the ground to keep the aerostat as light and simple as possible, carrying its solar panels and battery is preferred in most of remote Australia which suffers from unreliable or non-existent-of-mains power.

1.3.4 Scientific Research

To perform scientific research in the field of astronomy and astrophysics, the high-altitude aerial platforms have gain potential and have shown versatility in many projects. Some of them are listed in this section.

1.3.4.1 Ultra-Thin Balloons for High-Altitude Research

With the demand for high-altitude balloons to carry astronomical research as well as probe the stratosphere at TIFR, the balloon group had initiated research and development work on an indigenous balloon that can be deployed up to 42 km, unlike regular rubber balloons. By 1999, total indigenization of sounding balloon manufacture was accomplished.

The work on balloons with ultra-thin polyethene film (2.8–3.8 μm) commenced in 2011 to research in atmospheric sciences. In a successful trial, a 61,000 m^3 balloon made of 3.8 μm Antrix film reached the stratopause (48 km) for the first time in 2012. Further, fine-tuning of launch parameters was carried out to take the balloons to even higher mesospheric altitudes. Three successful flights with a total suspended load of 10 kg using 61,000 m^3 balloons were carried out in January 2014, and all three balloons crossed into the mesosphere reaching altitudes of over 51 km.

1.3.4.2 Development of Ultra-Thin Balloon Film

Over the years, scientists began to show interest in the study of atmospheric dynamics and meteorological parameters over the middle atmospheric region up to the mesosphere (~50 km) using balloon-borne experiments.

The study involved balloons with a payload-carrying capability of about 25 kg reaching the mesosphere. This led to the development of the basic raw material namely ultra-thin polyethene film with thickness ranging from 2.8 to 3.8 μm at TIFR balloon facility, Hyderabad. To reach mesosphere altitude required the following:

1. The balloon film must be ultra-thin to minimize weight compared to conventional scientific plastic balloons.
2. The ultra-thin film should have a high tensile strength to withstand large stresses during ground bubble inflation.
3. The balloon flight housekeeping electronics such as telemetry, telecommand, timer, GPS, etc., must be lightweight and consume very low power.

1.3.4.3 The First Trial on Ultra-Thin Balloon Film Extrusion

The ultra-thin film made out of linear low-density polyethene (LLDPE) resins with added metallocene catalysts proved to have outstanding toughness with improved tensile strength, impact resistance and puncture performance. The material properties of the ultra-thin film are discussed in reference [47]. The fabrication process of the balloon with these materials also has the advantage of lower heat-seal initiation temperatures and higher hot-tack strength for faster line speeds with excellent seal integrity.

Based on this valuable input, in 2008, a successful trial was performed for extruding 200 kg of 3.8 μm lay flat tubing with 1300 mm web width by gradually lowering the gauge thickness from 13 to 3.8 μm. During the trial, frequent bubble breaks in between and irregular web width were observed due to non-stability of the extrusion bubble. Laboratory (tensile) tests at room temperature gave encouraging results, qualifying the film as a candidate material for balloon fabrication.

1.3.4.4 Trial Balloon Fabrication Using 3.8 μm Film

In 2011, the first trial with ultra-thin film balloon (3.8 μm, 4077 m^3) was fabricated similarly to the regular balloon. The trial balloon was sealed on a curved table with only buffer strips of thickness 20 μm and width 33 mm. The balloon weighed less by 2 kg and reached a maximum altitude of 44.05 km compared to the regular one that could attain 43 km. This was the first time a TIFR balloon reached a record altitude in India. The performance of the balloon was normal during ascent, and after reaching the ceiling altitude, it initially descended and then started floating at the same altitude for a few minutes instead of bursting, which generally happens in a regular balloon. To overcome this, a micro-controller was installed based on a mini-timer for fusing the balloon's top portion using Ni-chrome wire. Based on the successful performance of the mini-timer and Nichrome wire flight termination system, it was decided to use the same for balloon termination for all sounding as well as balloons made using ultra-thin films.

1.3.4.5 Trial Balloon Fabrication Using 3.4 μm Film

In 2012, the second trial of a balloon of volume 5014 m^3 and fabricated using the 3.4 μm film was performed. Despite being slightly larger in volume than the regular-sounding balloon, this balloon weighed only 6.3 kg, which was lighter than the first trial balloon by 0.7 kg, and reached 45.03 km altitude. With the installed mini-timer, the balloon descended to the ground after reaching the maximum altitude on its own. Further, two more balloons of volume 5014 m^3 were fabricated using 3.2 μm thin film with the extraneous portion of seal trimmed for further reduction in weight. The balloons weighed 5.97 kg each.

1.3.4.6 Energy Harvesting and Power Generation

LTA crafts are attracting new potential applications compared to their typical market niches [48]. Also, more researchers have proposed diverse applications, ranging from high-altitude aerostats as astronomical platforms [49] to infrastructures for communication systems [50, 51]. One such application proves the feasibility for producing significant electric power by deploying a high-altitude tethered aerostat. A trial with a similar aerostat model was performed which showed its capability to deliver power to the ground over 95 kW. An estimation and costing were also carried out for the proposed concept.

To meet the power demands, regenerative fuel cells and solar panels can be potential power sources aboard the airship. During diurnal periods the power can be derived from the solar radiation using solar panels. A portion of the unused solar energy can be used to generate hydrogen and oxygen gases from the wastewater from the fuel cells or can be stored in rechargeable batteries. During nocturnal periods, the operation can be restricted on the harvested power.

Despite their potential in energy harvesting, these power sources suffer from various drawbacks besides a substantial maintenance expense. The rechargeable batteries are typically large and heavy, whereas the solar panels tend to be heavy and susceptible to various damages. Eventually, this negatively impacts the manoeuvrability of the airship, its ability to attain desired altitudes and travelling ranges, which significantly limit the overall performance. Thus, it creates a need for power-generating laminates that are integrated into an airship envelope, and are configured to harvest electrical power from the structural and thermal changes of the envelope even without occupying space within the payload.

U.S. Patent 7878453 (2011) describes an invention of an airship utilizing a piezoelectric and pyroelectric power-generating laminate following the concepts of the invention discussed earlier. This invention involves generating power based on structural and thermal changes, through piezoelectric/pyroelectric layers embedded into the airship envelope. This power-generating envelope comprises a base fabric layer, a barrier layer, an inner metal layer, a piezoelectric/pyroelectric layer, an outer metal layer and a cover layer that forms a gas-impervious envelope [52].

1.4 Functioning of LTA Systems

An aerostat or airship is made based on a 'balloon-within-a-balloon' concept where the envelope of the inner balloon is called the 'ballonet' and the outer envelope is called the 'hull'. Aerostats or airships fly based on Archimedes' principle of buoyancy, which states, "the upward buoyant

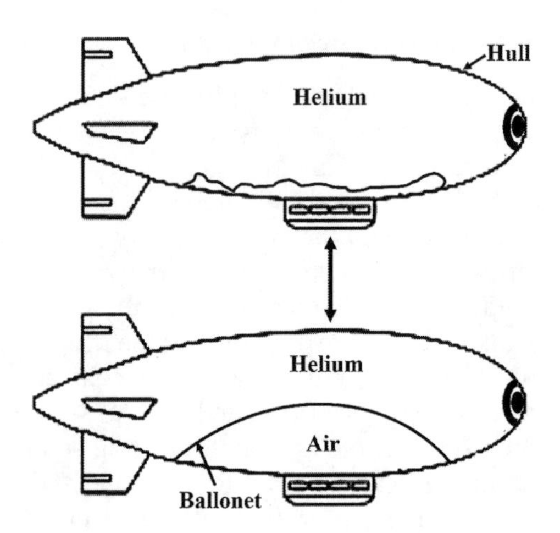

FIGURE 1.4
Lifting mechanism of LTA system.

force that is exerted on a body immersed in a fluid, whether fully or partially submerged, is equal to the weight of the fluid that the body displaces". During lifting of the aerostat, the outer balloon is filled with LTA gases such as helium gas (most commonly) or non-flammable hydrogen gas (sometimes), which tends to expand as the atmospheric pressure decreases with an increase in altitude. A ballonet is installed inside of the hull in order to maintain a constant pressure inside. The ballonet is deflated during lifting of the aerostat, and when it is totally empty, the aerostat reaches a maximum 'pressure height' and the height depends on the design and shape of the aerostat [53]. A reverse situation happens during descent – the ballonet is filled with atmospheric air, and helium is contracted (Figure 1.4). For the purpose of surveillance, weather monitoring and online transmission of pictures, the aerostats are equipped with a powerful telescope and other instruments such as solar array, radar system, propeller, elevator flap and gondola. However, a description of this equipment is beyond the scope of this chapter, which will focus mainly on the materials for hull and ballonet.

1.5 Working Atmosphere of LTA Systems

As already mentioned, on the basis of working altitude, LTA systems can be classified in two categories – low-altitude LTA aircrafts, i.e. aerostat (working altitude 2–5 Km), and high-altitude LTA aircrafts, i.e. airship (working altitude 17–55 Km). Due to this difference, the working atmosphere of these two types of LTA systems are slightly different, while the airship faces more severe weathering condition in comparison to the aerostat. In daytime, LTA systems are continuously exposed under intense UV radiation coming from sunlight, humidity variation, rain (when in troposphere level), temperature variation (+50 to −50°C) and varied pressure (5475 Pa–110 Pa) because of varying altitude, pollutants, etc. Additionally, HAA systems are exposed to concentrated ozone (in ozonosphere) and different cosmic radiations. Such a harsh atmosphere causes degradation of the envelope material of the aerostat and airship resulting in loss in strength, crack generation on envelope's surface and increase in helium gas diffusion through the envelope [1, 54]. At the ozonosphere and stratosphere levels, atmospheric temperature may reduce to as low as −50°C, in which condition most materials become brittle, provoking generation of cracks on the envelope surface and thus increase in helium gas permeability, resulting in loss in crores. Hence, material development for the envelope of the aerostat is very challenging, especially for HAA systems.

1.6 Material Requirements for the LTA Systems

1.6.1 Material Requirements for the Hull

The outer envelope or hull of aerostat/airship should fulfil the following requirements for providing a long service life:

1. Good resistance to environmental degradation, to sustain in harsh atmospheric condition containing UV and ozone, varying temperature and pressure, humidity, rain, wind, etc.
2. Low gas permeability to minimize loss of filled gas. Higher permeability of gas results in loss of operational capability and increased operational cost.
3. High strength: The strength of the material determines the maximum possible size of the Aerostat. The bigger the diameter, the more LTA systems face greater Hoope's stress, and hence higher strength is required.
4. Low weight: High strength-to-weight ratio is required to maximize payload capacity. Hence, with lower weight, payload capacity may increase.
5. High tear resistance: Tear strength is one of the most important parameters for inflatable applications, since even a small tear can lead to catastrophic failure.

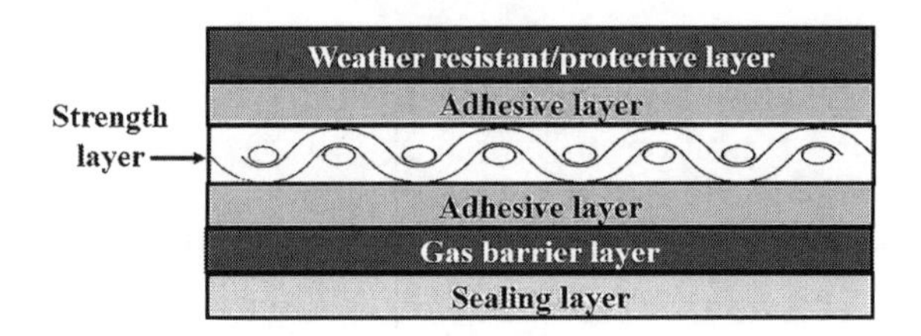

FIGURE 1.5
A typical multi-layered laminated structure for LTA systems. In case of coated structure, tie coats (with crosslinker) may be required in place of adhesive layers.

6. Flexibility even at sub-zero temperature to avoid brittleness of the material and loss of gas barrier property.
7. Adequate joint strength, required to produce strong and reliable joints.
8. Low creep to ensure that aerodynamic shape and desired properties are maintained throughout its life.
9. Good abrasion resistance and wear resistance for handling.
10. Long service life (expectedly > 3 years).

Any material having a single layer can't meet all these requirements, and hence generally, multi-layered coated or laminated textiles are used in the hull of LTA systems Figure 1.5 shows a typical multi-layered structure for LTA systems containing the following layers:

1. Strength layer
2. Protective or weather-resistive layer
3. Gas barrier layer
4. Adhesive layer (for laminates)
5. Sealing layer

1.6.1.1 Strength Layer

A strength layer is the base of any envelope material for LTA systems providing required strength to the structure. A single-layer textile fabric (mainly plain, twill, rip-stop weave) or nonwoven (sometimes) [55] are generally used in strength layer. The fibre used in strength layer may be high-strength commodity fibres such as Nylon and polyester or high-performance fibres such as Vectran, Zylon, UHMWPE (Spectra and Dyneema), Kevlar, etc. The main requirements for strength layers are as follows:

1. High tensile and shear strength to sustain the high pressure experienced by the system in its working condition
2. Good tear resistance to avoid catastrophic failure

3. High work of rapture
4. Light weight to carry higher pay loads
5. Good bondability or sealability with adjoining layers
6. Low creep to retain its properties for long run

1.6.1.2 Weather Resistance or Protective Layer

The weather resistance layer protects the envelope and thereby the whole LTA system from harsh atmospheric conditions like exposure under UV radiation, ozone, rain, humidity, sand particles, pollutants, etc. The main requirements for this layer are as follows:

1. Good weather resistance property reducing photo-oxidative degradation
2. Excellent low-temperature flexibility
3. Good bondability or sealability with adjoining layer

Generally, different fluoropolymers such as polyvinylidene fluoride (PVDF), polyvinyl fluoride (PVF, Tedlar), polytetrafluoro ethylene (PTFE), fluorinated ethylene–propylene copolymer (FEP) are frequently used in protective layer of LTA systems. Sometimes, fluoropolymers are top coated with a very thin aluminium coating for better weather resistance property by providing a reflection surface [1].

1.6.1.3 Gas Barrier Layer

This layer restricts the passage of helium from inside to outside of the aerostat system and keeps them floating at a particular altitude for a longer time. The main requirements for this layer are as follows:

1. High gas barrier property against helium or hydrogen gas
2. Excellent low-temperature flexibility
3. Good bondability or sealability with adjoining layer

Generally, a polymer having a good helium gas barrier property, such as biaxially oriented polyester (BoPET, Mylar®) polyvinylidene chloride (PVDC), and ethylene vinyl alcohol copolymer (EVOH), is used in this layer [1].

1.6.1.4 Sealing Layer

The coated or laminated fabric is finally sealed by either welding the edges of the fabric or by applying an adhesive to provide a specific shape to the

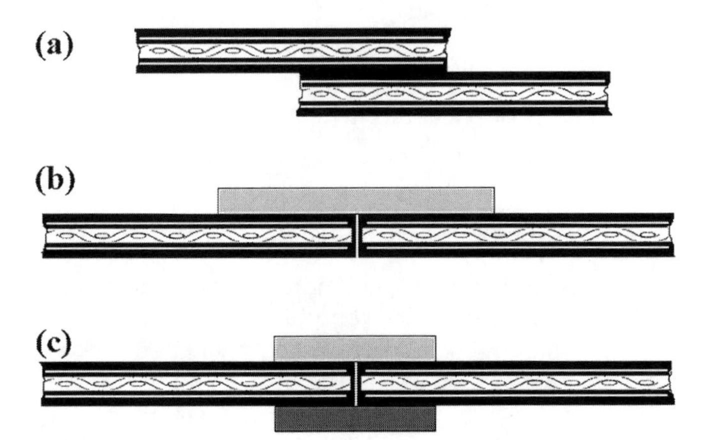

FIGURE 1.6
Different designs for sealing: (a) lap joint, (b) single-butt joint, (c) double-butt joint.

aerostat or airship. Sealing may be done by forming one of three types of basic joints [1, 11, 56], which are lap joint, single butt joint and double butt joint (Figure 1.6).

Generally, thermoplastic elastomers like polyurethanes are used in the sealing layer. The service life of any LTA system is strongly dependent on the joint strength, which again depends on the polymer used in the sealing layer, sealing technique (radio frequency welding, thermal welding, ultrasonic welding, laser welding, adhesive bonding, etc.), sealing parameters and joint type [1].

1.6.1.5 Adhesive Layer

In a laminated LTA envelope, the adhesive layer binds the strength layer with the protective and gas barrier layer, as shown in Figure 1.5. This layer is very important for proving enough bonding among different layers in a laminated structure. Selection of the proper adhesive is very crucial to maintaining a proper balance between flexibility and inter-layer adhesion, as well as avoiding delamination of layers in laminated structure. The adhesives used for this purpose are mainly polyurethane- or acrylic-based. Very frequently, various formulating agents like cross-linker, UV stabilizers are mixed with adhesive formulation to improve adhesion and whether resistance property of adhesive layer. The main properties required for this layer are:

1. Easy application of adhesive having suitable viscosity
2. Good compatibility with both substrates
3. Good bondability or adhesion power
4. Excellent low-temperature flexibility after drying/curing

Sometimes, the substrates (strength layer/protective layer/gas barrier layer) are functionalized by treating with plasma or corona or e-beam to improve adhesion. For example, the surface energy of Tedler film is 30 dynes/cm, which is very difficult to joint with other substrate, while after plasma treatment the surface energy becomes about 40 dynes/cm, providing better adhesion property [57].

1.6.2 Requirements for Ballonet Materials

The ballonet is another vital component of an LTA system such as an aerostat and airship. The ballonet is generally filled with air to enable the LTA system to descend from a higher altitude. Actually, it acts as an internal barrier, separating helium gas (present inside of the hull) and air (present inside of the ballonet). During ascending and descending of the aerostat or airship, the ballonet is continuously flexed as deflated and inflated, respectively. The main requirements for ballonet materials are:

1. Good air and gas barrier property to minimize lifting gas loss
2. Light weight
3. Excellent low-temperature flexibility
4. Good abrasion resistance
5. Good flex fatigue resistance
6. Good bondability or sealability

Ballonets are also made of multi-layered coated or laminated textiles. Generally, fine denier polyester/nylon or high performance fibre-based lightweight fabrics are used for making strength layer for ballonet. Both side of the fabric coated or laminated with a highly flexible polymer with good air and gas barrier property. In this context, a lightweight polyester or nylon fabric coated with thermoplastic polyurethane (TPU) based formulation having good air/gas barrier property, low-temperature flexibility and abrasion resistance are very common [1].

1.7 Challenges in Material Development for LTA Systems and Future Scopes

The envelope of LTA systems is prepared from a multi-layered coated structure. Due to the harsh working atmosphere, there are many challenges in developing suitable material for hull and ballonet material for

LTA systems, especially for stratospheric airships. The main challenges are summarized next.

Firstly, selection of right grade of polymer systems for different layers (gas barrier, weather resistant and adhesive) is very important, and a very challenging task considering all requirements for LTA systems such as lower degradation under UV/ozone, helium gas barrier, low weight, high strength, flexibility at sub-zero temperatures, etc. Different fluoropolymers such as PVF, PVDF, Teflon, etc., have a very good weather resistance property, and they are the potential materials for weather resistance layer. However, generally, they face an adhesion problem due to very low surface energy. On the other hand, Mylar (BoPET), EVOH, etc., have very good gas barrier properties and they might be suitable material for the gas barrier layer, but they also have some issues. Polyurethane is a unique material with versatile properties which can be used for the protective, gas barrier, adhesive and sealing layers. The only thing is that gas barrier and weather resistance properties need to be improved slightly. Development in the field of nanotechnology, nanomaterials and nanocomposites has a great potential for overcoming these issues to a large extent. However, many additional challenges are associated with nanotechnology, which will be discussed in detail in another chapter.

Secondly, selection of the right textile material (fibre type and fabric structure) for the strength layer is also very challenging, as there are a wide variety of fibres as raw materials including high strength commodity fibres such as nylon and polyester as well as high-performance fibres such as Vectran, M5, Zylon, UHMWPE and aramid fibres. Though use of most of the high-performance fibres are very advantageous in terms of strength and weight, we also have to consider other factors such also such as UV resistance, creep, fatigue, moisture resistance, and cost. In comparison to commodity fibres such as nylon or polyester, most of the high-performance fibres are very costly, have less extensibility, and not available in large quantity.

Thirdly, proper processing also has a great role in developing appropriate materials for LTA systems. Both coating and lamination processes have some merits and demerits. In coating, there is a greater chance of pinhole generation on the coated fabric surface resulting in a poor gas barrier property and very frequently with reduction of tear strength. Moreover, in most of the coatings, toxic solvents are used. In lamination these issues can be overcome, but this technique introduces some more issues such as delamination of layers under stress as well as harsh weathering conditions and lower flexibility compared to coated structures.

Fourthly, designing and development of proper structure is also very important for LTA systems. For obtaining good joint strength and adhesion property, proper joint design and selection of suitable seaming technology and adhesive are very important.

Fifthly, LTA systems are not very useful and economical without having a long service life. However, surface life estimation directly by experiment is very time-consuming because performance or life of an LTA aircraft-envelope material is analysed after long-time exposure under natural or accelerated weathering conditions. Therefore, models are very helpful in estimating service life of LTA systems, and lots of researchers are going on in this field. However, development of universal models is very challenging because of many variables in the systems starting from the raw material to the process, design and manufacturing of final structure for LTA systems.

1.8 Brief History of Development of LTA Systems

The history of airship invention was started in 1784, when Jean-Pierre Blanchard first recorded a flight of a non-rigid *dirigible* (French word for airship) and fitted a hand-powered propeller to a balloon for propulsion [58]. After that, in 1852, Henri Giffard was the first person who made steam-powered airship flight [53]. In 1863, Dr. Solomon Andrews devised first the fully steerable airship, named *Aereon*. In 1872, Paul Haenlein flew an airship powered by an internal combustion engine using coal gas. Charles Renard and Arthur Constantin Krebs first made a fully controllable free-flight airship (*La France*) in 1884. The 'Golden Age of Airships' began in 1900 when *Zeppelins*, the most successful airships of all time, were launched. Between two world wars, a number of nations operated airships. Britain, the United States and Germany mainly operated rigid airships, while Italy, the Soviet Union and Japan concentrated on semi-rigid-type airships. In 1923, the first noble helium was used in an American-built rigid airship, the *USS Shenandoah*. After that, there were many attempts in the twentieth century for improving the performance of the aerostat/airship and also for developing better structures for it. The twentieth century brought many international players such as Goodyear, Cargolifter AG, Aereon Corporation, and Skyhook International [9, 12]. Table 1.1 summarizes the various airship projects of recent years.

Currently, the research organizations in many developing countries such as the USA, Korea, UK, Germany, Canada, and Japan have launched various programs for design and development of unmanned stratospheric airships

TABLE 1.1

Summary of Aerostat/Airship Projects in Recent Years [9, 59–63]

Designer	Model	Unique Features	Maximum Speed (mph)	Maximum Altitude (ft)	Country
Lockheed Martin	HAA	Solar-powered high altitude unmanned, and un-tethered	28	60,000	USA
Lockheed Martin	HALE-D airship	Unmanned, solar-powered, first flights of new technologies like HALE-D	–	32,000	USA
X (formerly Google X)	Project Loon	Capable of providing Internet access to rural and remote areas	–	~60,000	USA
Techsphere Systems International	SA-60	Spherical shape and low altitude	35	10,000	USA
Southwest Research Institute	HiSentinel Airship	Stratospheric and solar-powered	–	74,000	USA
Ohio Airships Inc.	DynaLifter PSC-3	Winged hybrid, VTOL, and heavy lift	115	10,000	USA
AEROS	Aeroscraft ML866 model	Control of static heaviness, heavy lift, and VTOL	115	10,000	USA
World View Enterprises	Geostationary balloon satellites (GBS)	Powered with solar panels, Laser broadband internet access	–	60,000 to 70,000	USA
Skyhook–Boeing	SkyHook JHL-40	Heavy-lift four-rotor and 40-ton lifting capacity	80	–	Canada/USA
Millennium Airship Inc.	SkyFreighter	Hybrid, heavy lift, and VTOL	80	20,000	Canada/USA

21st Century Airship Inc.	–	Spherical shape	35	Low altitude	Canada
ATG/World SkyCat	Skycat-20	VTOL and cargo aircraft	97	10,000	UK
Zeppelin Luftschifftechnik Gmbh	Zeppelin LZ NT-07	Semi-rigid, internal rigid framework consisting of carbon fibre triangular frames and aluminium members	80.8	8203	Germany
European Commission, Aeronautics and Air Transport	MAAT FP7	Low-cost, flexible aerial transport solution conceptualized as a cruiser/feeder system, energetically self-sufficient	124	42,000–56,000	European Union

TABLE 1.2

Organization and Laboratories Working on Development of HAA

Name of the Organization/Laboratory	Country
National Aeronautics and Space Administration (NASA), Lockheed Martin Corporation, Defence Advanced Research Projects Agency (DARPA), Near Space Corporation, Lamart Corporation, Raven Aerostar, ILC Dover, Raytheon, Millennium Airship Inc., TCOM, L.P.	USA
Japan Aerospace Exploration Agency (JAXA), National Aerospace Laboratories (NAL), Kawasaki Heavy Industries Ltd., Taiyo Kogyo Corporation	Japan
Lindstrand Technologies, Airship Industries Skyship	UK
Korea Aerospace Research Institute (KARI)	Korea
RosAeroSystems International Ltd.	Russia
RT Aerostats Systems	Israel
Aerial Delivery Research and Development Establishment (ADRDE), Defence Research and Development Organization (DRDO), National Balloon Facility	India

for different communication-related applications. Table 1.2 provides information about different organizations and laboratories in various countries working on development of high-altitude airship (HAA) systems. Among these, the current key players dominating in the market of global LTA systems are TCOM, L.P. (USA), Lockheed Martin (USA), ILC Dover (USA), Raven Aerostar (USA), Raytheon (USA), RosAeroSystems (Russia) and RT (Israel) [64].

In India, Aerial Delivery Research and Development Establishment (ADRDE), Agra, which is a part of Defence Research and Development Organization (DRDO), has also launched two new aerostats in recent years, called 'Akashdeep' and 'Nakshatra'. In both cases, a polyurethane-coated multi-layered textile structure was used. Among these two aerostats, 'Nakshatra' has a bigger payload capacity with improved surveillance functionality [65, 66].

1.9 Conclusion

LTA systems have various technical applications starting from network monitoring to defence. However, there are significant challenges in material development for aerostat and airships, because of many critical requirements such as good weather resistance, excellent helium gas barrier property, high tensile and tear strength, low-temperature flexibility, low weight, low creep and long service life. Generally, multi-layered coated or laminated structures are used for these applications as any single-layered material can't fulfil all these requirements. However, there are significant challenges in selecting suitable materials (polymers, fibres and fabric structure) for specific LTA system fulfilling desired properties. Different high-performance fibres such as Vectran, Kevlar, Zylon and M5 have better potential in this regard, compared to conventional fibres like polyester and nylon. However, they also have the barrier of high cost and lower availability compared to the commodity fibres. On the other hand, improvement of polymer properties using the potential of nanocomposites is possible to a great extent, but it also has many constraints related to application of nanotechnology for large-scale production. However, extensive research work is going on in different labs and research organizations all over the world for developing better materials for different LTA systems.

References

1. H. Zhai, A. Euler, *Material Challenges for Lighter-Than-Air Systems in High Altitude Applications*, in: *AIAA 5th ATIO and 16th Light. Sys Tech. Balloon Syst. Conf.*, American Institute of Aeronautics and Astronautics, Reston, Virigina, 2005. doi:10.2514/6.2005-7488.
2. J.L. Hall, D. Fairbrother, T. Frederickson, V.V. Kerzhanovich, M. Said, C. Sandy, J. Ware, C. Willey, A.H. Yavrouian, Prototype design and testing of a Venus long duration, high altitude balloon, *Adv. Sp. Res.*, 2008. doi:10.1016/j.asr.2007.03.017.
3. G. Lin, H. Tan, *The influence of ozone on the property of envelop materials of near space vehicle*, in: *Proc. 2011 Int. Conf. Electron. Mech. Eng. Inf. Technol. EMEIT 2011*, 2011. doi:10.1109/EMEIT.2011.6022992.
4. U. Chatterjee, B.S. Butola, M. Joshi, Optimal designing of polyurethane-based nanocomposite system for aerostat envelope, *J. Appl. Polym. Sci.* 133, 2016. doi:10.1002/app.43529.
5. G.A. Khoury, *Airship Technology*, Vol. 10, Cambridge University Press, Cambridge, 2012.
6. X.L. Xia, D.F. Li, C. Sun, L.M. Ruan, Transient thermal behavior of stratospheric balloons at float conditions, *Adv. Sp. Res.*, 2010. doi:10.1016/j.asr.2010.06.016.
7. F.A. D'Oliveira, F.C.L. De Melo, T.C. Devezas, High-altitude platforms – Present situation and technology trends, *J. Aerosp. Technol. Manag.*, 2016. doi:10.5028/jatm.v8i3.699.
8. A. Alsayed, 2017, *Pitch and Altitude Control of an Unmanned Airship with Sliding Gondola*, PhD Thesis, University of Ottawa, Ottawa, Canada. https://ruor.uottawa.ca/handle/10393/36594
9. L. Liao, I. Pasternak, A review of airship structural research and development, *Prog. Aerosp. Sci.*, 2009. doi:10.1016/j.paerosci.2009.03.001.
10. W. Kang, Y. Suh, K. Woo, I. Lee, Mechanical property characterization of film-fabric laminate for stratospheric airship envelope, *Compos. Struct.*, 2006. doi:10.1016/j.compstruct.2006.04.060.

11. M. Praskovia, C. Nicolas, H. Patrick, *Fly Win, a H2-lifting gas airship demonstrator*, in: *Proc. Seventh Eur. Conf. Aeronaut. Sp. Sci.*, 2017.
12. C. Stockbridge, A. Ceruti, P. Marzocca, Airship research and development in the areas of design, structures, dynamics and energy systems, *Int. J. Aeronaut. Sp. Sci.*, 2012. doi:10.5139/IJASS.2012.13.2.170.
13. W.E. Symolon, *High-Altitude, Long-Endurance UAVs vs. Satellites: Potential Benefits for US Army Applications*, Massachusetts Institute of Technology, Cambridge, 2009.
14. L.P. Teisserenc de Bort, *Encyclopædia Britannica*, 1922. https://en.wikisource.org/wiki/1922_Encyclopædia_Britannica/Teisserenc_de_Bort,_Léon_Philippe.
15. S. Paul-Davies, *The Prisoner Handbook*, Pan Books, London, 2002.
16. J. Hall, *Venus Balloon Technology Summary*, Report of Technology Focus Group, n.d.
17. K. Baines, K.H. Hall, J.L. Balint, T. Kerzhanovich, V. Hunter, G. Atreya, ... , S.K. Zahnle, *Exploring Venus with balloons: science objectives and mission architectures for small and mediumclass missions*, in: *6th Int. Planet. Probe Work*, Atlanta, Georgia, 2008.
18. J.A. Cutts, D. Mimoum, D.J. Stevenson, *Probing the Interior Structure of Venus*, Reported by Keck Institute for Space Studies (KISS), Venus Seismology Study Team, Pasadena, CA, United States, 2015, 1–77. https://authors.library.caltech.edu/59019/1/2015_KISS_Venus_Final_Report.pdf (accessed October 3, 2021).
19. R.E. Grimm, A.C. Barr, K.P. Harrison, D.E. Stillman, K.L. Neal, M.A. Vincent, G.T. Delory, Aerial electromagnetic sounding of the lithosphere of Venus, Icarus, 2012. doi:10.1016/j.icarus.2011.07.021.
20. Raven Aerostar, Responsive aerostat systems, n.d. https://ravenaerostar.com/products/tethered-aerostats/responsive-aerostat-systems.
21. Lockheed Martin, Persistent surveillance systems – always there. Always on, n.d. https://www.lockheedmartin.com/content/dam/lockheed-martin/rms/photo/aerostats/LTA_Brochure.pdf.
22. Rosaerosystems, Augur RosAeroSystems, n.d. http://rosaerosystems.com/aero/.
23. Raytheon, Missiles and defense, n.d. https://www.raytheonmissilesanddefense.com/capabilities/missile-defense.
24. Harris, L3HARRIS Capabilities, n.d. https://www.harris.com/press-releases/2014/10/exelis-supports-customs-and-border-protection-with-upgrade-of-aerostat.
25. TCOM, *Air Surveillance*, n.d. https://tcomlp.com/air-surveillance-systems/.
26. RT, Products, n.d. https://www.rt.co.il/products/.
27. Stratospheric Balloons, *CNIM Air Sp.* n.d. https://cnim-air-space.com/en/STRATOSPHERIC-BALLOONS/#CONTACT.
28. STRATOBUS, *The Sentinel of The Stratosphere*, n.d. https://cnim-air-space.com/en/technologies-for-stratobus/.
29. SKYSTAR-180, n.d. https://www.rt.co.il/skystar-180.
30. G. Pfotzer, History of the use of balloons in scientific experiments, *Space Sci. Rev.*, 1972. doi:10.1007/BF00175313.
31. M. Pagitz, The future of scientific ballooning, *Philos. Trans. R. Soc. A Math. Phys. Eng. Sci.*, 2007. doi:10.1098/rsta.2007.0002.

32. V.F. Hess, Penetrating radiation in seven balloon flights, *Phys. Zeitschrift.*, 1912, 1084–1091. http://www.fisicateorica.me/repositorio/howto/artigoshistoricosordemcronologica/1912-HESS 1912 Conclusive evidence for the cosmic rays.pdf.
33. I.S. Smith, The NASA Balloon Program: Looking to the future, *Adv. Sp. Res.*, 2004. doi:10.1016/j.asr.2003.07.052.
34. E. Mack, Meet Google's "Project Loon": Balloon-powered Net access, 2013. https://www.cnet.com/news/meet-googles-project-loon-balloon-powered-net-access/.
35. J. Brodkin, Google flies Internet balloons in stratosphere for a "network in the sky", 2013. https://arstechnica.com/information-technology/2013/06/google-flies-internet-balloons-in-stratosphere-for-a-network-in-the-sky.
36. T. Koulopoulos, *The Moonshot to Create the Next Google*, 2018. https://www.inc.com/thomas-koulopoulos/the-moonshot-to-create-next-google.html.
37. NZ Her, Google launches Project Loon, 2013. http://www.nzherald.co.nz/internet/news/article.cfm?c_id=137&objectid=10890750.
38. F. Lardinois, Google X announces Project Loon: Balloon-powered Internet for rural, remote and underserved areas, 2013. https://techcrunch.com/2013/06/14/google-x-announces-project-loon-balloon-powered-internet-for-rural-remote-and-underserved-areas/.
39. Raven Aerostar, Loon: Raven Aerostar's Super Pressure Balloon, n.d. https://ravenaerostar.com/about/loon.
40. Raven Aerostar, Project loon: Raven aerostar: Google collaboration, 2013. https://web.archive.org/web/20130617063829/http://ravenaerostar.com/about/project-loon-raven-aerostar-google.
41. The Loon Flight System, n.d. https://www.loon.com/technology/flight-systems/.
42. L. Kelion, Google tests balloons to beam internet from near space, *BBC News*, 2013. https://www.bbc.com/news/technology-22905199.
43. The Hindu BusinessLine, Google releases Internet-beaming balloons, 2013. https://www.thehindubusinessline.com/info-tech/google-releases-internet-beaming-balloons/article20624236.ece1.
44. P. Gupta, R.S. Pant, *A methodology for initial sizing and conceptual design studies of aerostats*, in: *Int. Semin. Challenges Aviat. Technol. Integr. Oper. (CATIO-05), Tech. Sess. 57th Annu. Gen. Meet. Aeronaut. Soc.*, India, 2005.
45. Aerial Platforms Study - CORDIS, Absol. – Aer. Base Station. with Opportunistic Links Unexpected Tempor. Events, 2013. https://cordis.europa.eu/docs/projects/cnect/2/318632/080/deliverables/001-FP7ICT20118318632ABSOLUTED23v10isa.pdf.
46. Tethered Aerostat Systems Application Note, Syst. Assess. Valid. Emerg. Responders, 2013. https://www.dhs.gov/sites/default/files/publications/TetheredAerostat_AppN_0913-508.pdf.
47. T. Yamagami, Y. Saito, Y. Matsuzaka, M. Namiki, M. Toriumi, R. Yokota, H. Hirosawa, K. Matsushima, Development of the highest altitude balloon, *Adv. Sp. Res.* 33 2004, 1653–1659. doi:10.1016/j.asr.2003.09.047.
48. A. Colozza, J.L. Dolce, High-altitude, long-endurance airships for coastal surveillance, 2005. NASA Technical Report; NASA/TM-2005-213427.

49. R. Ashford, P. Bely, *High-altitude aerostats as astronomical platforms*, in: *11th Light. Syst. Technol. Conf*, 1995. doi:10.2514/6.1995-1602.
50. S.S. Badesha, *SPARCL: A high-altitude tethered balloon-based optical space-to-ground communication system, in: Free.* Laser Commun. Laser Imaging II, 2002. doi:10.1117/12.450641.
51. M. Mohorcic, D. Grace, G. Kandus, T. Tozer, *Broadband Communications from Aerial Platform Networks, 2009 Int. Conf. Instrumentation, Commun. Inf. Technol. Biomed. Eng.*, 2009.
52. P.E. Liggett, Piezoelectric and pyroelectric power-generating laminate for an airship envelope, U.S. Patent 7878453, 2011.
53. How airship is made, 2020. http://www.madehow.com/Volume-3/Airship.html (accessed January 30, 2020).
54. B. Adak, M. Joshi, Coated or Laminated Textiles for Aerostat and Stratospheric Airship, in: *Advanced Text Engineering Materials*, John Wiley & Sons, Inc., Hoboken, NJ, USA, 2018, 257–287. doi:10.1002/9781119488101.ch7.
55. K. McDaniels, R.J. Downs, H. Meldner, C. Beach, C. Adams, *High strength-to-weight ratio non-woven technical fabrics for aerospace applications*, in: *AIAA Balloon Syst. Conf.*, 2009. doi:10.2514/6.2009-2802.
56. E. Pasternak, I. Nemirovsky, M. Kouchak, Aerostructure for rigid body airship, U.S. patent No. 9266597, 2016.
57. W. Raza, G. Singh, S.B. Kumar, V.B. Thakare, Challenges in design & development of envelope materials for inflatable systems., *Int. J. Text. Fash. Technol.*, 2016, 2319.
58. Airship, 2017. http://enacademic.com/dic.nsf/enwiki/36668 (accessed January 30, 2020).
59. AXS Digit. Gr. LLC, Atmospheric satellite, 2010. https://en.wikipedia.org/wiki/High-altitude_platform_station#High-altitude_airship%0A (accessed January 30, 2020).
60. Lockheed Martin's HALE-D airship learns to fly, makes a crash landing, 2011. https://www.engadget.com/2011/07/28/lockheed-martins-hale-d-airship-learns-to-fly-makes-a-crash-la/ (accessed January 30, 2020).
61. Loon LLC, Proj. Loon., 2019. https://en.wikipedia.org/wiki/Project_Loon (accessed January 30, 2020).
62. High-altitude balloon, Geostationary Balloon Satell., 2020. https://en.wikipedia.org/wiki/Geostationary_balloon_satellite (accessed January 30, 2020).
63. The MAAT project – Multibody advanced airship for transport, 2013. http://airtn.eu/wp-content/uploads/maat-cranfield_airtn.pdf (accessed January 30, 2020).
64. Aerostat Systems Market worth 10.95 Billion USD by 2021, Aerostat Syst. Mark., 2016. https://www.marketsandmarkets.com/PressReleases/aerostat-systems.asp (accessed January 30, 2020).
65. The Times of India, City–Agra, Advanced indigenous aerial surveillance system to be deployed along Pak border, 2016. https://timesofindia.indiatimes.com/%0Acity/agra/Advanced-indigenous-aerial-surveillance-system-to-be-deployedalong-%0APak-border/articleshow/52238174.cms (accessed January 30, 2020).
66. DRDO working on New "Nakshatra" Aerostat, 2015. http://indiandefence.com/threads/drdo-working-on-new-nakshatraaerostat.%0A51754/.

2

Modern Technologies for Manufacturing Aerostats and Airships

B. S. Butola, Shikha Chouhan, and S. Wazed Ali
Indian Institute of Technology, New Delhi, India

CONTENTS

DOI: 10.1201/9780429432996-2

2.1 Introduction

In recent years, there has been increased interest in high altitude, lighter-than-air (LTA) systems operating above 20 km [1]. There are mainly two types of LTA systems: the aerostat (tethered) and the airship (untethered). An untethered high-altitude airship is an unmanned, powered, free-flying vehicle. Lift is provided by a combination of aerodynamics and lifting gas, such as helium or hydrogen that is contained in the envelope. The stratosphere is the layer of the atmosphere where the temperature starts to increase with height. Immediately after the troposphere, which has a constant temperature of about −60°C, the stratosphere starts at an altitude of 7 km at the poles and 18 km at the Equator, extending to around 50 km [2]. Airships can be classified on the basis of (i) hull configuration (non-rigid, semi-rigid and rigid), (ii) the way of producing vertical force (lighter-than-air, heavier-than-air and hybrid), and (iii) payload capability (heavy-lift and medium-lift). The payload of traditional buoyancy air-vehicles is usually less than 30 tons, while for heavy-lift air vehicles it can reach as high as 500 tons. Airships can also be divided into conventional types and unconventional types [3, 4]. Generally speaking, conventional airships have a streamlined axis symmetric body, generate aerostatic lift by a hull with enclosed gas, have a low payload capability, and use fuel as a power source. Figure 2.1 shows different types of conventional airships and their different components. In contrast, the tethered high-altitude aerostat is operated from a fixed location. Lift is provided solely by a lifting gas such as helium or hydrogen which is contained in the envelope. The tether provides a structural link to the mooring system, power for the aerostat and payload, and secure command and communications. The tethered high-altitude aerostat is very power efficient, requiring no power for station keeping or altitude control. It is also easily

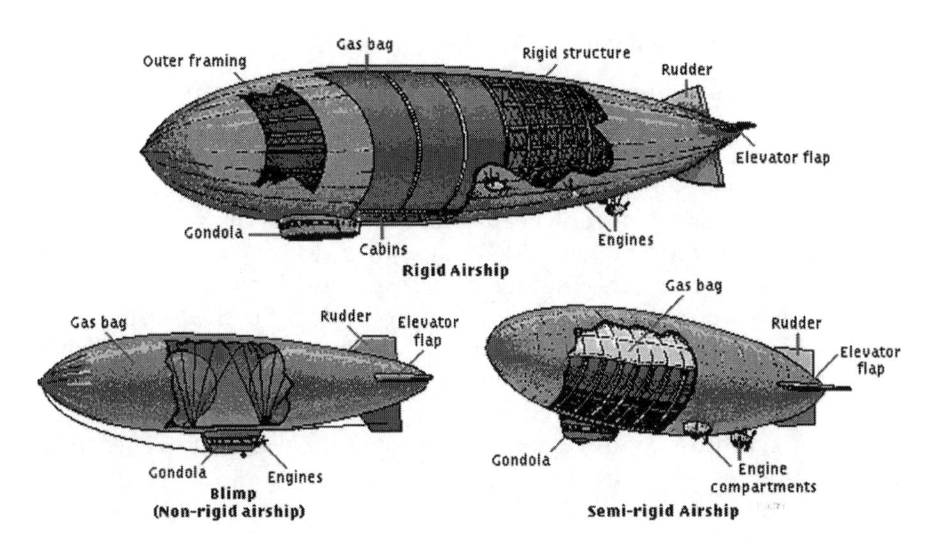

FIGURE 2.1
Types of airship [5].

recoverable for payload maintenance. The goal of both the un-tethered airship and the tethered aerostat is to remain on station for a year or more while supporting a sizable payload. The challenges for high-altitude LTA systems include:

1. Strong, lightweight LTA material
2. Efficient pressurization system
3. Altitude and altitude control (airship)
4. Icing and lightning
5. On-board power generation for long duration missions (airship)
6. Lightweight propulsion units (airship)
7. Lightweight tether (aerostat)

Selection of suitable material for the envelope is, thus, very important in order to deploy these systems effectively for a long time. Different types of envelope materials are used, such as single ply, coated and laminated/composite of two or more fabrics. In the early days, the envelope material was made using rubber-coated cotton fabric. This material was very heavy and gas permeability was also quite high. With the advent of synthetic fibre, materials having a light weight and low helium permeability were developed for airship envelope, e.g., thin film laminated polyester fabric, which has good gas retention capability, and is also able to sustain higher internal pressures [6].

2.1.1 Coating and Lamination in Textiles: Processes

Coating and lamination are increasingly important techniques for adding value to technical textiles. Coating and lamination enhance and extend the range of functional performance properties of textiles, and the use of these techniques is growing rapidly as the applications for technical textiles become more diverse. Cheaper fabric structures may be coated or laminated to provide higher added value to end-users and higher profit margins to manufacturers. The key to success in textile coating and lamination depends upon the application of appropriate technology using modern machinery [7–9]. Machine productivity is important, but flexibility in terms of production speed and the versatility of coating/lamination methods are important factors to consider, as well as a high level of process monitoring, process control and automation to satisfy demanding technical specifications.

2.1.2 Coating Technology

Coating is a process in which a polymeric layer is applied directly to one or both surfaces of the fabric. The polymer coating must adhere to the textile, and a blade or similar aperture controls the thickness of the viscous polymer. The coated fabric is heated and the polymer is cured (that is, polymerized). Where a thick coating is required, this may be built up by applying successive coating layers, layer on layer, Interlayer adhesion must therefore be high [8]. Finally, a thin top layer may be applied for aesthetic or technical enhancement of the coating. Depending upon the end-use requirements, heavy-duty technical textile coatings may be applied at high weight, while other end-uses for high-technology apparel may require coating weights very low. The chemical formulation of the coating, the coating thickness and weight, the number of layers, the form of the technical textile and the nature of any pre-treatment (such as to stabilize the fabric dimensions prior to coating) are of great importance. Traditionally, coating has been applied to woven technical textiles, but increasingly warp-knitted, raschel [9], weft-knitted and nonwoven fabrics must be coated on the same line. The machinery and method of application of the coating formulation must be versatile, minimize tensions on the fabric that may lead to distortion or stretch, and eliminate problems in knitted fabrics such as curling selvedges.

2.1.2.1 Types of Coating Techniques

There are several processes for the application of coating on the textile material depending upon the requirement of end product. Some of these processes are described as follows.

2.1.2.2 Knife or Blade Coating

Knife coating is a process where an excess of coating material is applied to the substrate and removed by a metering blade to achieve the desired coating thickness (Figure 2.2). There are different knife configurations, which are floating knife, knife-over-roll, knife-over-fixed-table and knife-over conveyor [10]. In a floating knife coating or a blade over air, the knife touches the surface of the fabric, which is supported only by tension between two rolls. The doctor blade pressures the upper surface of the fabric substrate and forces the coating formulations to penetrate into the substrate. Fabric tension and the depression of the knife mainly control the amount of coating applied and the depth of penetration, but sharpness of the blade and knife angle alignment also affect them [11]. This method is usually used for applying relatively small amounts of coating material to the fabric [12, 13]. A more common technique is the knife-over-roll coating, where the knife blade is suspended above a roller and does not touch the substrate directly. In this configuration, the coating layer thickness is controlled by a gap between the substrate and the knife. This enables precise thickness control, but variations in substrate thickness may result in non-uniform coating thickness [10–13]. Blade geometry (the shape and angle of the blade), blade flexibility, and the

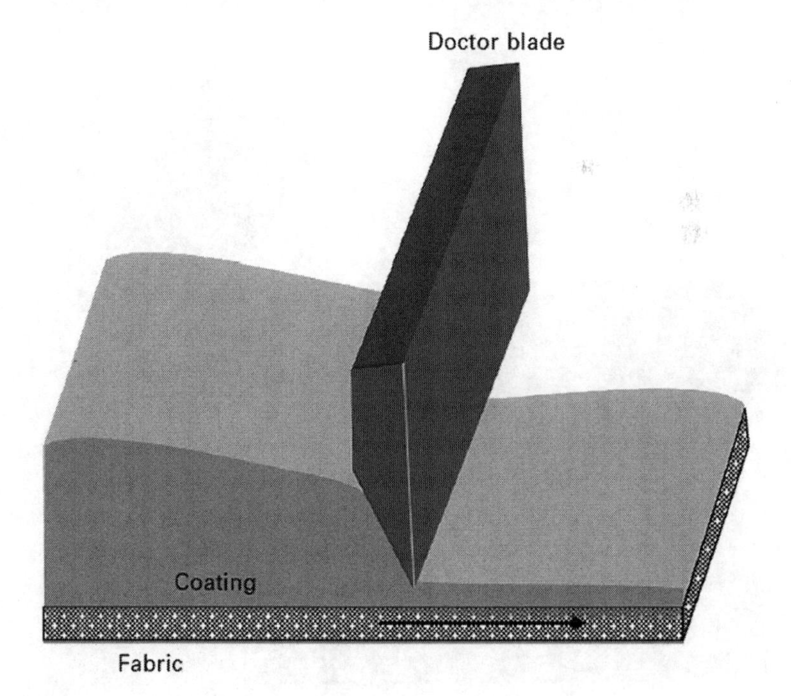

FIGURE 2.2
Knife coating.

rheology of the coating fluid also affect the amount of coating applied and the extent of penetration into the substrate [14]. Air knife coating uses an air jet to blow off excess coating materials and to control coating thickness instead of using a knife blade.

2.1.2.3 Roll Coating

Roll coating is a pre-metered coating and uses a series of rollers to meter and apply coating liquid on a substrate. A metered film of coating liquid is first formed on the roller surface before it is applied to the substrate, so the amount of coating material delivered to the substrate is nearly independent of the fabric properties and structures. Mostly determined by the rheology of the fluid and the relative speed of two rotation surfaces, precise control is possible [15–17]. Varying substrate thickness does not result in uneven coating thickness [13, 18]. The simplest roll coating set-up uses a single rotating roller. The bottom half of the roller is immersed in a coating liquid bath and the upper part of the roller is in contact with the fabric substrate (Figure 2.3). As it rotates, the coating liquid forms a film on the roller surface and part of the liquid film is transferred from the roller surface to a fabric substrate. The amount of coating on the substrate is governed by hydrodynamics. The rotation speed of the

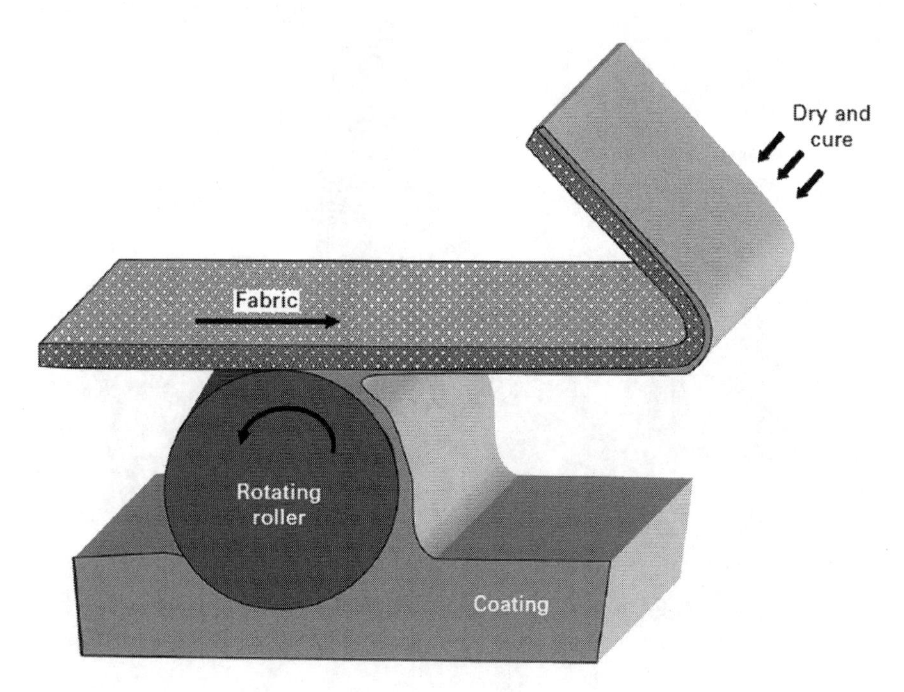

FIGURE 2.3
Roll coating (one roll reverse applications).

roller, the substrate speed and rheological properties of coating fluid (surface tension, viscosity, and density) are factors determining coating thickness [19]. In this set-up, one roller is a metering device as well as an application device. More precise control is achieved by adding more rollers.

Three-roll coating uses a metering roller, an applicator roller and a backup roller. Common three-roll configurations are nip feed coating and L-head coating. In three-roll nip feed coating (Figure 2.4(a)), the nip formed by a metering roller and an applicator roller is flooded with coating liquid and functions as a reservoir [16, 20]. The applicator roller picks up the coating liquid from the nip, and the amount of coating liquid delivered to the fabric

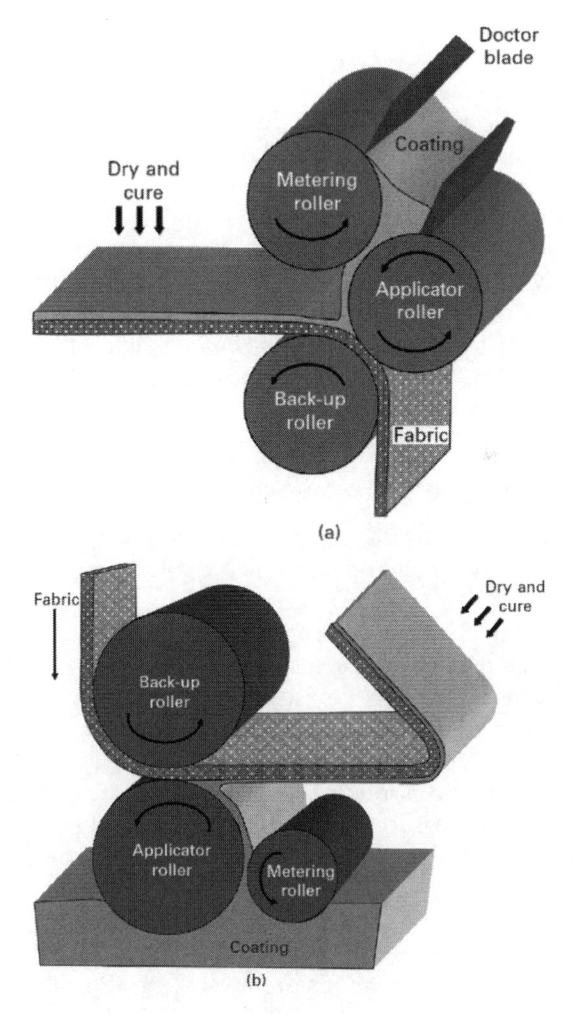

FIGURE 2.4
Three-roll coatings: (a) nip feed coating and (b) L-head coating.

substrate is metered by the metering roller rotating in a reverse direction to the applicator roller. After coating liquid transfers from the metering roller to the applicator roller, any coating liquid remaining on the metering roller surface is cleaned by a doctor blade; otherwise it will result in coating defects, such as streaks or film roughness. Film formed on the applicator roller is deposited onto the substrate surface supported by a back-up roller. This configuration needs only a minimal amount of coating fluid, but a drawback is possible coating fluid leakage. It can be problematic when the coating liquid has low viscosity. In a three-roll pan feed or L-head coating (Figure 2.4(b)), a liquid film is formed on the applicator roller rotating through the coating liquid, metered by a metering roller and deposited on the substrate fabric on a back-up roller. To increase coating speed, one more roller – a pick-up roller running at a reduced speed – can be added, and this system is called a four-roll pan-fed coating system [16, 20].

Configurations shown in Figure 2.4 are called reverse metering, since applicator and metering rollers rotate in opposite directions. When they rotate in the same direction, it is called forward metering [21]. Reverse metering produces a smoother film with better stability, while forward metering is prone to generate unstable, non-uniform films. Therefore, reverse roll coating is more commonly used [16, 20]. Roller coating can use water-based solutions, solvent-based coating materials as well as hot melts [22]. In hot melt roller coating, solid pellet is melted between the heated melt rollers, forming a melt film and deposited on a substrate. The substrate fabric is usually preheated before hot melt is applied [23].

2.1.2.4 Gravure or Engraved Roll Coating

Instead of a smooth roll surface, the coating roll surface can have an engraved pattern to carry the coating liquid. In engraved roll coating, coating liquid fills the engravings when it rotates through the coating reservoir and excess coating liquid remaining on the landing area is removed by a doctor blade before the roll presses against a substrate fabric (Figure 2.5) [20, 24]. The engraving on the roller acts as a metering device and the amount of coating delivered to the substrate is controlled mainly by the engravings. The land areas, cell opening, cell depths, cell volume, cell angles and cell space determine the amount of coating liquid applied to a fabric substrate [25]. Coating liquid viscosity, application pressure on the substrate and substrate structure may influence actual delivery amount. It can also create patterned coating [26].

2.1.2.5 Screen Coating

In screen coating, coating materials are applied to the substrate through a mesh screen by a squeezing motion (Figure 2.6) [27]. The amount of coating applied is determined by the screen mesh number, squeeze pressure, the

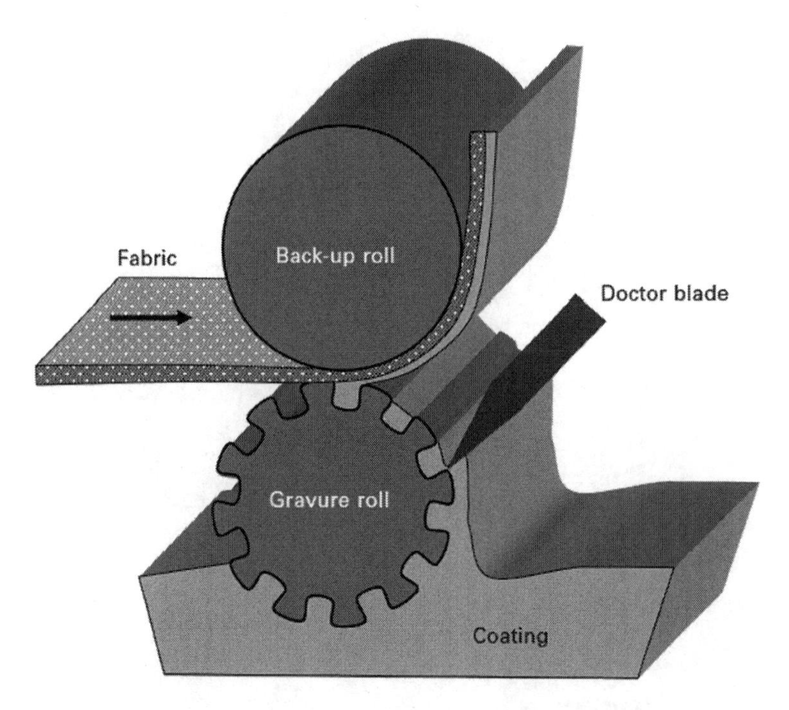

FIGURE 2.5
Gravure roll coating.

angle between the squeeze blade and the screen, and the viscosity of the coating liquid [28]. Screen coating is ideal for lightweight and delicate substrate, because little or no friction or tension is imposed on the substrate [28]. This method is also easily adapted to use hot melt and hot melt dot printing, or hot melt printing where pre-melted hot melt resin is deposited on the substrate through the screen. Advantages of using screen printing over engraved roll coating are relatively low cost of screen, shorter exchange times and lower friction [28] as well as good pattern definition. Since it is capable of providing a good pattern definition and precise coating thickness control, screen coating is also used for electrical applications, such as circuit board printing [29].

2.1.2.6 Transfer Coating

Direct application of coating material on a delicate substrate is challenging since tension required in the process often deforms or damages fabrics. It is also hard to prevent excessive penetration of coating liquid into a fabric substrate with very open structure. One solution to this problem is a sequential coating process where the coating material is first applied to a release paper and later transferred to the fabric substrate (Figure 2.7). This process is called

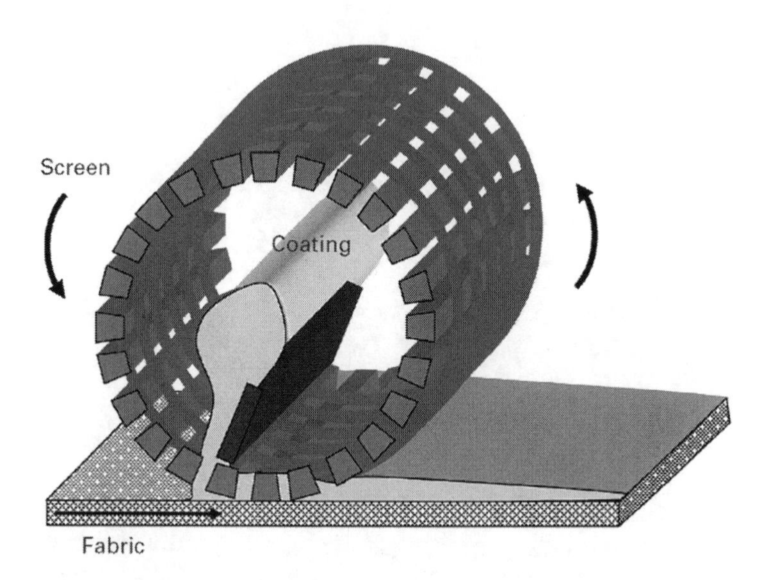

FIGURE 2.6
Rotary screen coating.

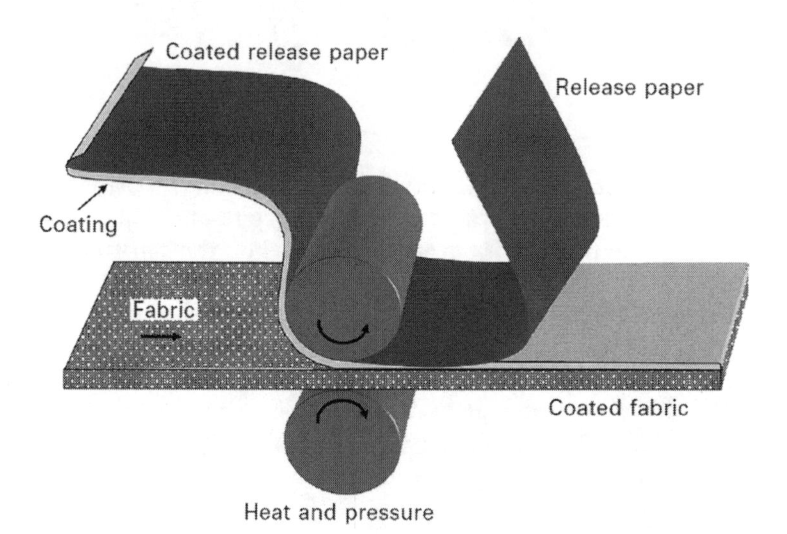

FIGURE 2.7
Transfer coating.

transfer coating [13, 30]. A desired amount of coating material is first applied to a silicone release paper through one or another direct coating method and then dried. Then, this coated silicone release paper and the fabric substrate pass through the lamination rollers under heat and pressure, so coating layers backed by a release paper form a bond with the substrate. After peeling off the silicone release paper, coating layers remain on the surface of the substrate fabric [13, 30]. It is more expensive than direct coating methods, but it can produce a flexible coated fabric because the process imposes little or no tension and penetration of coating formulations is low [30, 31].

2.1.2.7 Slot Die and Extrusion Coating

Slot die coating is the coating process where a plasticized coating compound or a coating liquid is pressed through a die and transferred directly onto the surface of the fabric substrate [23, 32]. The traditional slot die configuration is 'a closed system' where the die lip is in contact with a substrate backed by the roller, and coating materials are not exposed to air during the process [18]. In this configuration, the substrate is often under high shear pressure, so it must be able to withstand it. Some modern slot die systems can have a gap between a die lip and the substrate fabric to impart less tension. The die configuration, the lip shape and lip opening, and die-to-roller gap, flow rate and flow distribution influence coating quality. A balance between die pressure and substrate tension should be achieved for good-quality coating.

Extrusion coating mostly refers to a coating method that extrudes a polymer onto a substrate, and it mostly has a gap between a die tip and a substrate where extruded polymers are self-supporting. The configuration of common extrusion coating is shown in Figure 2.8 [33]. Molten polymers are extruded by an extruder pump through a die and deposited on the substrate – the substrate may be preheated – and the substrate and deposited coating layer are passed through pressure nip rollers to improve adhesion. When an additional substrate is inserted through the nip rollers with the coated substrate, a laminated fabric can be produced.

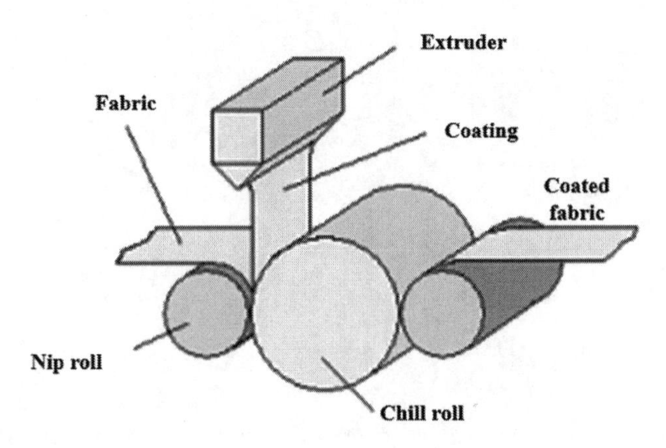

FIGURE 2.8
Extrusion coating [33].

In slot die and extrusion coating, coating weight is principally controlled by the pump delivery rate and the substrate line speed [18, 34]. One major application is hot melt coating since it can handle high viscosity of hot melt with relatively high line speed. Parameters that affect adhesion between the melt and the substrate are the compatibility of the coating materials and substrates, together with other processing factors such as extrusion melt temperature, air gap, film thickness, coating nip pressure, substrate construction, resin melt index, coating line speed and coating roller temperature [35]. Frequent maintenance of the die is required, and it makes extrusion coating an expensive option.

2.1.2.8 Spray Coating

In the spray coating technique, a spray gun system creates provision of passing heated air with a molten polymer adhesive in order to produce many continuous filaments through a nozzle (Figure 2.9). The filaments are applied to the moving fabric. Once the substrate is uniformly coated, a second substrate is introduced, and the two are permanently joined by pressure between two rolls. Chemicals which are relatively stable against heat and air are used for spray coating. Theoretically, all types of liquid adhesives – e.g., hot melt, solvent-based, water-based and high-solids versions – can be used but, in practice, hot melts need expensive apparatus to ensure that they do not solidify prematurely or char. Besides, the use of solvents having flammability and water-based adhesives which may not dry at commercial speeds is also excluded. Modular spray heads allow their placement at a variety of distances from the substrate. The non-contact nature of the spray systems has the advantage of working well with temperature sensitive substrates such as

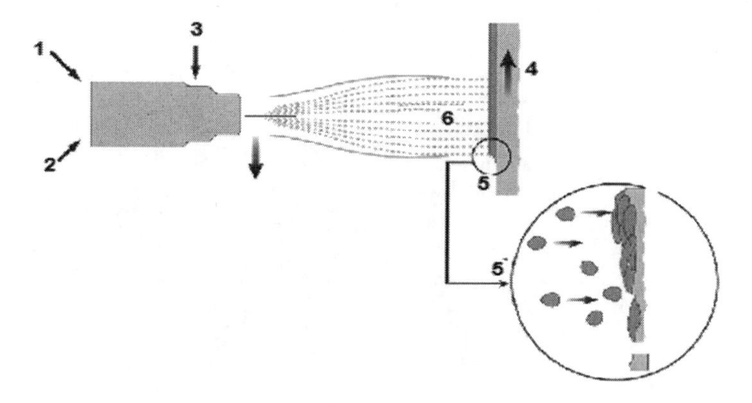

FIGURE 2.9
Spray coating: (1) energy source, (2) gas or other process burner, (3) coating material, (4) base fabric, (5) coated fabric surface (magnified view in zoom), (6) thermally spread multifilament cone.

polypropylene and the technique can typically be used with add-ons as low as 1 g/m [2]. The problems usually associated are uniformity in precision applications, penetration of material into the fabric, occasional blocking of the spray nozzle, continuous drying of the liquid by graded heating of the solvent to control blistering, and pollution.

2.1.3 Lamination Techniques

Lamination is the process that combines multiple substrates together and produces the stable multi-layer structures by adhesives, heat, pressure and mechanical bonding [36]. Good adhesion between layers is one of the essential requirements of the lamination process. Lamination can be done in a number of ways depending on the substrate type combinations and the number of layers. Some of the available lamination techniques are briefly described next.

2.1.3.1 Flame Lamination

Flame lamination is a simple and easy lamination process with high production rate and low cost. A preformed thin foam sheet is passed over an open flame to create a thin layer of molten polymer on the foam surface and subsequently bonded to fabric substrate by passing through the nip of a calender (Figure 2.10) [37]. Molten layers of foam surface act as adhesive and form stable bonds with the fabric as it cools. Polyurethane foam is the most frequently used material in flame lamination. Flame intensity should be optimally adjusted to ensure sufficient melting without burning or loss of excess foam layer thickness [38], and it is affected by gas type, flame height and

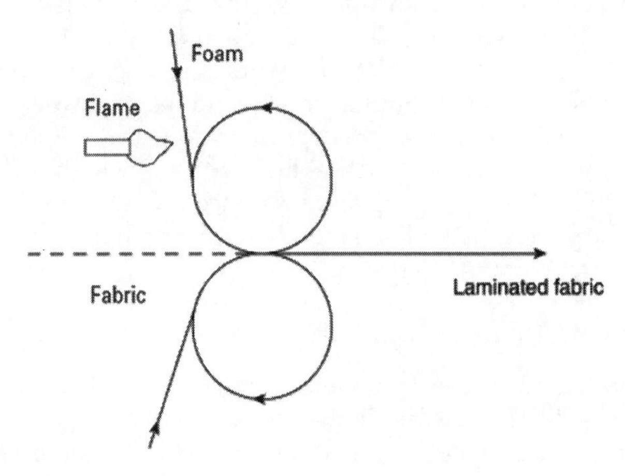

FIGURE 2.10
Flame lamination [37].

spread. Triple-ply lamination also can be produced using a dual burner or two-pass lamination [38]. As indicated earlier, this is an inexpensive process and does not require drying or a curing oven. But flame laminated fabrics are stiff since the process produces a solid bond layer with little porosity [38]. Another main drawback is environmental unfriendliness, as it generates harmful emissions during the process.

2.1.3.2 Ultrasonic Lamination

Ultrasonic lamination uses sound energy instead of adhesives to combine multiple layers of substrate (McIntyre, 2006) [39]. Ultrasonic is high-frequency sound waves, with frequencies of 20 kHz to 1 GHz. This range is beyond what humans can hear, but ultrasonic waves carry energy, which can generate bonding in different materials. Ultrasonic bonding or welding has been used for embossing, perforating, slitting, cutting and slicing, as well as lamination [40]. In the ultrasonic welding process, high-frequency electric signals produced by a generator are converted to mechanical oscillations in a weld horn or a sonotrode. This causes repeated compression and friction of the fibres in the substrate, resulting in melting and bonding [40, 41]. The most frequently used frequencies are 20, 30, 35 and 40 kHz. The amplitude of vibrations and the welding force, the forces that the sonotrode applies to the material, affect bond strength. Optimum amplitude depends on type of the material to be bonded and the welding force and needs to be adjusted according to desired bond strength, materials, welding time, substrate feeding speed and coverage area of the bonding pattern [40]. An engraved anvil drum can produce patterned bonding points. The percentage pressing area of the engraved roll, the gap between the ultrasonic head and the anvil roller, and the speed of substrate are important processing parameters [41]. No use of adhesive (mostly made from petroleum-based materials) and no heat application make this an environmentally friendly and safe process. Ultrasonic lamination is also very versatile, since multiple layers, up to 6 to 12, can be laminated simultaneously, and pattern variety can be used to improve appearance and tailored properties. Ultrasonic bonds provide textile-like hand, open characteristics, high loft, softness, breathability, and high absorption, without causing stiffness since the substrate is melted only at the point of pattern [42]. This method is used for multi-layer wipes, sorbents and medical products [40].

2.1.3.3 Wet Adhesive Lamination

Wet adhesives used in the lamination process are either water-based or solvent-based [43]. They are applied to one substrate surface in liquid form by conventional coating methods, such as gravure roll coating, spraying, roll coating and knife coating. Then, the adhesive-coated web is bonded with other substrates under pressure, and dried or cured in an oven [26, 44].

Adhesive performance, such as bond strength, durability and resistance to heat, depends on the chemistry of the adhesive as well as the lamination process. Crystallinity, viscosity, flow characteristics, chemical resistance and set times of adhesive used all affect performance [45]. Penetration of the adhesive into the substrate structure should be controlled. While some penetration is necessary to achieve good adhesion, deep penetration may generate stiff structures and loss of drapability, hand, tear strength, and breathability [26, 45]. A solvent-based adhesive has the advantage of being fast-drying, but it poses environmental risks, such as high VOC emission and hazardous waste generation. Owing to the environmental concerns and government regulations, conventional solvent adhesives are used only when coupled with expensive incineration or recovery equipment [37, 46]. Water-based adhesives do not use toxic solvents, but water is difficult to evaporate, so a drying unit is required, which has high energy requirements and occupies a large amount of floor space [37].

2.1.3.4 Hot Melt Lamination

Hot melt lamination uses hot melt adhesive as the bonding agent [47]. Hot melt adhesives are 100% solid and melt to the liquid phase in temperature ranges of 80 to 200°C, solidifying to form bonding when they cool [48]. Hot melt adhesives are mostly polymers and their compounds, specially formulated to satisfy lamination requirements [49]. They should have the appropriate melt viscosity, melting temperature, curing time, hardness, and good adhesion with the substrate to provide good peel, shear and tensile strength. Depending on the applications, breathability, lightness, porosity and washing durability may also be required [50]. The appropriate adhesive is selected to achieve the desired quality, coating flexibility and cost [47, 49]. There are two different hot melt adhesives classes; thermoplastic polymer-based systems and reactive hot melt adhesive systems [48]. Thermoplastics soften and solidify solely according to temperate changes and include ethylene vinyl acetate (EVA), polyamide (PA), polyethylene (PE), polyvinylchloride (PVC) and polyester-based compounds [45, 48]. These are sensitive against steam and water and, due to low softening points, the application ranges are limited. Reactive hot melt adhesives are fully cross-linked by the reaction with moisture, after being applied to the substrate. After crosslinking, heat does not soften the adhesive. They are highly durable and have good boiling and climate resistance, but are rather expensive. This group includes moisture crosslinking polyurethanes (PUR) [50]. Hot melt adhesive is normally melted in an off-line melting unit and delivered to a hot melt coating unit [37]. Many conventional coating processes are used to apply hot melt adhesives, including roll coating, gravure coating, slot die coating, screen coating, knife-over roll coating and spray coating. The coated substrate is subsequently bonded with another substrate to form a multi-layer laminate. Hot melt adhesives are replacing water- and solvent-based adhesives due to their environmentally friendly operating conditions [48]. They also have advantages over dry lamination for heat-sensitive substrate coating, since the substrate is not exposed to direct heat during the lamination process [51]. Nor do they require a drying step, where the energy requirement can be high [47, 48]. Lamination can be achieved at low coating levels, so it can produce soft handling and flexibility [47]. Bonds are formed almost instantly without drying, so operating speeds are not limited by the rate of drying and high productivity can be achieved [45]. The shelf-life of hot melt adhesive is longer than that of wet adhesives. However, the hot melt adhesive system has some drawbacks. Changing the adhesive type means that the entire pre-melt and application system must be cleaned, and these adhesives are relatively expensive [48].

2.1.3.5 Dry Heat Lamination

Dry adhesive is another type of 100% solid adhesive. Unlike hot melt adhesive, this type is applied to the substrate in solid form and later activated by heat and pressure to form a laminated product. Dry adhesives are powders, webs or films made from polyester, polyamide, EVA, polyethylene and thermoplastic polyurethane [37, 52]. High solid content without harmful emissions is advantageous from an environmental point of view; however, the energy requirement for melting dry adhesives and the high heat exposure of the sample are drawbacks [37]. Powder adhesives are cryogenically ground to a fine texture, with sizes from 1 to 500 μm. They can be applied by scatter coating, powder dot coating or paste dot coating. When melted, they form a discontinuous bond between two layers, imparting bonding while providing high softness, drape, and permeability. Even though they do not produce any harmful emissions, fi ne particles may generate airborne dust [37, 45]. Other dry adhesives come in the form of films and webs. The adhesive rolls are simply inserted between or over the substrate to be laminated [53]. These can be used to laminate textiles with open structures, where powder adhesives tend to fill the voids instead of staying on the surface [37, 45]. Lower add-on can also be achieved. However, they tend to be more expensive than powder adhesives, and it is difficult to adjust add-on level and application width. They also require the use of different rolls of adhesive to change add-on or application width [37]. One way to produce dry adhesive lamination is flat bed dry heat lamination. Dry adhesives are sandwiched between two substrates and transported through the heating tunnel by an upper and lower conveyer, where the adhesive melts and forms a bond between the two layers. Further pressure can be added by nip rollers at the end of the heating section as the materials cool (Figure 2.11) [54]. This is a high-volume production and can take place in a continuous operation. It also enables the use of dry adhesive at relatively low temperatures, thus reducing the problem of substrate heat exposure [55].

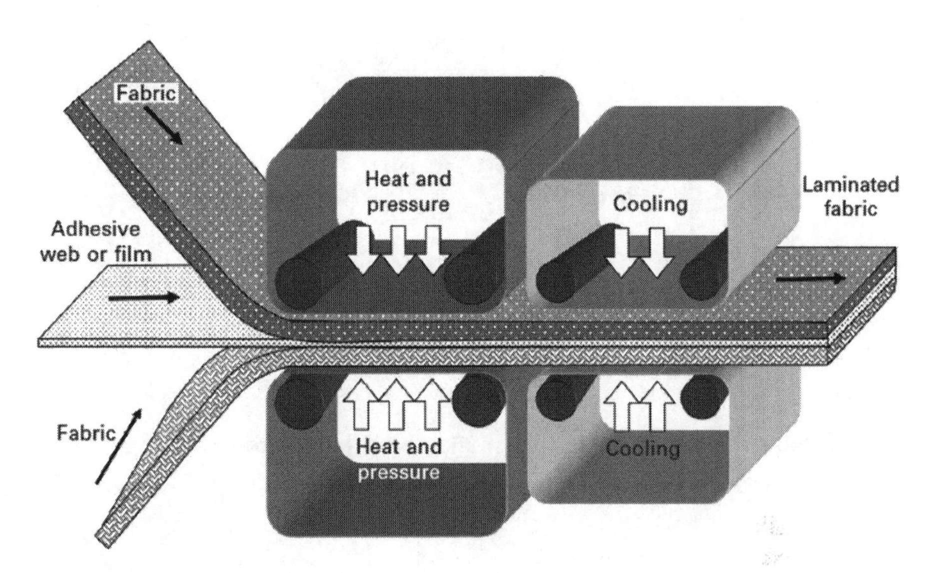

FIGURE 2.11
Flat-bed lamination [54].

2.1.4 Sealing Technique

The fabric panels used for airships, especially for meteorological/defence applications, are constructed by air-sealed seams between sheet materials composite fabrics consisting of different layers.

One of the layers in the composite sheet is invariably a woven textile fabric to provide high strength to the structure. This is required because at high altitudes, the helium gas contained in the airship expands due to the highly rarefied atmosphere and puts the structure under a lot of pressure. Other layers have different functions: for example, the outermost layer should provide adequate protection from highly degrading environment at high altitudes and an inner layer may provide good gas barrier properties to retain the gas in the airship.

As the surface area of an airship is quite large as compared to the width of the fabric that can be prepared on the widest looms, the starting material for airships are fabric panels of limited length and width. These fabric panels are first made into multi-layered composites by coating/lamination and then joined together to create the final shape of the airship. To create the final shape of the airship, the panels must be joined at the edges, resulting in the formation of seams. The seams can be created either by stitching the panels together, by use of adhesives or by welding. In this particular application, seams cannot be created by stitching the panels together as the holes created by needles will affect the gas barrier properties of the structure adversely. The seams need to be created in such a way that these are airtight. Hence,

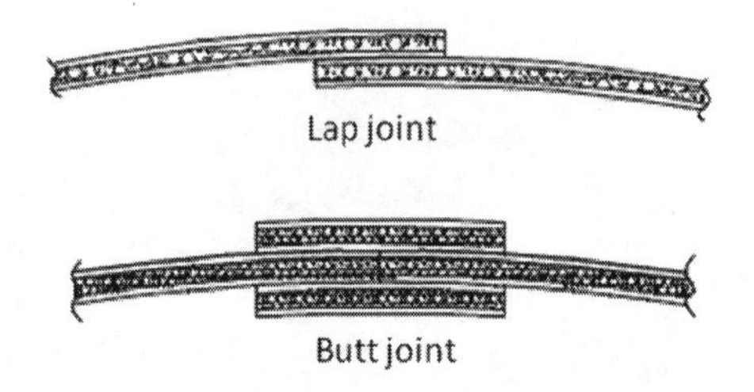

FIGURE 2.12
Different designs for sealing.

for inflatable structures, seams are created by either welding the edges of the panels or by applying an adhesive.

2.1.4.1 Type of Joints

There are two types of basic joints:

a. Homogeneous joint
b. Heterogeneous joint.

In the homogeneous type of joint, the joining substrate is same as the body substrate, whereas in the heterogeneous joint the two substrates are of different types.

There are two basic designs for making the joints (Figure 2.12)

1. Lap joint
2. Butt joint.

2.1.4.2 Welding Technology

Welding is a process in which materials (including fabrics) of the same type or class are joined together through the formation of primary (and occasionally, secondary) bonds under the combined action of heat and pressure. These bonds are the natural consequence of bringing two similar materials close enough together to allow intermolecular diffusion.

Techniques currently used for welding fabrics can be divided into three groups: those in which heat is generated by mechanical movement; those in which heat is conducted to the joint from an external heat source; and those using electromagnetic radiation directly.

It involves partial melting of the panel edges by application of heat to the surfaces of the panels in contact and then joining them by commingling of the melted portion. The heat can be applied by various means. Based on the mode of heat application, the welding can be done in the following ways:

2.1.4.3 Radio Frequency Welding

Radio frequency (RF) welding, also called high-frequency welding, HF welding or dielectric welding, uses high-frequency (normally 27.12 MHz) electromagnetic energy to generate heat in materials by virtue of their electrical properties; the material to be welded must be able to convert the alternating electric field into heat. Polar molecules in an electric field tend to orient in the field direction so that the positive (or negative) end of the dipole aligns to the negative (or positive) charges in the electric field. This process is called dipole polarization. Nonpolar molecules in an electric field displace electron clouds to align with the field (electronic polarization), so that centres of positive and negative charges no longer coincide. Electronic polarization is instantaneous and does not result in heat generation. Dipolar polarization, however, is not instantaneous at the high frequencies used in RF welding; as the dipoles try to align with the rapidly alternating electric field, orientation becomes out-of-phase. The imperfect alignment causes internal molecular friction and results in the generation of heat.

The delay between changes in electric field direction and changes in dipolar polarization is shown in Figure 2.13. An oscillating electric field E generates an oscillating current I within the polar material. At high frequencies, the two curves are out of phase by the phase angle θ; the loss angle is defined as $90 - \theta$, or δ. The amount of energy absorbed per cycle from the electric field is

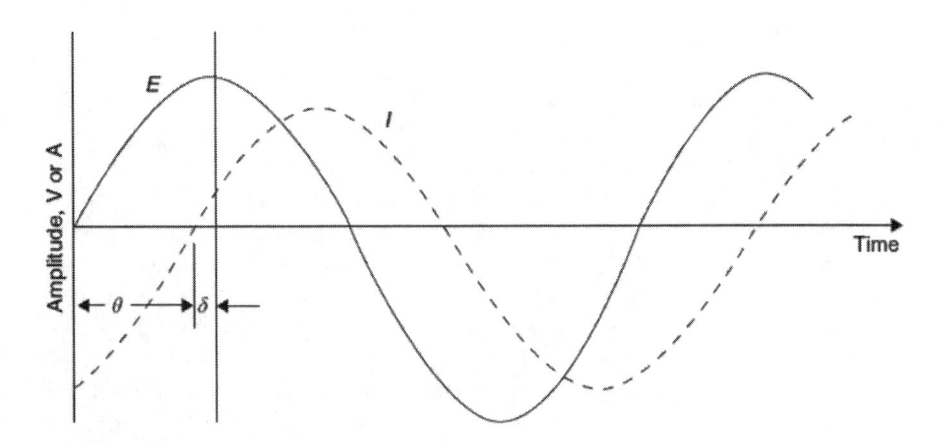

FIGURE 2.13
Electrical representation of dipole polarization resulting in heat generation due to imperfect alignment.

represented by the power factor and the dissipation factor; the power factor is defined as cos θ, and the dissipation factor or loss tangent is tan δ, a ratio of current dissipated into heat to current transmitted. The amount of dipole polarization is dependent on frequency and temperature. At low frequencies, the power lost by the electric field is low due to dipole alignment with the electric field. At high frequencies, field reversal is so rapid that dipole alignment becomes out of phase, and power losses increase. Eventually, a maximum in power loss is reached, so that further frequency increases result in decreased power loss. Dipole polarization is low at low temperatures but increases with increasing temperature [56].

Typically, an RF welding machine has two platens – an upper, moveable one and a lower, fixed one, also called a bed. The upper platen is connected to the RF generator, whilst the lower platen is connected to electrical earth. Tooling, known as the electrode or welding rule, is attached to the upper or lower platen and defines the shape and dimensions of the weld.

The parts to be welded are placed between the platens, and a press then lowers the upper platen so that a pre-set amount of pressure is applied to the joint area. Once the parts are clamped, the RF energy is applied for a pre-set time, called the heating time. Once the heating time has elapsed, the electrodes continue to clamp the parts until the weld solidifies. This is called the hold or cooling time. The press then opens and releases the welded assembly.

RF welding is normally used only for joining films and thin sheets. This is because a sufficiently strong electric field to produce material melting can be achieved only when the gap between the electrode and the opposing platen is less than around 1.5 mm (0.06 inches).

Radio-frequency welding is useful for sealing and welding plastics such as polyvinyl chloride (PVC), thermoplastic polyurethanes, nylons, cellulose acetate, cellulose butyrate, cellulose acetate butyrate, PET, EVA, polyvinylidene chloride (saran) and some ABS resins. Special grades of polyethylene and polypropylene films are now available which can be sealed using RF sealing.

Materials generally considered not compatible with this method of welding include PTFE, PC, acetal, PS and rigid materials with high melt temperatures. Standard grades of PP and PE are also not possible to join with this method. However, special grades are available that are RF weldable.

The most common application of RF welding and sealing machines is in factories. It is used in the fabric industries to seal polyurethane and PVC coated fabrics. The process is also commonly used in different kinds of industries where polymer films need sealing. Even sealing equipment and automation systems make use of RF welding. Some of the common products that are sealed using RF sealing include inflatable seat cushions, inflatable rafts, bladders of footballs, dust filters, automobile sun visors/tops/doors and sheet protectors.

2.1.4.4 Joint Design

Various types of joints can be produced by RF welding depending on the shape of the electrode, or welding rule, including plain seam welding, tear/seal welding and combined plain seam and tear/seal welding. In plain seam welding (Figure 2.14), a flat-ended welding rule is used. This type of weld is used to produce folds and can also be used to produce patterned welds, by machining the face of the rule. In tear/seal welding (Figure 2.15) a knife-edged welding rule is used to produce a very thin welded seam that can be torn after welding to enable the welded part to be separated from the surrounding material. For this application, a barrier material must be placed underneath the parts to be welded to prevent the welding rule from touching the lower platen, which would generate an arc and also blunt the rule. One of the problems with a tear/seal weld is that the integrity of the joint is not high due to the narrow weld width. Therefore, to produce a high strength weld that can be easily separated without a secondary cutting operation, electrodes

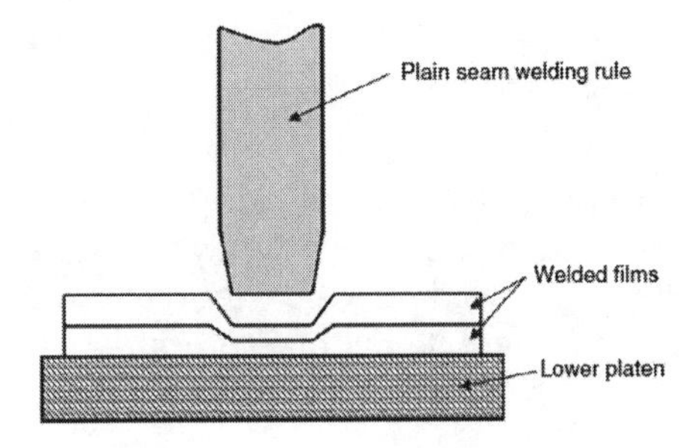

FIGURE 2.14
Plain seam welding.

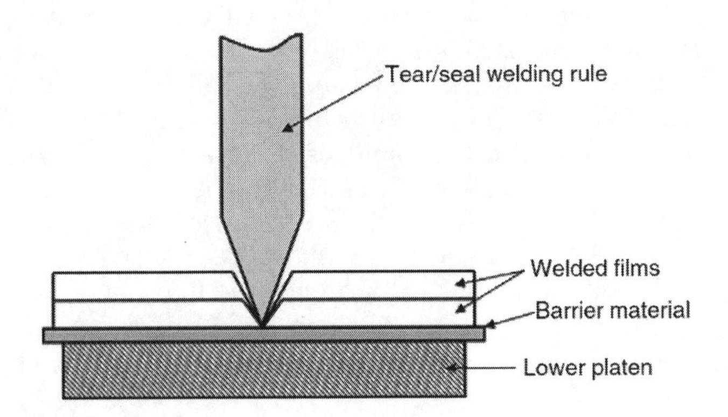

FIGURE 2.15
Tear/seal welding.

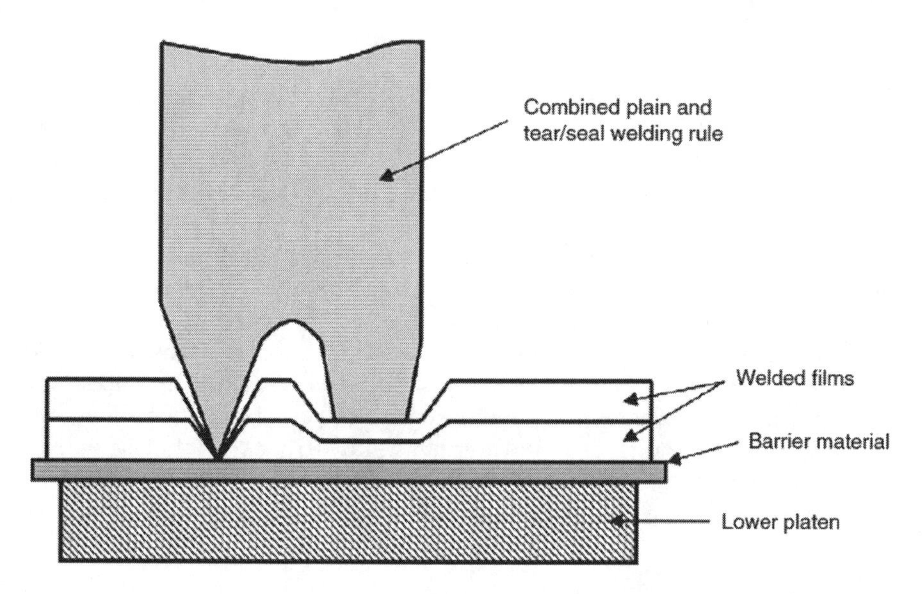

FIGURE 2.16
Combined plain and tear/seal welding.

can be designed to produce a tear seal next to the welded edge, by putting a knife edge a short distance from a flat-ended welding rule (Figure 2.16).

The main parameters in RF welding which decide the weld quality are power, heating time, cooling time, weld depth, pressure and platen temperature.

2.1.4.5 Heat Sealing

This technique uses an external heat source, either a heated bar, roller or plate. As the hot tool is pressed against the fabric layers, they are softened and a joint can be made. The heating is by conduction through one of the fabric layers. The process is rapid, taking a few seconds when a bar is used or at a rate of a few metres per minute when using rollers. The outer surface of the fabric is melted when this process is used for welding two similar layers together. However, it is more commonly used to fuse a hot melt adhesive film between two layers of fabric. The film layers can be melted before the fabric melts either due to use of polymer film with a lower melting point than the fabric or because of the insulating effect of the polymer layer.

The main advantages of using heat sealing over other joining processes are:

- Short weld times (typically a few seconds) or a few metres per minute
- Can be used for discreet parts or areas, or as a continuous process in lamination

- Simple technique; can be manually manipulated (heated tool or fabrics), and straightforward to put into automated production
- Flexible process applicable to films, fabrics and combinations of material and may be used for small or large parts
- Low cost of equipment and processing
- Low energy consumption, especially when impulse heating is used
- An impermeable seam can be achieved.

The main disadvantages of heat sealing are:

- It relies on conduction to heat the materials, therefore is not effective for thick or low thermal conductivity materials
- There can be heat marking on the outer surfaces
- There can be inconsistent heating or pressure at different parts of the roll, bar or plate
- There is potential to contaminate the outer surfaces of the materials if melt products stick to the heated tool surface.

2.1.4.6 Hot Press

A hot press consists of two metal plates, one or both of which are heated, which press and heat the fabrics to be joined. The bottom platen is sometimes covered with a sponge pad to aid even application of the pressure. With this type of equipment, pressure, temperature and time are accurately controlled.

A relatively high pressure is required, which is applied either pneumatically or hydraulically. It is typically used where two fabrics or a hem are sealed using a hot melt adhesive film between the two layers. The process can be applied to uncoated, coated or laminated fabrics to provide a sealed seam.

Double-sided heating plates can be used to improve the production rate or when joining thick materials. Single-sided heating can be used for thin material, or where one of the parts to be joined is heat sensitive.

2.1.4.7 Heated Belts or Bands

Heated belts can be used to provide a continuous sealing process in some thin material-joining applications. The belts are made of a thermally conducting material supported by driven rollers at either end. The belts move over fixed heating elements. They may be used single-sided with fabric moving with the belt, or as two moving belts with the fabric between them. They can form a stand-alone machine or part of a seaming machine with other forms of heating. In the latter case, the heated belt can provide part of the controlled cooling cycle, enabling consolidation and completion of the sealing whilst the parts are pressed together for a more extended period than can be achieved with a roller alone.

2.1.4.8 Heated Rollers/Laminator

Heated rollers may form part of a laminating machine typically with a smooth surface, a calendering machine where the surfaces are patterned to put a textured finish on a film or fabric, or as part of a product manufacturing machine where the rollers may be smooth or embossed to give a pattern to the joints in the product. The latter type of equipment has been proposed for making nappies from non-woven fabrics. The fabrics proposed contain a proportion of low melting point fibres which act as a fusing mechanism and can be melted without requiring the whole of the fabric to be melted. These are layered and positioned as required in the product and sealed in a pattern defined by the embossing on the rollers in a continuous process [57].

2.1.4.9 Continuous Fusing Press

The fabrics (potentially adhesive coated) and fusing interlayer films are laid up as a multiple-layer system. They enter a heating oven followed by a press to consolidate the seals and provide control over the cooling cycle. This type of equipment is typically used in clothing manufacturing, where the fusing press can be used to bond an interlining to an outer textile fabric by means of an adhesive, often previously applied to the interlining and which melts under certain conditions of temperature, time and pressure. They are also used for applying foam layers and for leather backing.

Fusible interlinings are textiles coated with an adhesive. The textile structure can be woven, knitted or non-woven. The resin is usually a polymeric material, which melts and flows within a defined temperature range. The adhesive can be applied to the textile in a variety of ways, e.g., scatter coating, dot printing, paste printing, lamination, melt coating, etc. Temperature, pressure and time are infinitely variable within the working ranges. Short pressing times give a relatively gentle process. Fusing can be achieved with a significantly lower temperature of the top fabric, which helps to avoid heat stress. Pressure and temperature are two critical parameters of the fusing process. The touchscreen control allows the adjustment and control of all required parameters.

2.1.4.10 Hot Air Welding

In the case of using a hot air stream, thermostatically controlled, the fabrics are heated without direct contact being made with the hot surfaces. The

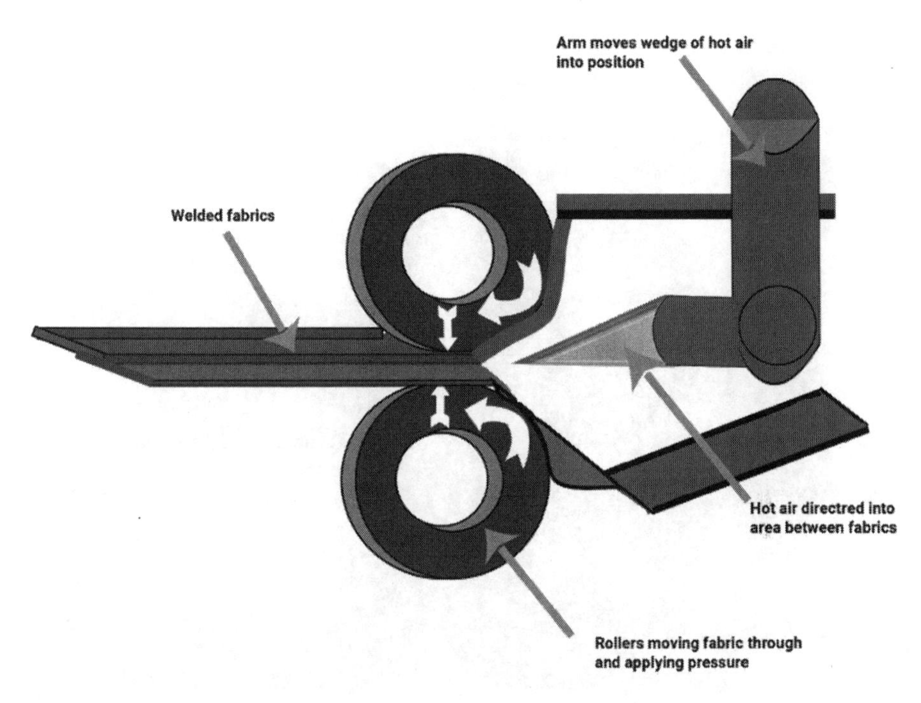

FIGURE 2.17
Hot air wedge welding equipment.

process will operate at a few m/min and can produce strong joints that do not melt the outer surfaces of the fabrics.

This equipment uses a shaped or perforated nozzle to deliver hot air to heat one or both of the fabric surfaces before they are pressed between driven rollers (Figure 2.17) to apply pressure and complete the weld between the two surfaces. The hot air method of welding makes use of a device consisting of a resistance heater, a blower and a temperature controller to blow hot air between the fabrics. The hot air wedges and nozzles are typically 10–40 mm wide. The same equipment can often be used for applying sealing tape to a stitched seam. A tape feed is added to deliver tape into the rollers over the previously stitched seam. The hot air stream is directed to heat the lower surface of the tape before it enters the rollers (Figure 2.18). Tapes in the range of 10–26 mm wide can typically be processed. The air nozzle and rollers are specified to be at least 3 mm wider than the tape to ensure effective sealing. The lower roller surface may be grooved for seam taping to allow the stitching thickness to be accommodated and present the flattest possible surface to the tape.

2.1.4.11 Hot Wedge Welding

Hot wedge welding uses an electrically heated wedge, which contains one or more cartridge heaters. The temperature is controlled by closed-loop

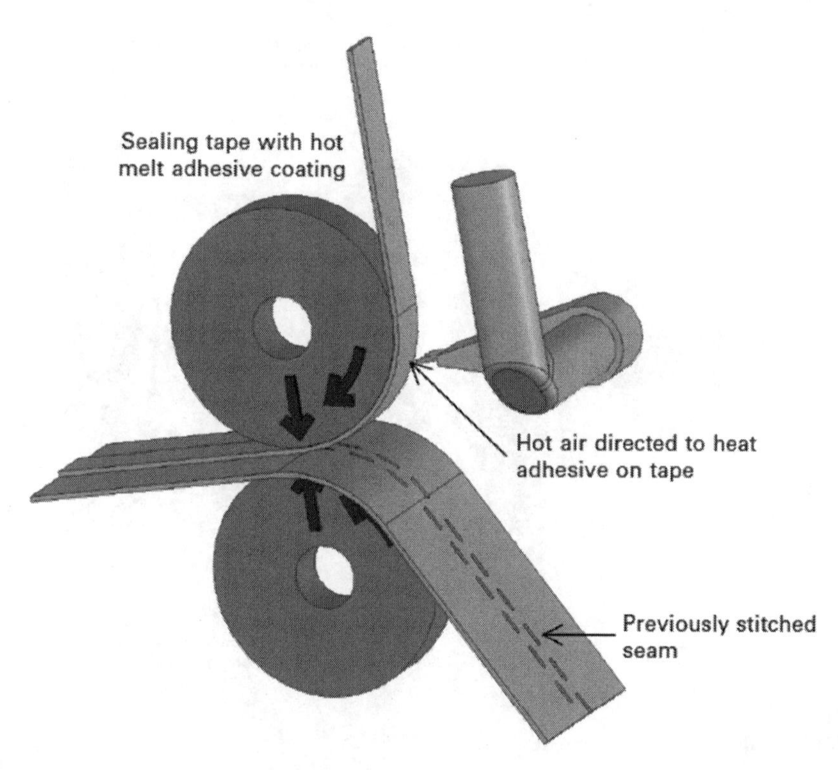

FIGURE 2.18
Hot air seam taping and sealing equipment.

feedback using a thermocouple positioned in the wedge. This maintains the wedge temperature typically to within ±10°C of the set temperature during welding. The principle of electrically heated wedge welding is that both fabric surfaces to be welded pass over, and are forced into intimate contact with, the hot surface of the heated wedge. In some systems, for heavy-duty applications, there are additional wedge rollers to hold the fabrics against the heated wedge. The drive rollers or consolidating rollers pull the fabric through the machine and press the heated surfaces together (Figure 2.19). The hot wedge is typically 7–50 mm wide for bench top machines and 25–75 mm wide for floor or ground-based machines.

Seams can be made as a single or double track. Double-track welding requires a heated wedge with a cartridge heater at each edge and a gap in the centre.

The main advantages of using hot air/wedge welding over other joining processes are:

- Applicable to thin and thick materials (including foams)
- Heat is applied directly at the joint line, enabling high-speed, energy-efficient joints to be made without affecting the outer surfaces of the fabrics

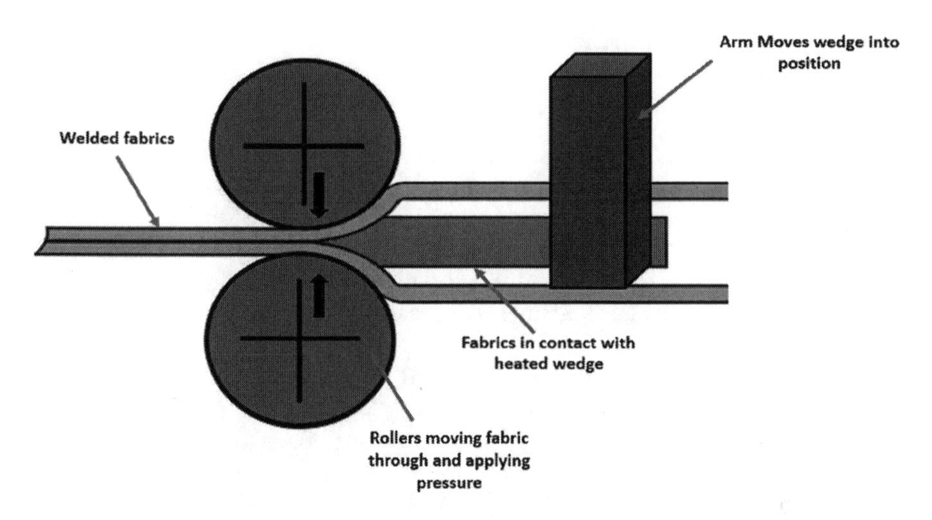

FIGURE 2.19
Hot wedge welding equipment.

- Sealed joints can be tested non-destructively when a dual-track machine is used
- Consistent performance once process is stabilized
- Relatively low cost (consumables such as needles, solvents, or adhesives are not required)
- An impermeable seam can be achieved
- A fusible film can be melted at the joint over a previously stitched seam to provide sealing
- Edges can be sealed and there are no stitch holes
- The air jet blows contaminants away from the joint
- The simple overlap seam provides an alternative to stitching that is lower in weight and less bulky. With suitable film adhesive it can be a soft, stretchy, flexible joint.

The main disadvantages of hot air/wedge welding are:

- The wedge or air temperature cannot be altered quickly, so care is needed when changing process speed to go round more complex parts of a seam to avoid melting through the fabrics
- Pre-placement of the fabrics with the joint in contact before welding is not possible
- The seam must be open at the end so that the wedge can be removed from between the fabrics; a heated press is sometimes used to close seams
- Seam designs are restricted to overlap, peel and taped.

2.1.4.12 Seam Design/Preparation

The seam designs used in wedge welding are restricted due to the need to insert the wedge between the fabric layers during welding. Most commonly, either an overlap seam or a taped seam is used, but it is also possible to use a peel seam or double tape seam for extra strength (Figure 2.20). The equipment may be fitted with guides to control the size of the overlap at the joint or the position of fabrics during taped seaming. Guides may also be fitted to fold the fabrics as they enter the welding zone, allowing hemming and other more complex features to be carried out (hem with rope, hem with pocket, welds with channels, etc.). Accessories for tube welding for sleeves or other applications are available. Taped seams can also be made without previously stitching the two pieces together, with precise position of the two fabric pieces and the overlying tape. Particularly when welding coated fabrics and plastic and when using adhesive films or tapes, it is important that the area of the seam to be welded is clean and free from dirt and dust particles. This includes the parts of the welding machine that come into contact with the fabric or the adhesive. Sometimes it may be necessary to use a water-based detergent solution to clean the materials. When using the hot air wedge, the air stream is effective in removing loose dust from the joint region. The gap at the rollers and between the wedge and the rollers is set to correctly accommodate the thickness of fabric being welded. If only one side of the joint is to be heated, such as in bonding with a pre-coated adhesive, the hot air stream is positioned accordingly (Figure 2.18).

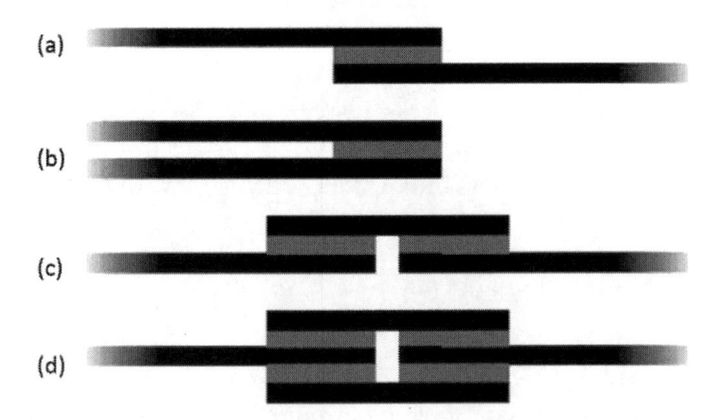

FIGURE 2.20
Some seam designs suitable for wedge welding. Various hemming types are also possible such as hems with a channel and hems with rope as well as plain hems: (a) ASTM D6193 designation LSa – Simple lap seam; (b) ASTM D6193 designation SSa – Simple peel seam; (c) ASTM D6193 designation LSp – Taped seam. This may also be over a previously stitched seam. (d) No ASTM D6193 designation – Double taped seam carried out in two stages.

Prior to welding, it is important to check that the heat source is centred in line with the rollers. The position of the hot air stream in relation to the fabric and the tape for seam taping is important. It should be positioned to heat the tape preferentially to allow the adhesive to be melted and then pressed into the fabric surface structure, and not to damage the sensitive backings of multi-layer waterproof fabrics.

2.1.4.13 Wedge or Air Temperature

The temperature of the air stream or solid wedge should be set to melt the material at the joint line. This is either the base fabric, plastic coating or an adhesive film at the joint. The temperature also controls the viscosity of the melt. Higher temperatures lead to greater flow of the melt into the surrounding fabric. Overheating causes excess melt flow and leads to holes melted through the fabric layers, or degradation or carburizing of the materials, severely reducing the joint strength. The temperature is measured using a sensor in the wedge or in the air stream. The temperature set here needs to be significantly higher (100–250°C) than the material melting point. Hot air systems typically allow operation at up to 600–750°C, hot wedge at up to 500°C. It will generally be necessary to carry out tests at a range of temperatures until the required joint performance is achieved. For hot air processing, the air flow rate and/or pressure must also be controlled to ensure suitable transfer of heat. It should also be noted that the ambient temperature of the working zone (such as working outdoors), or effects of the sun will affect the optimal welding parameters. If the ambient conditions are not controlled, then it is advisable to carry out test welds and assess the weld strength in advance of production.

2.1.4.14 Clamping Pressure

The pressure at the rollers is controlled by a pneumatic actuator that can typically be controlled by setting the air pressure in the range of 0–7 bar. The pressure and gap setting are varied depending on the thickness and resilience of the materials to be welded. Thin, tightly structured fabrics are processed with higher pressure than more open thicker materials. Very low pressures are used for foams.

2.1.4.15 Roller Gap

It is usually possible to set the roller gap independently of the roller pressure, otherwise called the depth stop. It should be set to less than the combined thickness of the materials and can be set to zero allowing the weld consolidation to be fully controlled by the pressure setting. The gap is opened to work on foam-backed material such as neoprene to prevent stretching.

2.1.4.16 Process Speed

The process speed is decided by the speed of drive rollers, which is generally set in the range 0–20 m/min. Typically, 0.7–4.0 m/min is used for thick materials and faster for thinner fabrics. If the speed is too low, or reduces as the operator is negotiating a corner in the product, there is a danger of melting through the fabrics or overheating the materials. Some equipment has automated adjustment of the heating power to compensate for changes in the processing speed. If the speed is too high, the fabric or adhesive materials will be insufficiently softened, resulting in a low strength joint. Some welding equipment allow differential adjustment of the welding speed between the upper and lower rollers. This helps to control puckering at the seam.

After welding is completed, in the case of coated fabrics or a plastic sheet, it is common to see a small melt bead at the weld edge. This is a reliable indication that proper seaming procedures have been used. Excessive melt beads or wrinkling of the sheets suggests that excessive heating or pressure may have been applied. When double-track welding is used, in geomembrane welding for example, the unwelded centre channel can be non-destructively tested to ensure that the welds are sealed. It is also possible to cut mechanical test specimens from the surplus at the ends of welds.

2.1.4.17 Ultrasonic Welding

Ultrasonic welding is a process that uses mechanical vibrations to soften or melt a thermoplastic material at the joint line. The fabrics to be joined are held together under pressure and subjected to ultrasonic vibrations, usually at a frequency of 20–40 kHz. The mechanical energy is converted to thermal energy due to intermolecular and surface friction. The mechanisms responsible for generating heat at the joint line are not fully understood. However, Benatar and Gutowski [58] have summarized the key steps as:

- The mechanics and vibration of the parts
- The viscoelastic heating of the thermoplastics
- Heat transfer
- Flow and wetting
- Intermolecular diffusion

Studies suggest that the ultrasonic bonding mechanism is a physical process and not a chemical process, i.e. when the process is carried out within the normal operating range, no new chemical groups or bonds are formed. Other studies of the mechanisms of ultrasonic welding include Tolunay [59], Shi and Little [60] and Mao and Goswami [61]. The ultrasonic process can be used for seaming, cutting, slitting, trimming, tacking, embossing, quilting or simultaneously cutting and sealing (Figure 2.21).

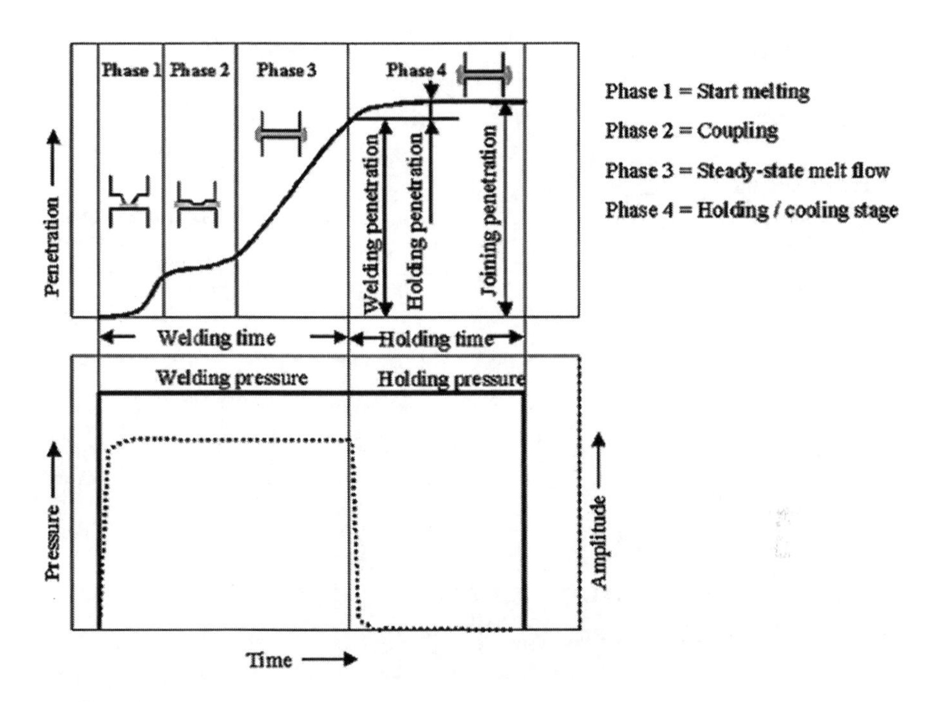

FIGURE 2.21
Process diagram of the ultrasonic welding process using an energy director.

The ability to ultrasonically weld textiles depends on their thermoplastics content and the desired end results. As a minimum, the material must have uniform thickness and thermoplastic content of 65%. Yarn density, tightness of weave, elasticity of material and style of knit are all factors which can influence the weld ability.

Benefits of the process include energy efficiency, high productivity with low costs and ease of automated assembly line production. The main limitation of the process is that the maximum component length that can be welded by a single horn is approximately 250 mm. This is due to limitation in the power output capability of a single transducer, the inability of the horns to transmit very high power, and amplitude control difficulties due to the fact that joints of this length are comparable to the wavelength of the ultrasound (Figure 2.22).

2.1.4.18 Vibration Welding

Vibration welding uses heat generated by friction at the interface of two materials to produce melting in the interfacial area. The molten materials flow together under pressure, forming a weld upon cooling. Vibration welding can be accomplished in a short time (1–10 seconds cycle time) and is

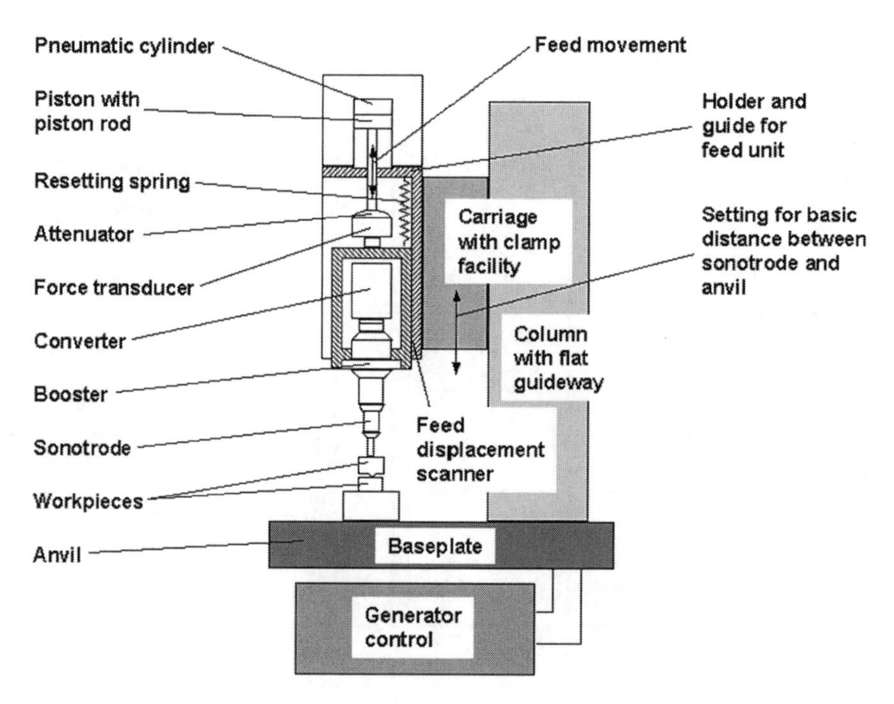

FIGURE 2.22
Basic construction of a pneumatic ultrasonic welding machine.

applicable to a variety of thermoplastic parts with planar or slightly curved surfaces.

The two main types of vibration welding are: linear, in which friction is generated by a linear, reciprocating motion (Figure 2.23); and orbital, in which the upper part to be joined is vibrated in a circular motion. Linear vibration welding is most commonly used, but orbital vibration welding makes the welding of irregularly shaped plastic parts possible.

Process parameters are the amplitude and frequency of vibration, weld pressure and weld time, all of which affect the strength of the resulting weld.

The welding process consists of four phases (Figure 2.24). In Phase I, the heat generated through friction raises the temperature of the interfacial area to the glass transition temperature of amorphous thermoplastics or the melting point of semi-crystalline plastics. Since the material is still in the solid state, there is no displacement (penetration) – the distance through which the parts approach each other during welding due to lateral flow of molten material – in this phase. In Phase II, material at the interface begins to melt and flow in a lateral direction, causing an increase in weld displacement. In Phase III, the rate of melt generation equals the rate of melt displacement, which therefore increases linearly with time. At the end of Phase III,

FIGURE 2.23
Linear vibration welding.

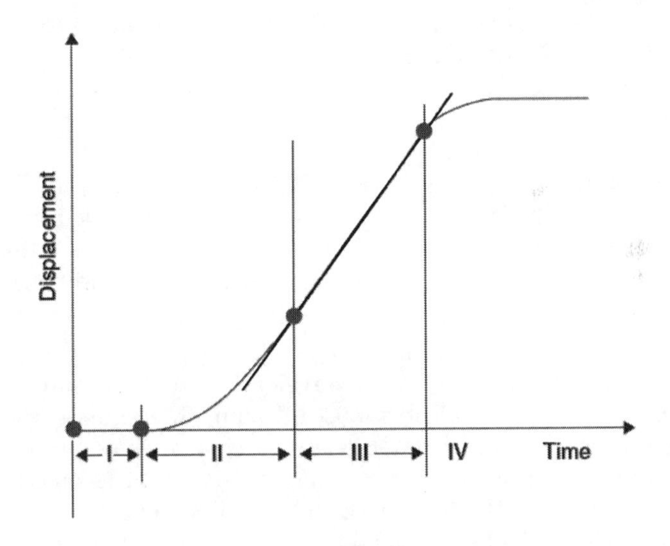

FIGURE 2.24
Displacement versus time curve showing the four phases of vibration welding.

the vibratory motion is stopped, and during Phase IV, the weld penetration increases slightly as the molten film solidifies under pressure.

Advantages of vibration welding include relatively short cycle times, energy efficiency, capability of welding large parts, and insensitivity to surface preparation. Because of the smearing action due to friction, welds can be obtained on surfaces that have been vacuum metalized, painted or contaminated. No additional materials are introduced as in the implant welding techniques or adhesive bonding, so the weld interface is composed of the same material as the welded parts [62].

A drawback of vibration welding is the initial high capital cost of the equipment and tooling compared to other processes such as hot plate or ultrasonic welding. However, this must be judged against the ability to weld larger parts at one go, with a faster processing cycle.

Vibration welding is commonly used on large parts, although smaller parts can be welded economically in multiple cavity tooling. Typical part sizes range from 3 × 3 inches (76.2 × 76.2 mm) to 24 × 60 inches (61.0 × 152.4 cm). The technique is used when strong, leak-proof pressure or vacuum joints are necessary.

Almost any thermoplastic can be vibration welded: crystalline, amorphous, filled, foamed, and reinforced. The only polymers that can be difficult to weld are the fluoro polymers, due to their low coefficient of friction

2.1.4.19 Laser Welding

Laser welding uses electromagnetic radiation directly for creating the weld. The methods for laser welding of fabrics fall into two distinct groups defined by the type and, more specifically, the wavelength of the laser being used for the process:

- Direct welding using CO_2 lasers (10,600 nm wavelength). Welding occurs when the laser beam is absorbed directly at the interface by the top surface of the fabric. The laser is often directed at the nip between two rollers so that heating and clamping can be achieved almost simultaneously.
- Transmission laser welding using Nd: YAG, diode or fibre lasers that, unlike CO_2 lasers, have wavelengths that transmit through unpigmented thermoplastics (400–1500 nm). To use these lasers successfully for joining fabrics, the laser energy must be properly harnessed to create the heating and intermolecular diffusion critical for weld formation. The location of the weld is controlled by placing an absorber of the laser radiation at the joint interface. The absorber heats to generate melting locally.

2.1.4.19.1 Direct Laser Welding

In direct laser welding (Figure 2.25), the materials are heated from the outer surface, and generally an additional absorber applied to the fabrics is not required. A laser source is chosen that has a wavelength that is absorbed by the materials. Laser sources from 0.8 to 10.6 mm wavelength have been used. At 10.6 mm (CO_2 laser), radiation is strongly absorbed by plastic and fabric

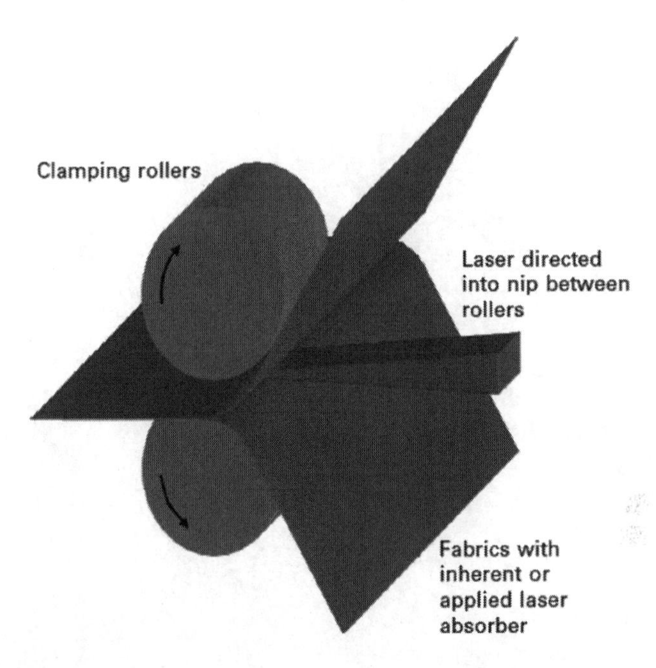

FIGURE 2.25
Direct laser welding.

surfaces independently of any colouration or additive, allowing high-speed joints to be made. If diode or fibre lasers in the 0.8–1.1 mm wavelength range are used, then the fabrics must either have a dark carbon-based pigmentation or an absorptive coating is applied as used in transmission laser welding. A diagram of direct laser welding is shown in Figure 2.25.

2.1.4.19.2 Through-Transmission Laser Welding (TTLW)

Through-transmission laser welding of fabrics involves the use of laser energy to produce the heat necessary to generate a weld. Traditionally, this was accomplished by using an additive in one of the substrates to absorb the laser energy. Most thermoplastic fabrics and some natural fabrics transmit near infrared laser energy. Therefore, one substrate must remain in its natural state, while the second substrate is altered, typically by adding carbon black, in order to make it absorbent to the laser energy. The laser beam transmits through the top surface until it reaches the bottom substrate where it is absorbed. Under favourable welding conditions, the heating conducts into the top substrate, allowing the two materials to melt and flow together, forming a weld. The main advantages of the laser welding process are that it does not involve vibration, is non-contact, does not impart mechanical damage and is a clean process. The main disadvantage is the use of conventional laser absorbing materials such as carbon black, as it virtually rules out using

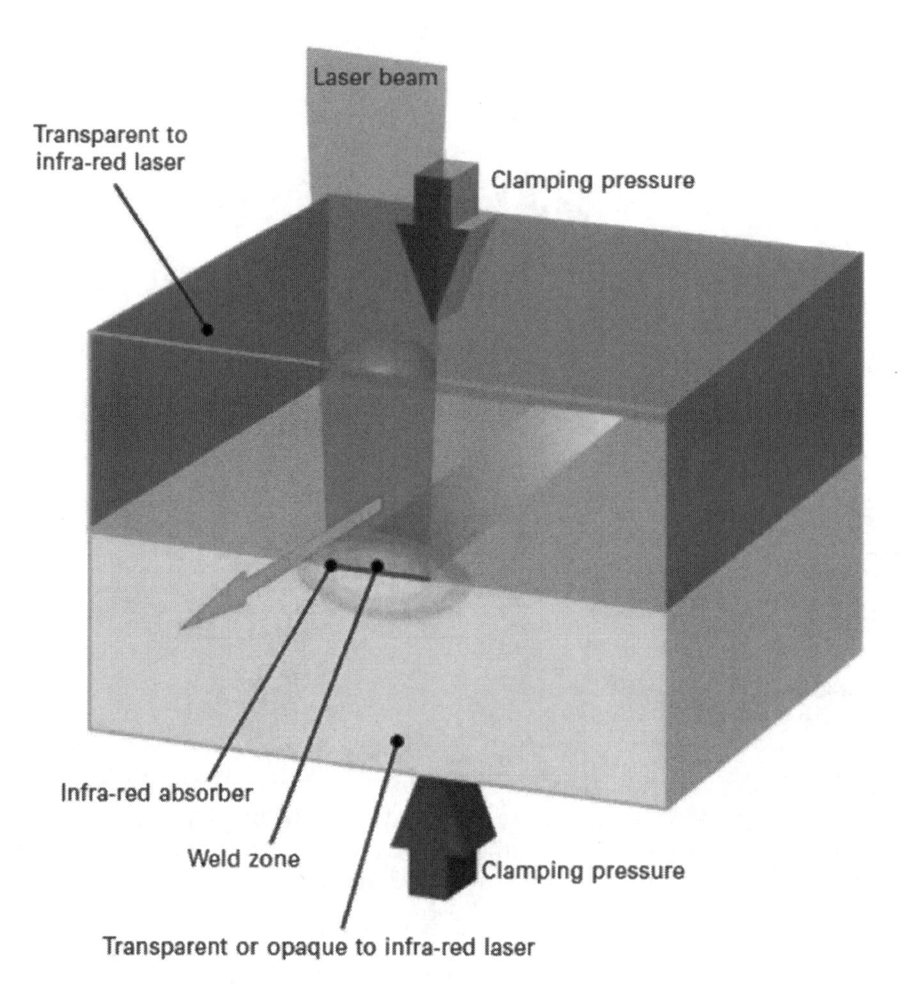

FIGURE 2.26
Transmission laser welding.

this process for any applications in which aesthetics and lack of colour are important. A diagram of transmission laser welding is shown in Figure 2.26.

2.1.4.20 Colour Independence

A significant process advancement to through-transmission laser welding has been the development of an innovative process called the Clearweld process, which offers all the advantages of conventional laser welding without the use of opaque materials or addition of unwanted colour [63, 64]. The process uses near-infrared (NIR) absorbing material systems to convert laser energy into heat. A thin layer of these materials applied at the interface of two pieces of fabrics to be joined absorbs light, acting as a focal point for the laser.

Localized heating of the substrates occurs at the joint interface, producing clean, optically clear joints with no particulates and visible colour. The Clearweld process was invented by TWI Ltd; a UK-based industrial research and development organization, and has been developed for commercial use by Gentex Corporation [65]. Clearweld consists of a series of materials capable of powerful absorption in the near infrared spectrum, while remaining virtually colourless.

2.1.4.21 Induction Welding

Induction welding, also called electromagnetic or EMA welding, uses induction heating from radio frequency (typically 2–10 MHz) alternating current to magnetically excite an implant placed at the joint interface of the two parts being welded. This implant, or gasket, is normally a composite of the polymer to be welded with either metal fibres or ferromagnetic particles. The heat generated melts and fuses the implant with the surrounding material. It is a reliable and rapid technique, ranging from fractions of a second for small parts to 30–60 seconds for parts with long (400 cm; 157 inches) joint lines, and results in structural, hermetic, or high-pressure welds [66, 67].

The two most commonly encountered mechanisms by which heat can be generated by an induction field are eddy current heating and heating due to hysteresis losses. In eddy current heating, a copper induction coil (work coil), which is connected to a high-frequency power supply, is placed in close proximity to the joint (Figure 2.27). As electric current at a high frequency

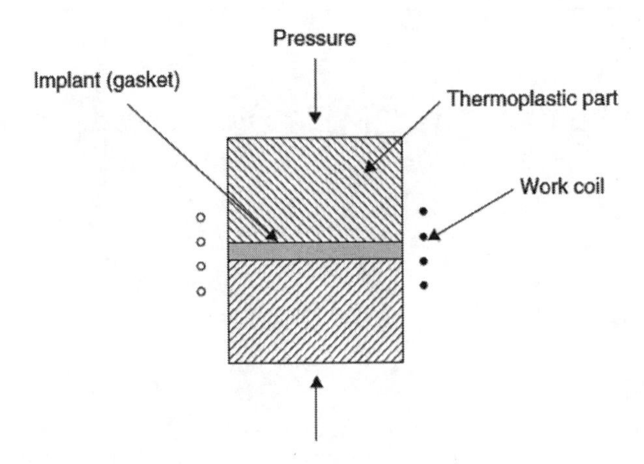

FIGURE 2.27
Set-up for induction welding.

passes through the work coil, a dynamic magnetic field is generated whose flux links the implant. Electric currents are induced in the implant, and when these are sufficiently high to heat the conducting material, the surrounding thermoplastic parts soften and melt. If pressure is applied to the joint, this aids wetting of the molten thermoplastics, and a weld forms as the joint cools [68, 69].

Ferromagnetic materials (such as stainless steel and iron) also heat up in a dynamic magnetic field due to hysteresis losses. As the magnetic field (B) increases, the ferromagnetic material becomes magnetized, and the magnetic intensity (H) increases. As the magnetic field decreases, so does the magnetic intensity. However, there is a lag between the two, which results in a hysteresis loop (Figure 2.28). This phenomenon results in dissipation of energy in the form of heat. The induction welding process can be divided into four main steps (Figure 2.29):

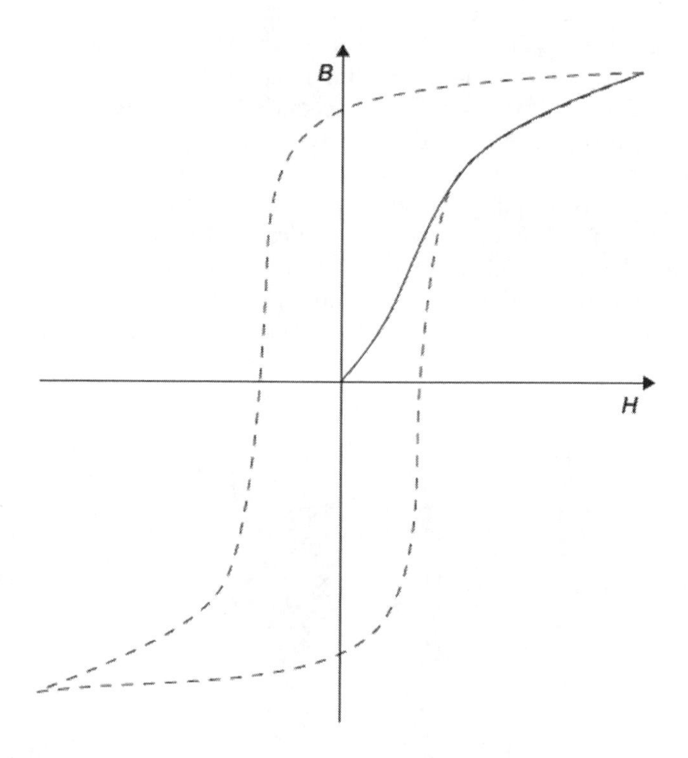

FIGURE 2.28
Graph of the magnetic field (*B*) in a ferromagnetic material and the magnetic intensity (*H*) for an alternating magnetic field. The magnetization curve is the full line and the hysteresis curve is the broken line. The area enclosed by the hysteresis loop is equal to the energy dissipated in one circuit.

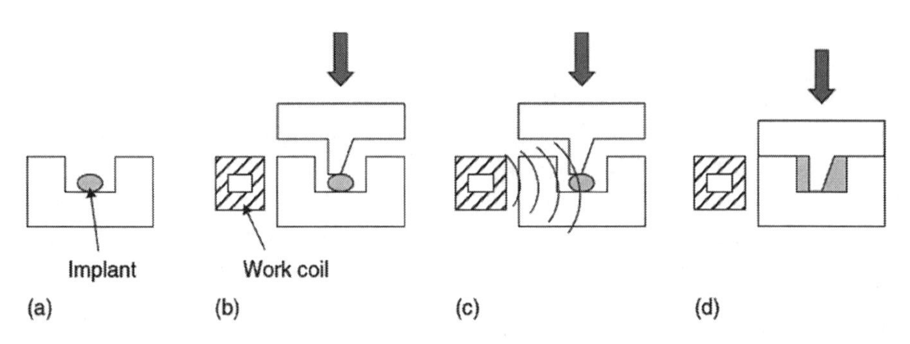

FIGURE 2.29
The induction welding process: (a) placement of the implant, (b) application of pressure, (c) induction heating and (d) cooling under pressure.

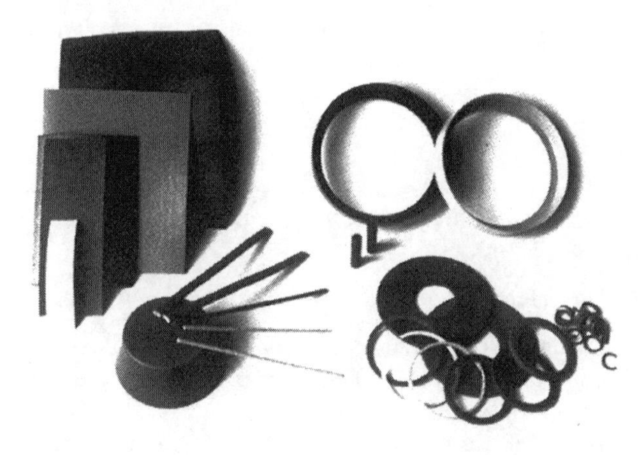

FIGURE 2.30
Various forms of induction welding implants.

Step 1: ***Placement of the Implant***
Implants are available in many forms, such as sheets, extruded profiles, injection-moulded parts, tapes and strands (Figure 2.30), depending on the size and shape of the parts to be welded, and on the position of the work coil. They can be positioned at the joint line, either by hand or by using an automated system [70].

Step 2: ***Application of Pressure***
Pressure can be applied to the parts by placing them in fixtures attached to a pneumatic cylinder, or the work coil may be embedded in a PTFE or ceramic block, which applies the pressure.

Step 3: ***Induction Heating***
Power is applied to the work coil, creating the electromagnetic field that heats the implant, which in turn heats and melts the surrounding

thermoplastic by thermal conduction. Electromagnetic fields become exponentially weaker as the distance from the work coil increases, so that joints placed as close as possible to the coil maximize the heating of the implant. During heating, the implant flows to fill the gap between the parts.

Step 4: ***Cooling and Removal of the Parts***

After a pre-set time, the power is switched off and the parts are allowed to cool under pressure for a pre-set time. The welded assembly is then removed and the cycle is repeated for creating other joints.

Weld strength is proportional to the surface area of contact; however, it is important that the molten flow be contained within the joint area. The amount of molten material needed to fill the joint cavity can be calculated using the cross-sectional area of the implant (A_E), the cross-sectional area of the cavity between the parts being joined (A_V) and a constant (k) with values ranging from 1.02 to 1.05, depending on the pressure used in welding and the material being welded using the following relation:

$$A_E = kA_V$$

The constant (k) allows for overfilling the cavity and ensures that any surface irregularities are smoothed out during welding [67, 68, 71].

2.1.4.22 Hot Plate (Heated Tool or Thermal) Welding

Heated tool welding, also known as hot plate, mirror, platen, butt or butt fusion welding, is a widely used technique for joining injection moulded components or extruded profiles.

The process uses a heated metal plate, known as the hot tool, hot plate, or heating platen, to heat and melt the interface surfaces of the thermoplastic parts. Once the interfaces are sufficiently melted or softened, the hot plate is removed and the components are brought together under pressure to form the weld. An axial load is applied to the components during both the heating and the joining/cooling phases of the welding process.

Welding can be performed in either of two ways: welding by pressure, or welding by distance. Both processes consist of four phases, shown in the pressure versus time diagram in Figure 2.31.

In welding by pressure, the parts are brought into contact with the hot tool in Phase I, and a relatively high pressure is used to ensure complete matching of the part and tool surfaces. Heat is transferred from the hot tool to the parts by conduction, resulting in a local temperature increase over time. When the melting temperature of the plastic is reached, molten material begins to flow. This melting removes surface imperfections, warps, and sinks at the joint interface and produces a smooth edge. In Phase II, the melt pressure is reduced, allowing further heat to soak into the material and the

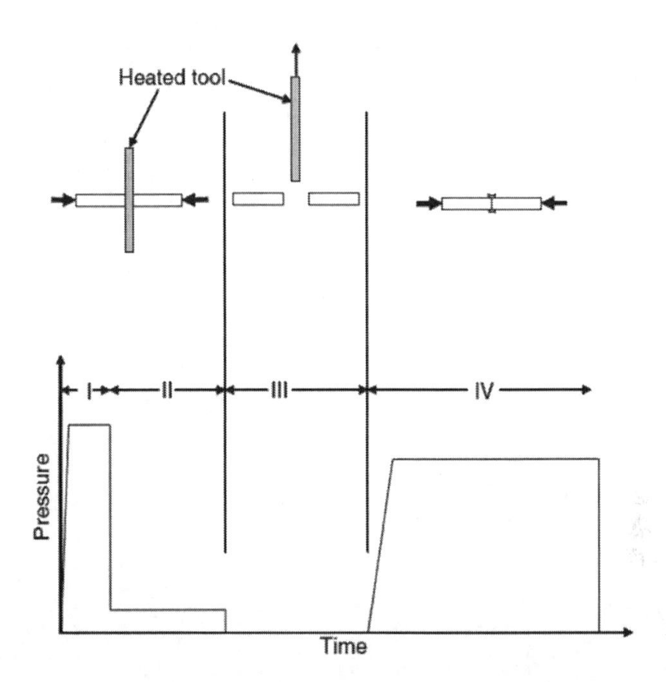

FIGURE 2.31
Pressure vs. time curve showing the four phases of heated tool welding [72].

molten layer to thicken; the rate at which the thickness increases is determined by the heat conduction through the molten layer. Thickness increases with heating time – the time that the part is in contact with the hot tool.

When a sufficient melt thickness has been achieved, the part and hot tool are separated. This is Phase III, the changeover phase, in which the pressure and surface temperature drop as the tool is removed. The duration of this phase should be as short as possible (ideally, less than 3 seconds) to prevent premature cooling of the molten material. A thin, solid "skin" may form on the joint interface if the changeover time is too long, affecting the weld quality.

In Phase IV, the parts are joined under pressure, causing the molten material to flow outward laterally while cooling and solidifying. Intermolecular diffusion during this phase creates polymeric chain entanglements that determine joint strength. Because the final molecular structure and any residual stresses are formed during cooling, it is important to maintain pressure throughout the cooling phase in order to prevent warping. Joint microstructure, which affects the chemical resistance and mechanical properties of the joint, develops during this phase [73, 74].

Welding by pressure requires equipment in which the applied pressure can be accurately controlled. A drawback of this technique is that the final part dimensions cannot be controlled directly; variations in the melt thickness and sensitivity of the melt viscosities of thermoplastics to small temperature changes can result in unacceptable variations in part dimensions.

In welding by distance, also called displacement-controlled welding, the process described earlier is modified by using rigid mechanical stops to control the welding process and the part dimensions. Figure 2.32 shows the process steps.

In Step 1, the parts are aligned in holding fixtures; tooling and melt stops are set at specified distances on the holding fixture and hot tool (heating platen), respectively.

The hot tool is inserted between the parts in Step 2, and the parts are pressed against it in Step 3. Phase I, as described for welding by pressure, then takes place. The material melts and flows out of the joint interface, decreasing part

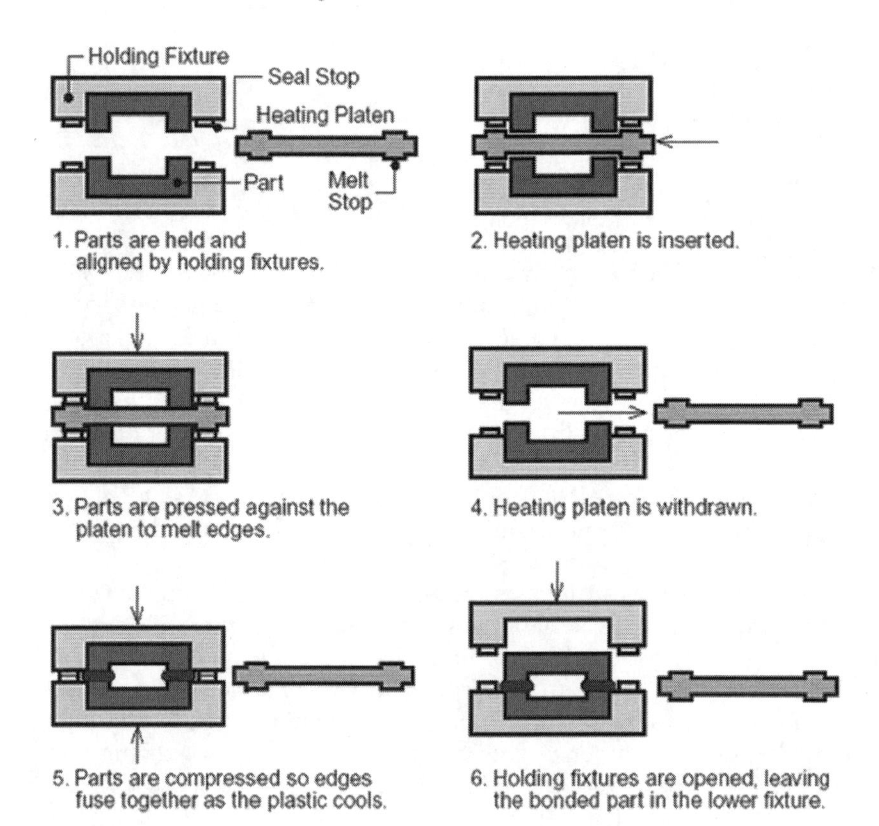

FIGURE 2.32
The heated tool (welding by distance) process [75].

length until, in this case, the melt stops meet the tooling stops. Melt thickness then increases (Phase II) until the hot plate is removed in Step 4, the change-over phase (Phase III). The parts are then pressed together in Step 5 (Phase IV), forming a weld as the plastic cools; tooling stops inhibit melt flow. The welded part is then removed in Step 6.

Heated tool welding is suitable for almost any thermoplastic but is most often used for softer, semi-crystalline thermoplastics such as PP and PE. It is usually not suitable for nylon or high molecular weight materials. The temperature of the molten film can be controlled by regulating the hot tool temperature so that plastics that undergo degradation at temperatures only slightly above the melting temperature can be welded.

2.1.4.23 Seaming/Sealing Technology Used in HAA

The Korea Aerospace Research Institute (KARI) has designed and fabricated a highly efficient high-performance seaming machine for huge Airship Hull construction (Figure 2.33)

NASA has used Gore sealing technology for development of their recently developed high-altitude balloon for a mission to Venus [76] (Figure 2.34).

These sealing/seaming technologies used for high-altitude airships are mostly guarded technologies, and very little information is available regarding them. However, no organization has disclosed the technology and technical details.

FIGURE 2.33
High performance seaming machine.

FIGURE 2.34
Gore sealing technology [76].

References

1. Zhai, H., & Euler, A. 2005. *Material challenges for lighter-than-air systems in high altitude applications.* In *AIAA 5th ATIO and 16th Lighter-Than-Air Sys Tech. and Balloon Systems Conferences* (p. 7488).
2. d'Oliveira, F. A. 2016. High-altitude platforms – Present situation and technology trends. *Journal of Aerospace Technology and Management* 8: 249–262.
3. Liao, L. 2009. A review of airship structural research and development. *Progress in Aerospace Sciences* 45: 83–96.
4. Kim, D. M. 2003. *Korea stratospheric airship program and current results.* In *AIAA's 3rd Annual Aviation Technology, Integration, and Operations (ATIO) Forum* (p. 6782).
5. http://en.wikipedia.org/wiki/Airship, accessed on March 18, 2009.
6. McDaniels, K., & Downs, R. J. 2009. *High strength-to-weight ratio non-woven technical fabrics for aerospace applications.* In *AIAA Balloon Systems Conference* (p. 2802).
7. Ian, H. 2003.*Coating and Lamination Enhance Textile Performance.* Technical Textiles International, UK: International Newsletters LTD.
8. Mondal, S. 2008. Phase change materials for smart textiles – An overview. *Applied Thermal Engineering*, 28: 1536–1550.
9. Giraud, S., Bourbigot, S., & Rochery, M. 2005. Flame retarded polyurea with microencapsulated ammonium phosphate for textile coating. *Polymer Degradation and Stability* 88: 106–113.
10. Herrera, A. 1991. Versatility in Coating Operations—Knife Coatings. *Journal of Coated Fabrics* 20: 289–301.
11. Woodruff, F. A. 1992. Environmentally Friendly Coating and Laminating: New Processes and Techniques. *Journal of Coated Fabrics* 21: 240–259.
12. Fung, W., & Hardcastle, J. M. 2000. *Textiles in Automotive Engineering.* Elsevier.

13. Scott, R. A. 1995. Coated and laminated fabrics. In *Chemistry of the Textiles Industry* (pp. 210–248). Springer, Dordrecht.
14. Schindler, W. D., & Hauser, P. J. 2004. Finishing with enzymes: bio-finishes for cellulose. In *Chemical Finishing of Textiles* (pp. 181–188). Cambridge, England: Woodhead Publishing Ltd.
15. Alonso, S., Bertrand, F., & Tanguy, P. A. 2003. A torque-based analysis of the reverse roll coating process. *Chemical Engineering Science* 58: 1831–1837.
16. Greer, R. 1995. The reverse roll coater. *Journal of Coated Fabrics* 24: 287–297.
17. Abbott, N. J. 1972. A study of tearing in coated cotton fabrics: Part III: The influence of coating application techniques. *Journal of Coated Fibrous Materials* 1: 130–149.
18. Glawe, A., Reuscher, R., & Koppe, R. 2003. Hot-melt application for functional compounds on technical textiles. *Journal of Industrial Textiles* 33: 85–92.
19. Wright, E. L. 1981. Low wet pick-up finishing: A review of commercial experience. *Textile Research Journal* 51: 251–255.
20. Grant, R. R. 1978. Coating techniques for the manufacture of continuous web products. *Journal of Coated Fabrics* 8: 171–182.
21. Hannachi, A., & Mitsoulis, E. 1990. Numerical analysis of multilayer forward roll coating. *Journal of Plastic Film and Sheeting* 6: 170–190.
22. Grant, R. R. 1981. Coating and laminating applied to new product development. *Journal of Coated Fabrics* 10: 232–253.
23. Zickler, D. 1978. Coating industrial fabrics by means of the melt roll process. *Journal of Coated Fabrics* 8: 121–143.
24. Hewson, R. W., & Gaskell, P. H. 2010. Tri-helical gravure roll coating. *Chemical Engineering Science*, 65: 1311–1321.
25. Robinson, J., & Marrick, J. 2006. Printing techniques for enhancing nonwovens. *Nonwovens World* 15: 36–44.
26. Stukenbrock, K.H. 2003. Chapter 8 Chemical finishing, in Albrecht, W., Fuchs, H., Kittelmann, W., & Lünenschloss J., *Nonwoven Fabrics* (pp. 435–454). Weinheim, Wiley-VCH.
27. Licari, J. J. 2003. *Coating Materials for Electronic Applications: Polymers, Processing, Reliability, Testing*. William Andrew.
28. Goossens, A. 2001. Rotary screen coating. in Satas, D., & Tracton, A. A., *Coatings Technology Handbook* (pp. 179–183). New York: Marcel Dekker.
29. Matthews, A., & Eskildsen, S. S. 1994. Engineering applications for diamond-like carbon. *Diamond and Related Materials* 3: 902–911.
30. Fung, W. 2002. *Coated and Laminated Textiles* (Vol. 23). Woodhead Publishing.
31. Keeley, V. E. 1991. Transfer coating with polyurethanes. *Journal of Coated Fabrics*, 20: 176–187.
32. Whiteman, R. 1993. On-line gauging, control, and benefits for the coating/laminating process. *Journal of Coated Fabrics*, 23: 87–104.
33. Giles, H. F., Jr, Mount III, E. M., & Wagner Jr, J. R. 2004. *Extrusion: The Definitive Processing Guide and Handbook*. William Andrew.
34. Lippert, H. G. 2001. Slot die coating for low viscosity fluids, in Satas, D., & Tracton, A. A., *Coatings Technology Handbook* (pp. 139–155). New York: Marcel Dekker.
35. Mamish, A. L. 1990. Co-extrusion coating of woven fabrics. *Journal of Coated Fabrics*, 20: 108–125.

36. Nair, G. P. 2006. Spotlight on laminating machines. *Colourage* 53: 96–112.
37. Mansfield, R. G. 2003. Combining nonwovens by lamination and other methods. *Textile World*, 153: 22–25.
38. Gillessen, G. 2000. Flame, dry or hot-melt. *International Dyer*, 185: 34.
39. McIntyre, K. B. 2006. Coatings and laminates move nonwovens forward. *Nonwovens Industry*, 37: 54–61.
40. Brieger, T. 2006. Consideration factors for ultrasonic bonding vs. adhesive use in nonwovens products. *Nonwovens World*, 15: 60.
41. Knorre, K. 2001. Laminating by ultrasonics. *Nonwovens, Industrial Textiles*, 47: 51.
42. Gil, G. A. 1999. Ultrasonic bonding – new possibilities and opportunities. *Nonwovens Industry*, 30: 46.
43. Walker, R. 1994. Thermoplastic powder adhesive for lamination. *Technical Textiles International*, 3: 28.
44. Swedberg, J. 1998. Sticking to it: Learning about laminates. *Industrial Fabric Products Review*, 75: 30.
45. Crabtree, A. 1999. Hot-melt adhesives for textile laminates. *Technical Textiles International*, 8: 11.
46. Halbmaier, J. 1992. Overview of hot melt adhesives application equipment for coating and laminating full-width fabrics. *Journal of Coated Fabrics*, 21: 301–310.
47. Anon. 2002a. Laminating and coating: Flexible future. *Textile Month*, March, 24.
48. Glawe, A., & Reuscher, R., 2003. Hot-melt application for functional compounds on technical textiles. *Journal of Industrial Textiles*, 33: 85–92.
49. Nussli, R. 2001. The hot-melt coating and laminating of industrial textiles. *Nonwovens, Industrial Textiles*, 47: 14.
50. Mansfield, R. G. 2003. Combining nonwovens by lamination and other methods. *Textile World*, 153: 22–25.
51. Woodruff, F. A. 2002a. Laminating and coating: Advanced composites. *Textile Month*, March 23.
52. Anon. 2002c. Lamination and coating. *Textile Month*, March 22.
53. Jarrell, C. 1992. A new process for coating and laminating face-finished fabrics. *Journal of Coated Fabrics*, 21: 212–221.
54. Field, I. 2000. Why is the industry turning to dry heat lamination technology? *Technical Textiles International*, 9: 21–24.
55. Field, I. 2001. Dry heat lamination technology. *International Dyer*, 186: 39–43.
56. Litman, A. M. 1988. Fowler NE: Electrical properties. *Engineered Materials Handbook. Volume 2: Engineering Plastics*, Reference book, ASM International.
57. Mukai, H. 2009. Stretchable non-woven fabric, absorbent article and method of producing absorbent article. *European Patent EP2 090 684 A1*.
58. Benatar, A., & Gutowski, T. G. 1989. Ultrasonic welding of peek graphite APC-2 composites. *Polymer Engineering and Science*, 29: 1705–1721.
59. Tolunay, M. N. 1982. An experimental study of the bonding mechanism in the ultrasonic welding of thermoplastics. Cornell University, Aug.
60. Shi, W. and Little, T. 2000. Mechanisms of ultrasonic joining of textile materials, *International Journal of Clothing Science and Technology*, Vol. 12 No. 5, pp. 331–350.

61. Mao, Z., & Goswami, B. C. 2001. Studies on the process of ultrasonic bonding of nonwovens: Part 1 – Theoretical analysis. *International Nonwovens Journal*, (2), 1558925001OS-01000210.
62. Stokes, V. K. 1995. *Toward a weld-strength data base for vibration welding of thermoplastics*. In *The 53rd Annual Technical Conference. Part 1 (of 3)* (pp. 1280–1284).
63. Jones, I. 1999. *Use of infrared dyes for transmission laser welding of plastics. ICALEO'99*, Nov.
64. Hilton, P. A., Jones, I. A., & Kennish, Y. (2003, March). *Transmission laser welding of plastics*. In *First International Symposium on High-Power Laser Macroprocessing* (Vol. 4831, pp. 44–52). International Society for Optics and Photonics.
65. Jones, I. A., & Wise, R. J. 1998. Welding method. Patent WO 00/20157, October 1.
66. Sanders, P. 1987. Electromagnetic welding: an advance in thermoplastics assembly. *Materials & Design*, 8: 41–45.
67. Emaweld. 1995. *Electromagnetic Welding System for Assembling Thermoplastic Parts*, Supplier technical report, Ashland Chemical Company.
68. Davies, J., & Simpson, P. 1979. *Induction Heating Handbook* (Vol. 124). London: McGraw-Hill.
69. Harry, J. E. 1971. *Plastics Fabrication and Electro Technology* (Reference book), Heydon & Sons.
70. Stokes, V. K. 1989. Joining methods for plastics and plastic composites: An overview. *Polymer Engineering and Science*, 29: 1310–1324.
71. Thompson, R. 1987. Assembly of fabricated parts. *Modern Plastics Encyclopedia 1988*, Reference book (M603.1), McGraw-Hill.
72. https://www.wikiwand.com/en/Hot_plate_welding
73. Nieh, J. Y., & Lee, L. J. (1993). Morphology characterization of the heat-affected zone (HAZ) in hot plate welding. In *Technical Papers of The Annual Technical Conference-Society of Plastics Engineers Incorporated* (pp. 388–388). Society of Plastics Engineers Inc.
74. http://www.mapeng.net/templets/map/Plastic-Part-Design-Checklist/hot_plate.html
75. Poslinski, A. J., & Stokes, V. K. 1992. Analysis of the hot-tool welding process. *ANTEC 92 – Shaping the Future*. 1: 1228–1233.
76. Beers, D. E., & Ramirez, E. J. 1990. Vectran: High-performance fibre. *Journal of the Textile Institute*, 81: 561–574.

3

Fibres and Fabrics for Aerostats and Airships

Abhijit Majumdar and Unsanhame Mawkhlieng
Indian Institute of Technology Delhi, New Delhi, India

CONTENTS

3.1 Introduction

An aerostat, in a broader sense, is a lighter-than-air (LTA) structure that is used primarily for ground surveillance, for communications and as a radar platform. In its simplest form, an aerostat consists of three segments: hull, fins, and ballonet. The hull is the main ballooned fabric envelope that is filled with gas, the fins are the wings that stabilize the structure during wind disturbances and the ballonet is the air bag that is used to adjust the lift. An aerostat achieves its rise or lift via the use of a buoyant gas such as helium or, less preferably, hydrogen. Since the principle of operation of an aerostat is based on buoyancy and its lift is governed by the amount of air it displaces, it is important to reduce the weight of the structure as much as possible. Appropriately apportioning the hull as the main part of the aerostat, its weight and its construction strategy are of utmost importance, and these aspects will be addressed in this chapter.

DOI: 10.1201/9780429432996-3

Aerostats can either be tethered or free-flying and are meant to stay afloat for a long duration of time, only to be brought down for gas refilling and maintenance. Such applications inevitably subject the aerostat to unfavourable and harsh climatic conditions and ultraviolet (UV) radiations. Additionally, because of the vastness of the structure, handling the aerostat at ground level and lifting it to the air can expose the structure to possible cuts and rips. Further, the aerostat often carries heavy payloads, yet another external influence to withstand. Thus, material selection for the construction of aerostats, particularly the hull, is a crucial manoeuvre towards achieving an efficient product. The base material of choice should possess high tensile strength and modulus as well as excellent creep resistance. More inclusively, the construction assembly as a whole should additionally entail high tear resistance, low rigidity, high UV resistance and excellent barrier properties to prevent leakage of the buoyant gas. To give examples of a typical aerostat, Table 3.1 summarizes the performance properties of two different sizes of aerostats (Model numbers Mark VII and Mark VII-S) being utilized by TCOM Corporation [1].

The special laminate used for the aforementioned aerostat hull envelope provides a predicted life expectancy of more than 15 years. Such stringent requirements cannot be expected to unfold from one single material, unavoidably imposing the use of a multiple-layered assembly of different materials. Typically, the multi-layered construction consists of an interior load-bearing layer, usually a high-performance fabric, a gas barrier layer and, finally, an outer protective layer that is UV resistant, as shown in Figure 3.1. The layers can be adhered mechanically, chemically or with the help of a polymeric adhesive. For example, a patented multi-layered laminate of such construction invented by Mater and Kinnel consisted of a fabric layer, a thermoplastic polyester film and a thermoplastic polyurethane film [2].

The ballonet is another important part of the LTA system, an internal compartment that serves as an air bag. When inflated, the ballonet expansion causes the buoyant gas in the hull to compress, making it denser. Additionally, as the ballonet is air filled, the expansion makes the system heavier and thus brings down the lift of the aerostat. Likewise, deflating the ballonet increases the aerostat's lift. This very purpose of the ballonet subjects it to continuous flexing as it inflates and deflates, thus mandating the use of a brilliant bending resistant material that can perform excellently even in extreme temperature variations. Again, the requirements of light weight and excellent gas impermeability are not to be compromised to minimize lifting gas loss and purity decay.

The scope of this chapter is limited to the construction of the hull, with special attention towards the strength or load-bearing layer and the protective/gas barrier layer. The high-performance fibres used for the strength layer are addressed in detail in Section 3.2. The materials for the strength layer are presented in Section 3.2.1, whereas Section 3.2.2 deals with the materials used

TABLE 3.1

Performance Summary of Two Different Models of Aerostats [1]

Properties	Model Number	
	Mark VII	**Mark VII-S**
Tensile strength	Basic hull material strength requirement is a safety factor of 2 in a 70-knot wind at 120°F	
Shear strength	Minimum shear modulus of 200 lb/in	
Tear strength	Minor damage of 1 in or less should not lead to catastrophic failure	
Ply adhesion strength	Minimum required peel value is 10 lb/in	
Flex life	Hull material to be capable of surviving hundreds of crease cycles	
Helium permeability requirement limit	0.5 l/m²/day	
Weight	5000 lb	6400 lb
Dimensions		
Hull volume	267,000 ft³	375,000 ft³
Overall length	175 ft	215 ft
Hull length	148 ft	188 ft
Hull diameter	56.8 ft	56.8 ft
Fin span	81.5 ft	81.5 ft (horizontal) 99.0 ft (vertical)
Payload enclosure width	25 ft	25 ft
Payload enclosure height	15 ft	15 ft
Operational performance		
Wind speed @ MSL	90 knots	90 knots
Wind speed @ 10,000 ft	105 knots	105 knots
Ceiling altitude above mean sea level (MSL) (std day)	15,000 ft	15,000 ft
Maximum load (payload, power plant, fuel) (std day)		
@ 10,000 ft	4000 lb	8000 lb
@ 15,000 ft	8000 lb	3700 lb

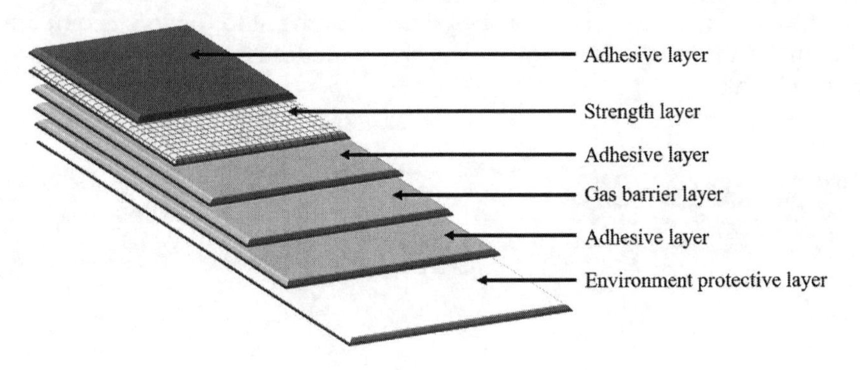

FIGURE 3.1
Laminated structure [2].

for the protective and gas barrier layers. Section 3.3 concludes the chapter with some futuristic advancements that may overtake the market as far as aerostat construction is concerned.

3.2 Hull Material

A multifunctional layer is inevitable for the construction of a hull for reasons discussed previously. As such, a multi-layered laminated or coated structure is used as shown in Figure 3.2(a) and (b), where different layers serve diverse purposes.

3.2.1 Strength Layer

The requirement of flexibility to form the hull's shape imposes the use of flexible material like textile fibres and fabrics. The advent of high-performance fibres has enabled the production of very efficient aerostats, with superior strength-to-weight ratio, which would have not been achievable with conventional commodity fibres. The purpose of the strength layer demands the use of such high-performance fibres of high tenacity and modulus, coupled with low density. Some of the potential fibres are ultrahigh molecular weight polyethylene or UHMWPE (Spectra, Dyneema, etc.), aromatic polyester (Vectran), poly *p*-phenylene-benzobisoxazole (Zylon), etc. With a superior strength-to-weight ratio, these fibres exhibit good tensile properties because of the internal molecular chain arrangement, generally characterized by high crystallinity and crystal orientation. The chemical structure and nature of the polymer also dictate the manner of longitudinal and transverse interactions among the different molecular chains. Furthermore, the length of the polymeric chain, defined in terms of molecular weight, decides the number of chain-ends and the overlapping length among neighbouring chains, which is ultimately reflected in the tensile property of the fibre. Table 3.2 compares the different fibres in terms of their density, modulus, tenacity, and elongation at break.

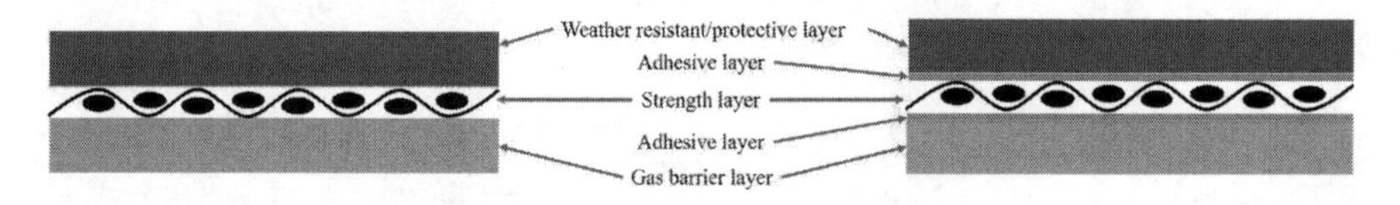

FIGURE 3.2
Different layers of a typical (a) multi-layered coated and (b) laminated structures for LTA systems.

TABLE 3.2

Properties of Some High-Performance Fibre

Fibre Class	Specific Fibre Type	Density (g.cm^{-3})	Modulus (GPa)	Tensile Strength (GPa)	Elongation at Break (%)
Para-aramid	Technora	1.39	70	3.00	4.4
	Twaron	1.45	121	3.10	2.0
	Kevlar 29	1.44	70	2.96	4.2
	Kevlar 129	1.44	96	3.39	3.5
	Kevlar 49	1.44	113	2.96	2.6
	Kevlar KM2	1.44	70	3.30	4.0
	Taparan 629	1.44	96	3.30	3.4
UHMWPE	Spectra 900	0.97	73	2.40	2.8
	Spectra 1000	0.97	103	2.83	2.8
	Spectra 2000	0.97	124	3.84	3.0
	Dyneema	0.97	87	2.60	3.5
PBO	Zylon	1.56	270	5.80	2.5
High-performance polyester	Vectran HT	1.40	75	3.40	2.8
PIPD	M5	1.70	271	3.96	1.4

It is not surprising that aerostats or balloon envelopes were conventionally made of cotton fabrics. A patent by Junius D. Edwards revealed a gas-retaining balloon using two or three plies of cotton fabric having rubber films in between the plies and being rubber-coated on the outside and inside of the fabric assembly [3]. The films of rubber in between the plies were incorporated to arrest the gas from exuding out of the fabric, while the outwardly coated rubber was meant for protective and aesthetic purposes. Later, with the dawn of high-performance fibres, the use of nylon and polyester became more common due to their better tensile properties. The tenacity of nylon and polyester can reach up to 10 g/den and breaking extension over 10%, resulting in high work of rupture [4]. Polyester has been a choice of material for a long time due to its high strength-to-weight ratio, low creep, enhanced hydrolysis resistance coupled with low moisture regain.

In addition to the fibres used, the construction of the base fabric out of these high-performance fibres is also crucial. Keeping in mind the requirement of the strength layer, the fabric structure can be altered to meet the need of tear and flex resistance as well. This aspect is achievable by exploring the merits of different weaves and constructional parameters to maximize the utilization of both fibre and fabric properties.

3.2.1.1 Fibres

3.2.1.1.1 Aramid

Aramid (aromatic polyamide) is one of the popular high-performance fibres available today. It is a rigid chain polymer and is known for its high strength, modulus, and thermal resistance. According to the US Federal Trade Commission, aramids are defined as 'long-chain synthetic polyamides in which at least 85% of the amide linkages (–CO–NH–) are attached directly to two aromatic rings'. The fibre gained attention quickly after DuPont first launched Nomex (meta-aramid) in the 1960s. Shortly after, Kevlar, a para-aramid fibre, came into existence. Nomex is generally known for its excellent flame retardancy, while Kevlar is known for its tenacity and modulus. Kevlar possesses such strength due to the molecular chains being neatly stacked and highly oriented. Transversely, the chains are held by hydrogen bonds, while longitudinally, they are held by strong covalent bonds. Additionally, the aromatic ring contributes to the superior modulus due to increased chain rigidity and enhanced chemical resistance due to resonance. Kevlar comes in different grades suitable for different applications, including ballistic protection, and has a tensile strength ranging from 2.96 to 3.3 GPa and an elastic modulus from 70 to 113 GPa. Apart from Kevlar, other *p*-aramid fibres are also available with different tradenames such as Technora, Twaron and Taparan, having comparable mechanical properties as shown in Table 3.1. The chemical structure of Kevlar is shown in Figure 3.3.

Although *p*-aramid (Kevlar) is a strong potential fibre with good resiliency, commercially, it is not used for the construction of the hull. A 1977 report by Sheldahl Inc. for NASA explored the possibility of using Kevlar as an alternative to the then used material, Dacron, as a strength material for inflatable structures [5]. The report showed that the Kevlar-based laminates and coated structures used were superior in strength; however, the tensile strength loss after creasing was significant. Also, the Kevlar-laminated structure showed more extensive yarn failure at 24,000 flex test cycles. In addition, the process of handling and packaging used then presented a number of difficulties. Presently, Kevlar is normally used as tether cables in

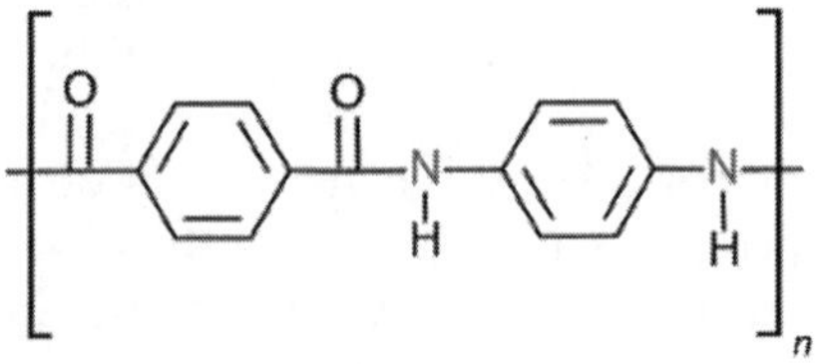

FIGURE 3.3
Chemical structure of *p*-aramid fibre Kevlar.

commercial aerostats to carry both power supply and fibre optics for communications. Briefly, in a typical tether cable design, Kevlar serves both as a strength component and as a protective cover due to its non-conductive property. It is used to surround a central core conducting component and is itself protected from environmental attacks with the help of a protective weathering barrier [6, 7].

3.2.1.1.2 Ultra-High Molecular Weight Polyethylene

Gel-spun ultra-high molecular weight polyethylene (UHMWPE) fibres are known for their excellent tensile properties and very low density. Commercially, UHMWPE is available under different trade names such as Dyneema (DSM, Netherlands) and Spectra (Honeywell, USA). UHMWPE yarns come in twistless multifilament form with filament linear density spanning from 0.3 to 10 denier per filament (dpf). UHMWPE is also known for its excellent UV resistance which is a necessity for aerostat applications. It is reported that the tensile strength of UHMWPE (Dyneema) drops to only about 5% when exposed to UV rays for a period of two days. However, aramid fibre (Kevlar) loses its strength up to 25% when subjected to similar conditions [8, 9]. The properties of UHMWPE fibres commercially available are given in Table 3.3.

Although UHMWPE stands as a potential strength material for the envelop or hull construction, it has not been explored extensively. The higher cost of the fibre may be one of the hindrances. In a recent doctoral thesis from the University of Southampton [11], CT155HB UHMWPE, a polyethylene laminate manufactured by CubicTech Corporation, acquired by DSM Dyneema was explored as an envelope material. This fibre has a potential of becoming a material of choice due to its low density and excellent tenacity.

TABLE 3.3

Properties of UHMWPE Fibres [10]

Fibre Type	Dpf	Tenacity (GPa)	Modulus (GPa)	Elongation to Break (%)
DSM				
Dyneema SK60	1	2.8	91	3.5
Dyneema SK65	1	3.1	97	3.6
Dyneema SK75	2	3.5	110	3.8
Dyneema SK76	2	3.7	120	3.8
Honeywell				
Spectra 900	10	2.6	75	3.6
Spectra 1000	5	3.2	110	3.3
Spectra 2000	3.5	3.4	120	2.9

3.2.1.1.3 *High-Performance Polyester*

Vectran is melt-spun fully aromatic polyester that possesses very high tenacity and modulus. It was originally created by Celanese Corporation in the 1990s and is currently manufactured by Kuraray Co. Ltd., Japan. Produced from thermotropic liquid crystal polymer (exhibit liquid crystallinity in melt state) by polycondensation of 4-hydroxybenzoic acid and 6-hydroxynaphthalene-2-carboxylic acid, Vectran is characterized by high molecular orientation. The chemical structure of Vectran is shown in Figure 3.4.

The tensile strength and modulus of Vectran are at par with those of aramids. It has a tensile strength of 24–27 gpd (3–3.4 GPa) and a modulus of 600–838 gpd (74–104 GPa). Additionally, it is rip resistant and is dimensionally stable. Above all, it is exceptionally resistant to cut and shear, making it a suitable choice as a protective material. Vectran absorbs minimal moisture (0.1% moisture regain) and thus, retains its mechanical properties at low temperature or when wet. With a high decomposition temperature in the range of 400°C, Vectran is able to withstand high temperature and heat, rendering additional advantages for aerostat applications. The ability to withstand higher temperatures is even more pronounced when wet.

Owing to these properties, Vectran is currently one of the choices for airship fabric and is readily available on the market. In fact, it is one of the fibres that have been explored more extensively in research as well [13–15]. Additionally, the choice of Vectran is because of its ability to tolerate steady-state super-pressure and transient deployment forces.

3.2.1.1.4 *Poly p-Phenylene Benzobisoxazole*

Poly *p*-phenylene-benzobisoxazole (PBO) is a lyotropic liquid crystal polymer and is synthesized using 2,5-diamino-1,3-benzenediol (DABDO) and terephthalic acid (TA) or TA derivative (e.g., terephthaloyl chloride) in poly(phosphoric acid) (PPA) [16]. Invented in the 1980s by SRI International, PBO was trademarked as Zylon. Presently, PBO is produced by Toyobo Corporation, Japan. The chemical structure of PBO is shown in Figure 3.5.

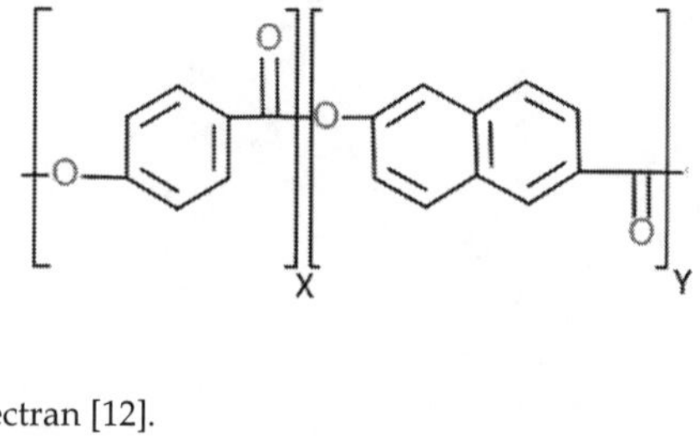

FIGURE 3.4
Chemical structure of Vectran [12].

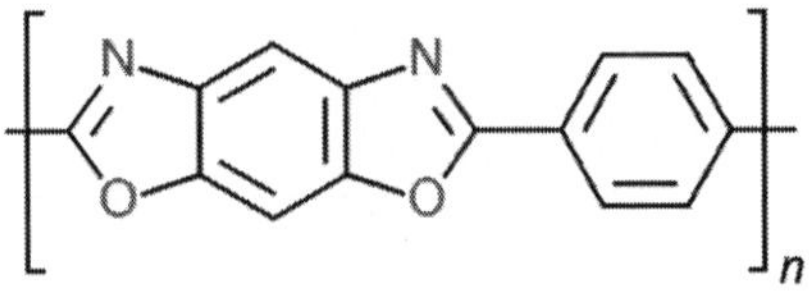

FIGURE 3.5
Chemical structure of PBO (Zylon) [17].

Probably the strongest fibre, Zylon is reported to have a tensile strength of 5.8 GPa and a modulus of 270 GPa. Its density is 1.56 g/cm^3, which is slightly higher than that of aramids. The fibre possesses outstanding flame resistance (in a vertical flame test, char length is almost zero), excellent thermal stability as well as good creep and chemical resistance. However, Zylon is sensitive to UV light and moisture. In fact, Zylon also has to be protected from visible light. When exposed to daylight, its residual strength falls to 35%, after exposure duration of only six months. This property necessitates the use of a protective covering material wherever Zylon is used outdoors. The effect of moisture is also substantial. When dry, Zylon can withstand temperatures as high as 400°C with strength retention of 75%. However, in the presence of saturated steam, the strength falls sharply (20% of the strength tested at standard conditions) even at 250°C [16]. Zylon also possesses low compressive strength, and special attention has to be given where the application involves flex fatigue.

Zylon has been studied for its applicability in stratospheric airship application, and the results clearly showed its superiority over other fibres in terms of its mechanical properties, both at high and low temperature as well as in terms of creep resistance [18–20]. However, its inability to retain these properties in wet and humid environments makes it a less preferred choice than Vectran.

3.2.1.1.5 Polyhydroquinone-Diimidazopyridine

Polyhydroquinone-diimidazopyridine (PIPD) fibre or M5 is a rigid-rod-based polymeric fibre characterized by very high strength and modulus, developed by the Dutch chemical firm Akzo Nobel. It is formed by condensation polymerization of tetraaminopyridine and dihydroxy-terephthalic acid in the presence of diphosphorus pentoxide as a dehydrating agent [21, 22]. Its chemical structure is shown in Figure 3.6.

Apart from its high modulus (271 GPa) and tenacity (3.96 GPa), M5 is known for its stability at high temperature and humidity, showing a strength retention of almost 100% when subjected to a temperature of 82.2°C, at humid conditions (85% RH) for a span of 11 days [21]. Additionally, it also shows superior UV stability to Zylon. It is able to retain almost 100% strength even after 100 hours of exposure to Xenon lamp. Ideally, these properties of M5 make it a

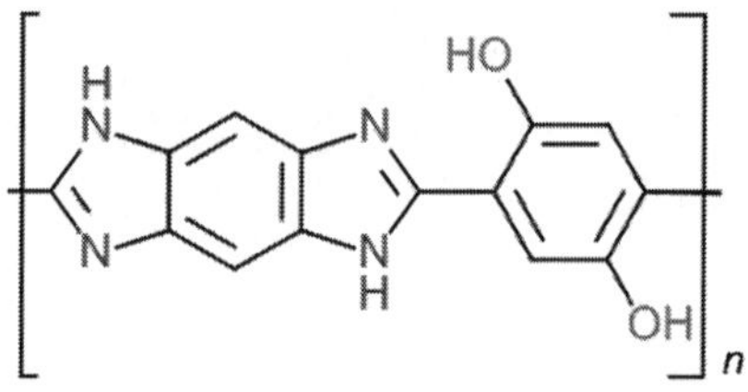

FIGURE 3.6
Chemical structure of M5 fibre [23].

potential choice for the strength layer of hull construction. Unfortunately, there is no or little evidence of M5 being used in practice as a hull material.

3.2.1.2 Fabrics

It is understood that for the fibre's excellent mechanical properties to be fully realized in the final product, the fabric has to be judiciously constructed. In other words, the structure should have the potential to take the load and to distribute the same among the individual yarns. Other than the necessary requirement of strength, the fabric used for aerostat application is deemed suitable if it also possesses excellent tear resistance, excellent flexibility, and good creep resistance. Commercially, it is normal to have a single layered fabric as the strength layer in medium-strength hull materials. The constructional parameters of the fabric are therefore of utmost importance and are different for varying fabric structures such as woven, unidirectional, and knitted. For a woven fabric, these parameters include fabric sett (i.e., number of yarns per unit area) and weave (i.e., manner of interlacement of warps and wefts). Fabric sett decides the load-bearing capacity as well as the openness of the structure. Weave affects several parameters such as strength, firmness and, more importantly, rip or tear resistance. For a unidirectional fabric, layer orientation and thickness are two of the influential parameters. For a knitted structure, the constructional parameters are wales and courses per unit length (or more commonly, stitch density) and the presence as well as distribution of tucks and floats. However, the very nature of the knitted structure hinders its use in aerostat or similar applications.

3.2.1.2.1 Two-Dimensional Woven Fabrics

Two-dimensional (2D) woven fabrics are flexible structures with very low thickness. In its simplest term, it is a structure that is formed by the interlacement of two sets of yarns called warp and weft which are aligned in longitudinal and transverse directions, respectively, as shown in Figure 3.7(a). Likewise, when three sets of yarns are used, the alignment of the sets is generally 60°

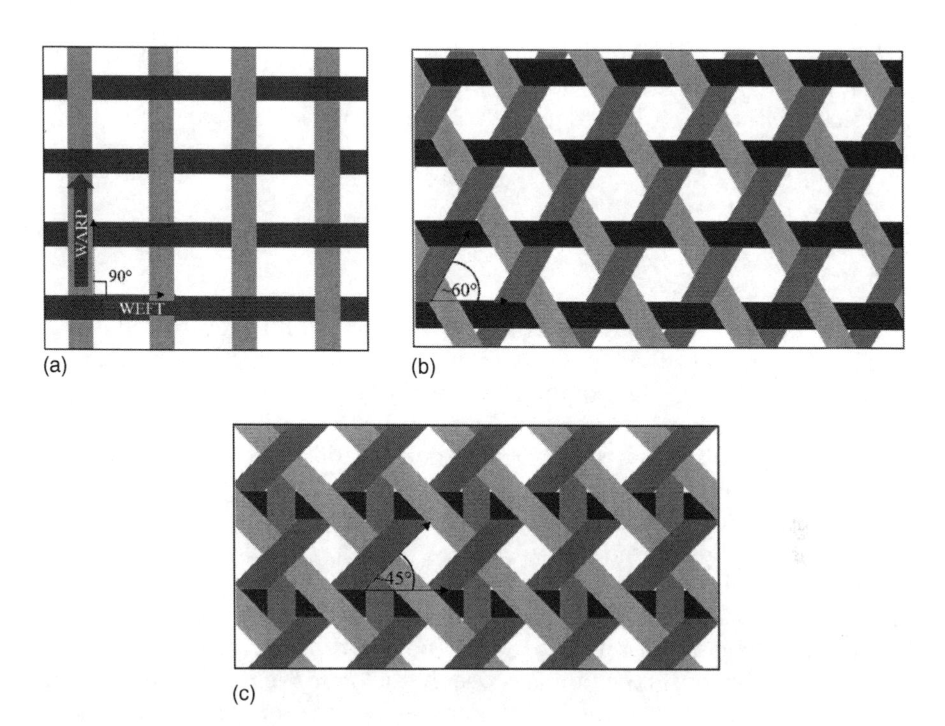

FIGURE 3.7
Biaxial (a), triaxial (b) and multiaxial (c) fabrics.

to each other, as shown in Figure 3.7(b). Such a fabric is called triaxial fabric. Similarly, the number of yarn sets can be higher than three, as shown in Figure 3.7(c), in which case, the term multiaxial fabric is generally used. Such fabrics are characterized by superior mechanical properties in the biased directions as well (i.e., the directions other than along the length or width).

The structure of woven fabrics is mainly characterized by yarn linear density, fabric sett, crimp, and weave. Yarn linear density influences the fabric thickness and areal density (g/m²). Fabric sett refers to the number of yarns present per unit length along the length and width of the fabric. The terms 'ends per inch (EPI)' or 'end per centimetre (EPcm)' and 'picks per inch (PPI)' or 'picks per centimetre (PPcm)' are commonly used to denote the number of warps and wefts used, respectively. Crimp is the measure of the degree of waviness present in the yarns inside a woven fabric. If L_y is length of yarn and L_f is corresponding length of fabric, then the yarn crimp can be calculated by using Equation (3.1).

$$\text{Crimp}(\%) = \frac{L_y - L_f}{L_f} \times 100 \tag{3.1}$$

Woven fabrics are normally specified by their areal density. If warp yarn count (tex) is T_1, weft yarn count (tex) is T_2, ends per unit length (EPcm) is N_1, picks per unit length (PPcm) is N_2, crimp % in warp is C_1 and crimp % in weft is C_2, then areal density of fabric can be calculated using Equation (3.2).

$$\text{Areal density}\left(g/m^2\right) = \frac{1}{10}\left[N_1T_1\left(1+\frac{C_1}{100}\right) + N_2T_2\left(1+\frac{C_2}{100}\right)\right] \tag{3.2}$$

Fabric sett has substantial effects on the properties of the fabric, and it governs the firmness of the structure. In fact, it would not be an exaggeration to state that the properties of the fabric are largely dependent on this single parameter alone. Some of these include tensile, tearing, bending, shearing, permeability and transmissibility properties. The dependency of these properties on the fabric sett is not always linear, and therefore, a higher sett does not necessarily lead to better performance. This aspect of relationship is dealt with in more detail subsequently.

Other than fabric sett, weave is another parameter than can alter the fabric property considerably. Weave is the pattern in which the yarns interlace among each other in forming the fabric. The unit cell of a weave is known as repeat unit. The most common is probably the plain weave, in which every yarn runs alternatively over and under a series of sequentially laid transverse yarns. Hence, the design repeats over two ends and picks. It is the most dimensionally stable weave due to the highest number of interlacement or crossover points. Other common weaves are basket or matt, twill, satin, ripstop, etc. Basket weave is an extension of the plain weave in which two or more adjacent warp and weft yarns follow the same interlacement pattern. In twill weave, the warp yarns float over or under a certain number of weft yarns and vice versa. It is characterized by the presence of prominent lines running diagonally on the fabric surface. There are multiple derivatives of a twill weave with different names such as pointed, broken, combined, herringbone, step, etc. Satin (warp-faced) or sateen (weft-faced) are weaves in which the length of float is sufficiently large so that only either the warp or the weft yarns predominate on the face side of the fabric. Hence, this weave produces a smooth and lustrous surface. The various simple weave types are shown in Figure 3.8(a)–(d). Ripstop weave, as the name suggests, is a special kind of elaborated compound weave (mixture of two or more weaves) that can prevent a rip or tear from propagating. This is because of the presence of a crosshatch pattern of coarser yarns along both the length and width of the fabric. The coarse yarns can be either be inherently of higher linear density or can be formed by combining two or more yarns of the same linear density (as the rest of the yarns in the fabric) but processed as one. Figure 3.9 shows one repeat of this weave kind. As in the case of sett, the fabric mechanical

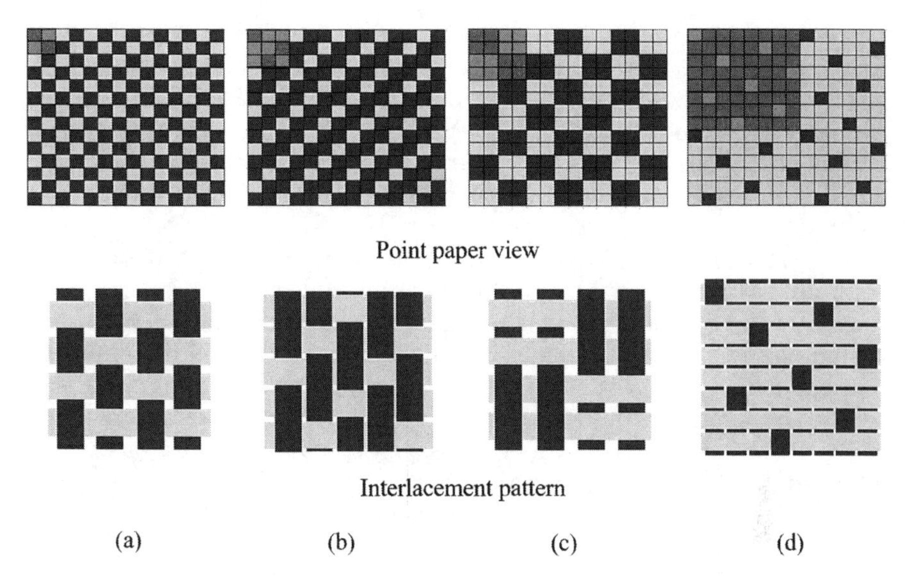

FIGURE 3.8
Different simple weaves: (a) plain; (b) twill; (c) basket; (d) sateen. Top: point paper view. Bottom: interlacement pattern. The yellow highlighted portion in the top left represents the repeat unit.

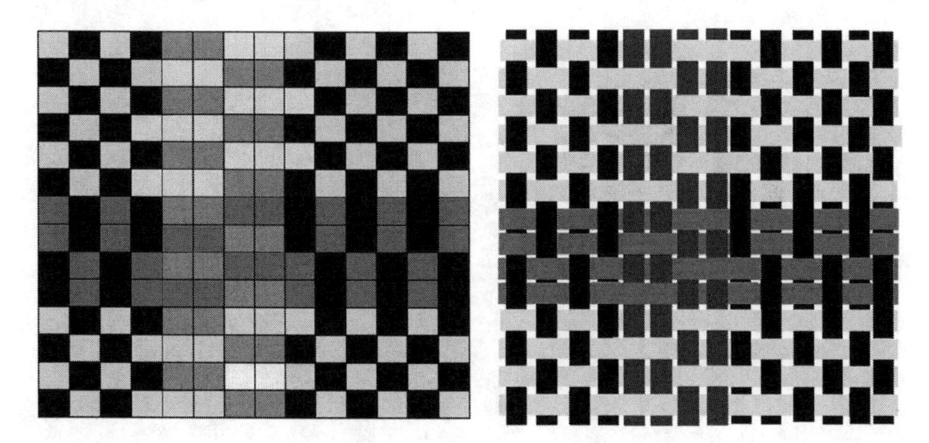

FIGURE 3.9
Ripstop weave.

and transmissivity properties are highly influenced by the weave. Hence, the response of a fabric towards a certain type of load can be tuned by varying the weave. The effect of fabric sett and weave on some of the important properties of the fabric, as far as aerostat application is concerned, is briefed as follows.

Leno is another potential weave for aerostat and airship applications. This weave produces a lightweight, open, and stable fabric structure. The warp yarns are twisted around the weft, as shown in Figure 3.10, to give excellent frictional contacts among the yarns. Thus, the structure, though porous, maintains its integrity under deformation. When combined with films, the composite material becomes an excellent candidate for structural applications.

Tensile properties: Tensile property is of utmost importance particularly for applications like the strength layer in the hull laminate. As far as fabric sett is concerned, the higher the value, the higher the number of yarns participating in sharing the load, hence, the higher the tensile strength. Furthermore, for a square fabric, the proportionately increasing number of transverse yarns causes an increase in the fabric strength *via* a phenomenon termed fabric assistance [24]. However, it can be envisaged that there is an optimum level of thread density that is beneficial to the strength, a further increase beyond which will only cause the tensile strength and modulus to fall due to excessive crimp (yarn path undulation).

Similarly, different weaves exhibit different response to tensile loading. For a given thread density, plain weave is known to be the strongest due to the highest number of interlacement points as discussed earlier, where each crossover point generates an additional source of frictional resistance

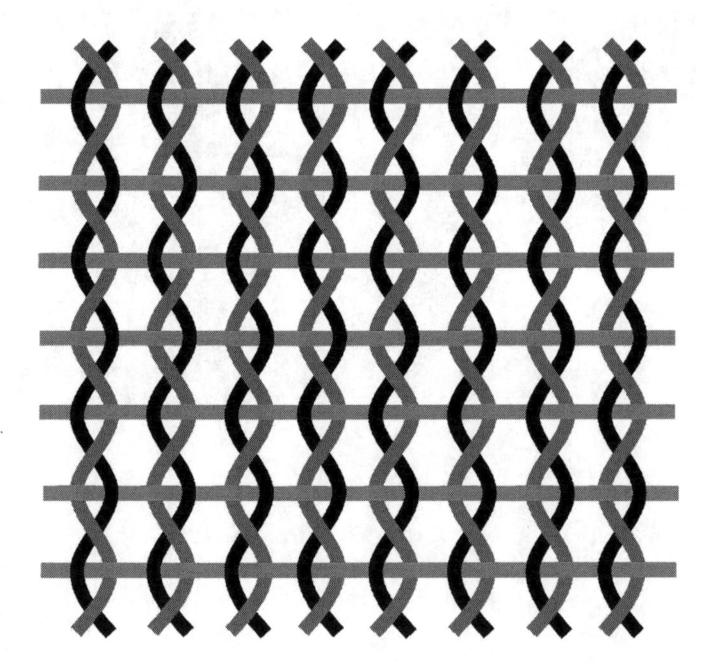

FIGURE 3.10
Leno weave.

that needs to be overcome. A research paper from way back in 1933 shows that the tensile strength of a fabric is basically dependent upon how closely woven the structure is [25]. In other words, a structure with a higher number of interlacement points and shorter floats gives higher tensile strength. However, a higher number of interlacements leads to higher crimp which may adversely affect fabric strength due to the obliquity effect.

Tearing strength: For a given yarn strength, the tearing strength of a fabric structure largely depends on the mobility of yarns during the act of tearing. This is because when being torn, movement of the yarns allow them to form a group or bundle which will collectively share the tearing stress. This means that a fabric with lower sett and loose weave offers higher tearing resistance [26]. Since twill and satin are weaves that have longer floats as compared to plain weave, they offer higher resistance to tearing while compromising on the tensile strength. Hence, such structures are not suitable for aerostat applications where the load-bearing capacity is a primary requirement. In order to achieve the optimal design, employing different weaves into the same structure is advantageous, and this is attained with the ripstop weave. The presence of the plain-woven section adds to the strength, whereas the existence of the rib segment aids in resisting tear and the propagation of tear. This concept of ripstop construction was adapted by Durney [27] in which a ripstop webbing seam network system was sewn externally on to a test bag (aerostat envelope) to enhance the tearing resistance.

Bending and shearing properties: The hull is periodically subjected to deflation and inflation, subjecting the strength layer to continuous flexing and shearing throughout its service life. For a given fibre, bending and shear resistance of fabric will depend on yarn linear density, fabric sett and weave. Higher yarn linear density and fabric sett increases the bending as well as shearing rigidity. This is because for a given area, higher sett means a higher number of crossover points between warp and weft and a higher angle (and area) of yarn contact at the points of interlacement. Therefore, the resistance of the fabric against bending or shearing forces is higher.

Figure 3.11 shows the bending rigidity of fabrics, having a sett of 70×70 inch^{-1}, woven with 30 Ne polyester spun yarns [28]. It is observed that plain woven fabric shows the highest bending rigidity, followed by twill and matt. Satin weave has the minimum bending rigidity. Plain woven fabric has the maximum number of interlacements between warp and weft yarns. Therefore, the length of yarn segment that undergoes bending deformation as cantilever is the minimum for plain weave. Thus, the created bending moment is also less.

Figure 3.12 depicts the shear rigidity of fabrics, having a set of 70×70 inch^{-1}, woven with 30 Ne polyester spun yarns [28]. It is observed that plain woven fabric shows the highest shear rigidity. In comparison to plain weave, the shear rigidity becomes almost half in 2/1 twill woven fabric. Fabric shear rigidity reduces further for 3/1 twill, 2/2 twill and 2×2 matt weaves, and

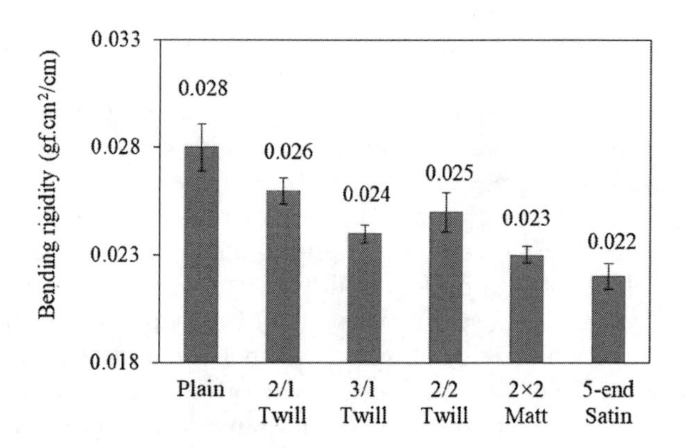

FIGURE 3.11
Bending rigidity of polyester fabrics [28].

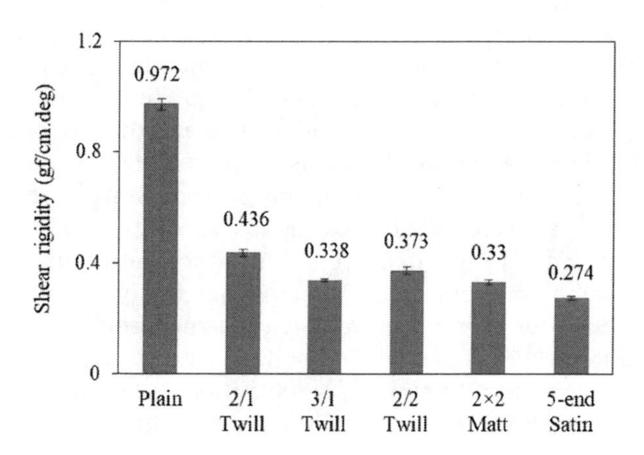

FIGURE 3.12
Shear rigidity of polyester fabrics [28].

it becomes the minimum for 5-end satin. Though the trends of bending and shear rigidities with respect to weave are same, the latter is influenced to a much greater extent by weave. The shear deformation in woven fabric is associated with the following things in the sequence they are mentioned.

- Bending of yarns
- Rotation of yarns at the crossover points
- Jamming of yarns

Bending of yarns is difficult in plain woven fabric, as explained earlier, and the number of crossover points is also the maximum. The latter makes the rotation of yarns difficult as reaction forces act between the warp and weft yarns at the crossover points. As the weaves become less firm, the bending as well as yarn rotation becomes easier. Therefore, the shear rigidity reduces.

3.2.1.2.2 Unidirectional Fabrics

In its true sense, a unidirectional fabric (UD) is one in which only one set of yarn is used, held transversely by fine yarns widely spaced apart only to keep the structure intact. The longitudinal yarns are responsible for taking the load while the transverse yarns maintain the structural integrity. Another definition that is used often in literature is that of a laminate. Here, a UD fabric is one in which the twist less multifilaments are just laid down one above the other, held together by a thermoplastic resin or through lamination. The multifilament yarns are usually aligned at different angles: (0°, 90°, 0°, 90°); (0°, ±45°, 90°); etc. Naturally, the use of different layouts and number of layers result in varying properties.

In contrast to woven fabrics, UD fabrics or sheets, particularly used as a composite of many layers, have an added advantage of being mechanically tailorable as required to obtain varying properties in different directions. Additionally, fibre properties such as strength and creep resistance can be fully utilized in the fabric due to the absence of crimp, and the fabric can be formed using high denier tows, which ultimately reduces the cost of manufacturing. Modelling and prediction of properties also become easier in case of UD fabrics. With this understanding, Cubic Tech Corporation developed composite UHMWPE fabrics, an alternative to nylon and polyester woven materials. It is claimed that a Dyneema laminate CT1.5 is 80% stronger than a 44 g/m^2 silicon coated woven nylon of comparable weight and thickness. Moreover, the laminate has modulus 10 times as high and tear strength 4 times as high. Such UD laminates can also be tuned to exhibit quasi-isotropic properties. These laminates have been explored for their suitability in aerostat applications and have shown promising results [29].

3.2.2 Barrier Layer

As the name suggests, the protective layer is intended to serve two crucial functions achievable through different layers:

1. An exterior-most layer, providing protection against extreme and harsh environmental conditions such as intense visible and UV radiations, ozone, temperature variation, rain, humidity, and hazardous environmental conditions, to name a few. The layer is also known as a weather-resistant layer, and the application requirement is to possess excellent low-temperature flexibility and remarkably high weather resistance property.

2. A barrier, preventing the leakage of helium gas from inside the hull and retaining it for a longer duration. Separately termed as the gas barrier layer, it should retain gas to the maximum capacity and, as such, must have superiorly high flex and shear stiffness to withstand the inflation and deflation. Also, the barrier layer should be stable at the aerostat's operational temperature.

Despite serving two important and different functions at a high altitude, the wholesome purpose of the protective layer, be it as a weather-resistant or gas barrier, is to protect the entire system from all the unwanted circumstances, thereby increasing the overall service life of the aerostat. Structurally, these layers are constructed by coating or laminating a polymeric film, while ensuring high bondability with the adjacent layers. Hence, this calls for a requirement of the protective layer to own bondability or sealability to its excellence.

3.2.2.1 Different Materials Used in Protective/Gas Barrier Layer

In general, the enormously vast size aerostat that floats approximately at an altitude of 20 km above sea level, against the harsh stratospheric environmental conditions, must possess prolonged endurance properties [30]. Due to temperature variations during the operational conditions of the LTA system, the materials used to construct the airship become brittle, resulting in loss of flexibility. In addition, the material is prone to strength reduction and permeability due to the constant exposure to intense UV radiations and ozone concentration. Hence, the protective layer must contain the following necessary characteristics to provide the required shield to the LTA systems:

- Weatherability
- Gas retention
- Thermal reflective/emissive properties
- Shear stiffness
- Bondability
- Low temperature flexibility

Technically, different materials having a unique set of properties are employed in order to meet the requirements of the two said layers (weather-resistant and gas barrier). For instance, polyvinylidene fluoride (PVDF), often with an aluminized top coating with better weather resistant property is used for the construction of a weather-resistant layer [13, 15]; whereas polyvinylidene chloride (PVDC), having good helium gas barrier properties, is used for fabricating the gas barrier layers [4, 13]. However, there are polymers that can be used for both purposes because of their outstanding properties as being weather resistant as well as having a high gas barrier capacity. These materials must meet the important criteria of least permeability to gases; high strength-to-weight ratio; high resistance to degradation to intense UV and visible light, hydrolysis, abrasion and other environmental factors; and ease of fabrication [31]. The following briefly delineates some of the major derivatives of fluoro-carbon-based polymers that are being employed in the construction of LTA hull protective layers.

3.2.2.1.1 PVF Tedlar

Polyvinyl fluoride (PVF) is a thermoplastic fluoropolymer with repeating units of vinyl fluoride and is structurally similar to polyvinyl chloride. Its chemical structure is shown in Figure 3.13. Commercially known by the trade name of Tedlar, from DuPont, since 1961, it is typically chosen as the outermost layer in laminated fabric construction [32, 33]. It holds excellent UV resistance and mechanical properties, gas retention properties, toughness, flexibility, good abrasion resistance and strength in lower mass. The fluorocarbon property of Tedlar provides the base for its excellent durability and inertness toward chemical, solvent, stain, and graffiti. Being resistant to sunlight degradation, Tedlar stands up well to the stratospheric pollutants and can combat mildew and acid rain attack. Since Tedlar is impermeable to grease and oils, the atmospheric airborne dirt does not adhere to it. This polymeric film can retain its properties over a wide range of temperature (–72°C to 107°C), thereby making it suitable for high altitude applications [32, 33]. However, Tedlar shows inferior adherable properties with the available adhesives due to its lower critical surface tension (~30 dynes/cm). The adhesion between PET and Tedlar films can be significantly enhanced by increasing the critical surface tension of the film to 44–48 dynes/cm after the application of corona treatment [33].

Tedlar polymeric film is available either as a near-colourless, transparent, or pigmented film. Coating of the outermost layer of Tedlar films with Aluminium or UV absorber additives (Aluminized Tedlar or

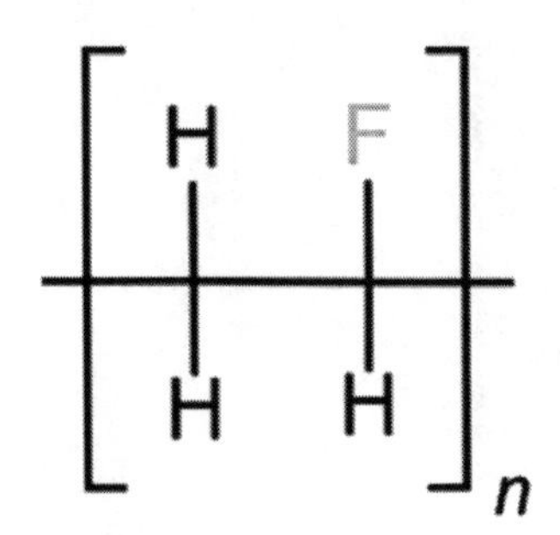

FIGURE 3.13
Chemical structure of Tedlar.

Pigmented Tedlar) enhances its barrier property, thereby offering the highest level of protection by blocking 99% of the UV rays in the wavelength range of 290–350 nm [15, 34]. This suggests that the materials underneath Tedlar film will not be exposed to the high-energy, destructive light. The long-term weathering effect of pigmented Tedlar was studied by Nakadate et al. [35] by incorporating it with Zylon woven fabric and polyurethane. The aluminium evaporated Tedlar film demonstrated improved resistance to the UV radiation and could also provide enhanced barrier to the environmental moisture as well as the prolonged outdoor exposure to high humidity. Similarly, the resistance of Tedlar to intense ozone was confirmed by Maekawa et al. [36] where ozone at 50 ± 5 ppm (5 times higher than that at high altitude) was applied to the test specimen. In the 24-hour test, Tedlar film protected critical damage to the interior structure of the laminate by preventing the penetration of ozone and its resultant erosion to the underneath layers. Hence, these studies confirmed the protective nature of Tedlar against the harsh environmental conditions, thereby validating its efficacy as a suitable material for hull construction for decades.

3.2.2.1.2 PTFE Teflon

Teflon, chemically known as polytetrafluoroethylene, is a fluorocarbon solid characterized by high molecular weight. It was discovered by Roy J. Plunkett in 1938 and is a synthetic fluoropolymer based on a chain of carbon atoms, surrounded by fluorine. It has multiple strongly bonded carbon-fluorine chains that make it highly resistant to solvents, acids, and bases. Thus, the unique properties of Teflon could be attributed to its tightly bonded carbon and fluorine structure, where fluorine shields the exposed carbon chain. The chemical structure is shown in Figure 3.14.

Being one of the inert materials, Teflon holds the necessary properties, making it a suitable candidate for high-altitude LTA applications. Being thermoplastic, it can easily be thermoformed, heat-sealed, vacuum-formed, welded, heat-bonded, metalized and laminated to other materials.

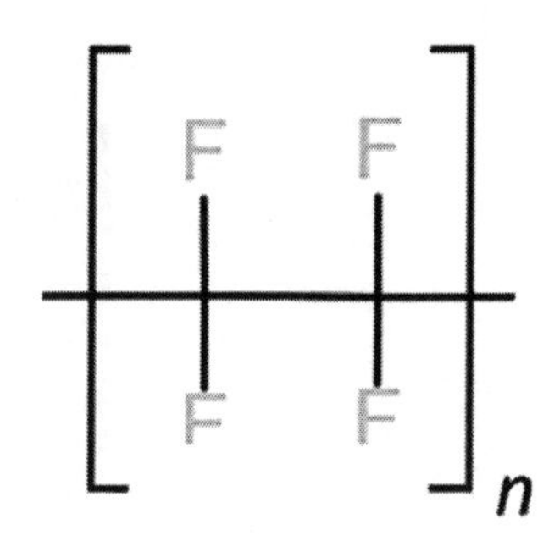

FIGURE 3.14
Chemical structure of Teflon.

In addition, it possesses low permeability to gases and has low temperature toughness ranging from −240°C to 205°C. Because of its superior properties as a true melt-processible fluorocarbon, it can also be applied as an excellent hot-melt adhesive. A multi-layer, multi-axial Teflon laminate was reported to be developed successfully for an inflatable military application with a tensile strength exceeding 2000 lb./in. in four directions (warp, fill and both 45° axes), thereby proving its excellent mechanical properties [4].

3.2.2.1.3 Mylar Polyester Film

Commercially available under the trade name of Mylar from DuPont, it is a biaxially oriented polyester film developed in the mid-1950s. It is made from the stretched polyethylene terephthalate (PET) whose chemical structure is shown in Figure 3.15. It is commonly used as a gas barrier material and possesses high tensile strength, shear stiffness, toughness, insulating properties, chemical and dimensional stability, high temperature resistance, durability and unique optical properties and low coefficient of friction [32, 37]. For instance, a laminate comprising of layers of Tedlar/Mylar/Dacron was fabricated to protect against environmental conditions, for gas retention and for load-bearing purposes. This special laminate is used by TCOM, an airborne surveillance systems provider, for the construction of its hull envelop and was reported to have a predicted life expectancy of more than 15 years [1]. In a work for super-pressure balloon application, Hall et al. [13] showed the fabrication of a full-scale prototype balloon using multi-component laminated structure encompassing layers of bonded aluminized Teflon film, aluminized Mylar film, Vectran fabric and polyurethane coating. The structural configuration yielded low gas permeability, resistance to sulfuric acid and high strength for operating under super-pressure conditions, thereby confirming the suitability of the balloon to be used in high-altitude flight at Venus for a long-duration mission.

3.2.2.1.4 Polyimide Kapton

Kapton is a polyimide film developed by DuPont in the late 1960s and is produced by the condensation of pyromellitic dianhydride and 4,4'-oxydiphenylamine. Its chemical name is poly (4,4'-oxydiphenylene-pyromellitimide),

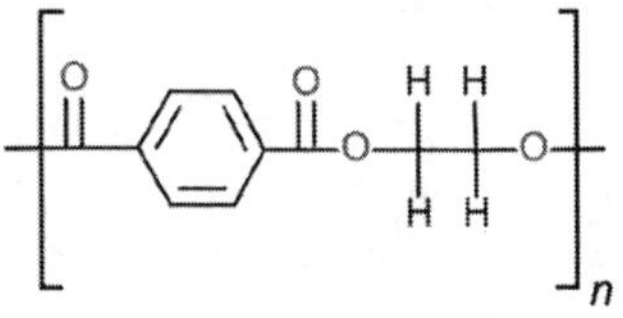

FIGURE 3.15
Chemical structure of Mylar.

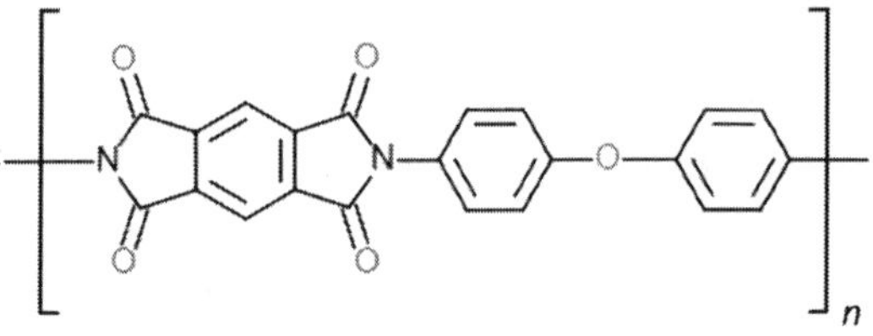

FIGURE 3.16
Chemical structure of Kapton.

and its structure is shown in Figure 3.16. It is characterized by its excellent capacity to hold the inflating gas within the hull structure, and thus, it is used as a gas barrier layer [32, 33]. In addition, Kapton shows stability in a wide range of temperatures (−273°C to 400°C). This could be attributed to its excellent dielectric constant that helps in diffusing sources of radiation such as laser light [32]. Lavan et al. [38] claimed that fabricating an airship hull material with Kapton as a gas barrier layer and PVF as protective weathering component worked well at an altitude of ~21.3 km (70,000 ft) having temperature ranging from −100°C to +60°C. However, the polyimide Kapton layer tend to degrade after prolonged exposure to UV radiations and, hence, coating with an UV stable polymeric layer is recommended [32]. Abide by this concept, Gouzman et al. [39] demonstrated that a protective coating with liquid phase-deposited titania (100–300 nm) resulted in enhanced efficiency of Kapton to overcome the problems associated with electrostatic discharge and reduction of atomic oxygen (AO)-induced surface erosion.

3.3 Conclusions

This chapter has presented a compendium of potential fibres and fabrics for aerostat and airship applications. The stringent performance requirements make the engineering design of these fabrics quite challenging. A monolithic fabric structure cannot fulfil the multiple requirements in terms of tensile strength, tear resistance, gas barrier and weather resistance while keeping the areal density within the specified limit. Therefore, multi-layered fabric structure is used in practice. Development of new-generation high-tenacity fibres and protective films are providing a necessary impetus to material development for aerostat and airships. Optimization of fabric structure in terms of weave and constructional parameters are also crucial to meet the requirement. Therefore, concerted efforts of materialscientists and textile engineers are needed to cope with the impending challenges of aerostat and airship design and development.

References

1. Arnold, E.M. 1977. Tethered aerostats used in TCOM systems. *Journal of Aircraft* 14: 1239–1243.
2. Mater, C.E., and Kinnel, M.J. 1992. Laminate material particularly adapted for hull of aerostats. US Patent 5118558, assigned to ILC Dover Incorporation.
3. Edwards, J.D. 1930. Balloon envelope. US Patent 1749474, assigned to Aluminum Company of America.
4. Zhai, H., and Euler, A. 2005. *Material challenges for lighter-than-air systems in high altitude applications*. In *Proceedings of the AIAA 5th Aviation Technology, Integration, and Operations (ATIO) Conference: AIAA 16th Lighter-than-Air Systems Technology Conference and Balloon Systems Conference*, Arlington, Virginia.
5. Niccum, R.J., Munson, J.B., and Rueter, L.L. 1977. Investigation of Kevlar fabric based materials for use with inflatable structures. NASA Contractor Report, Northfield.
6. Beach, G.R., Wheeler, M.S., and Jakubowski, P.R. 1989. Lightning hardened tether cable and an aerostat tethered to a mooring system therewith. US Patent 4842221, assigned to Westinghouse Electric Corporation.
7. Lindstrand technologies. n.d. Aerostats, https://www.lindstrandtech.com/what-we-do/aerostats (accessed April 14, 2020).
8. Gupta, A. 2005. *Improving UV resistance of high strength fibers*. MS thesis, North Carolina State University.
9. AG. Material Comparison Kevlar & Dyneema, n.d. https://static1.squarespace.com/static/54bd7b6de4b08f92b17133e1/t/54da70efe4b046b7e737b6e3/1423601903111/AG_Material+Comparison_Kevlar+%26+Dyneema_B.pdf (accessed April 14, 2020).
10. van Dingenen, J.L.J. 2001. Gel-spun high-performance polyethylene fibres. In *High-Performance Fibres*, ed. J.W.S. Hearle, 62–92. England: Woodhead Publishing Limited.
11. Greenhalgh, D. 2017. *Aerostat for electric power generation*. PhD diss., University of Southampton.
12. Vectran, n.d. https://en.wikipedia.org/wiki/Vectran (accessed April 14, 2020).
13. Hall, J.L., Fairbrother, D., Frederickson, T., et al. 2008. Prototype design and testing of a Venus long duration, high altitude balloon. *Advances in Space Research* 42: 1648–1655.
14. Liu, L.B., Cao, S., and Zhu, M. 2015. Mechanical characteristics of stratospheric airship envelope of Vectran fibre-reinforced-laminated composite. *Material Research Innovations* 19: S5606–S5612.
15. Kang, W., Suh, Y., Woo, K., et al. 2006. Mechanical property characterization of film-fabric laminate for stratospheric airship envelope. *Composite Structures* 75: 151–155.
16. Chae, H.G., and Kumar, S. 2006. Rigid-rod polymeric fibers. *Journal of Applied Polymer Science* 100: 791–802.
17. Zylon, n.d. https://en.wikipedia.org/wiki/Zylon (accessed April 14, 2020).
18. Komatsu, K., Sano, M.A., and Kakuta, Y. 2003. *Development of high-specific-strength envelope materials*. In *Proceedings of the AIAA Third Annual Aviation Technology, Integration and Operations Technology Forum*, AIAA, Denver, Colorado.

19. Li, A., Vallabh, R., Bradford, P.D., et al. 2019. Textile laminates for high-altitude airship hull materials – a review. *Journal of Textile and Apparel, Technology and Management* 11: 1–22.
20. Li, A. 2018. *Evaluation of laminated hull material for high altitude airship*. PhD diss., North Carolina State University.
21. Cunniff, P.M., Auerbach, M.A., Vetter, E., et al. 2002. *High performance "M5" fiber for ballistics/structural composites*. In *23rd Army Science Conference*, Orlando, Florida.
22. Lammers, M., Klop, E.A., Northolt, M.G., et al. 1998. Mechanical properties and structural transitions in new rigid-rod polymer fibre PIPD (M5) during the manufacturing process. *Polymer* 39: 5999–6005.
23. M5 fibre, n.d. https://en.wikipedia.org/wiki/M5_fiber (accessed April 14, 2020).
24. Rengasamy, R.S., Ishtiaque, S.M., Das, B.R., et al. 2008. Fabric assistance in woven structures made from different spun yarns. *Indian Journal of Fibre and Textile* 33: 377–382.
25. Schiefer, H.F., Cleveland, R.S., Porter, J.W., et al. 1933. Effect of weave on the properties of cloth. *Bureau of Standards Journal of Research* 11: 441–451.
26. Sharma, I.C., Malu, S., Bhowan, P., et al. 1983. Influence of yarn and fabric properties on tearing strength of woven fabrics. *Indian Journal of Fibre and Textile* 8: 105.
27. Durney, G.P. 1980. *Concept for prevention of catastrophic failure in large aerostats*. In *International Meeting and Technical Display on Global Technology 2000*, Baltimore, Maryland.
28. Alam, Md. S. 2019. *Studies on bending and shear behaviours of woven fabrics*. PhD diss., Indian Institute of Technology Delhi.
29. Adams, C. n.d. Light weight, spread filament, multilayer composite technology. *DSM Dyneema*, https://www.dsm.com/content/dam/dsm/dyneema/en_GB/Downloads/Presentations/Light Weight Spread Filament Multilayer Composite Technology by Chris Adams.pdf. (accessed April 14, 2020).
30. Adak, B., and Joshi, M. 2018. Coated or laminated textiles for aerostat and stratospheric airship. In *Advanced Textile Engineering Materials*, ed. Shahid Ul-Islam, and B.S. Butola, 257–287. Beverly: Scrivener Publishing LLC.
31. Joshi, M., Banerjee, K., Prasanth, R., et al. 2006. Polymer/clay nanocomposite based coatings for enhanced gas barrier property. *Indian Journal of Fibre and Textile Research* 31: 202–214.
32. Dasaradhan, B., Das, B.R., Sinha, M.K., et al. 2018. A brief review of technology and materials for aerostat application. *Asian Journal of Textile* 8: 1–12.
33. Raza, W., Singh, G., Kumar, S.B., et al. 2016. Challenges in design and development of envelope materials for inflatable systems. *International Journal of Textile and Fashion Technology* 6: 27–40.
34. DuPont. 2014. DuPont™ Tedlar® polyvinyl fluoride (PVF) films. https://www.dupont.com/solar-photovoltaic-materials/technical-resources/dupont-tedlar-polyvinyl-fluoride-pvf-films-general-properties.html (accessed October 22, 2021).
35. Nakadate, M., Maekawa, S., Kurose, T., et al. 2011. *Investigation of long term weathering characteristics on high strength and light weight envelope material Zylon*. In *Proceedings of the AIAA 11th Aviation Technology, Integration, and Operations (ATIO) Conference: AIAA 19th Lighter-than-Air Systems Technology Conference and Balloon Systems Conference*, Virginia Beach, Virginia.

36. Maekawa, S., Shibasaki, K., Kurose, T., et al. 2008. Tear propagation of a high-performance airship envelope material. *Journal of Aircraft* 45: 1546–1553.
37. Goedtke, P., Mathes, H., and Schaefer, W. 1978. Plastics for Aerospace Applications. *Kunststoffe, German Plastics* 68: 58–60.
38. Lavan, C.K., and Kelly, D.J. 2005. Flexible material for lighter-than-air vehicles. US Patent 6979479 B2, assigned to Lockheed Martin Corporation.
39. Gouzman, I., Girshevitz, O., Grossman, E., et al. 2010. Thin film oxide barrier layers: Protection of Kapton from space environment by liquid phase deposition of titanium oxide. *ACS Applied Materials and Interfaces* 2: 1835–1843.

4

Coated Textiles for the Envelope of Lighter-than-Air (LTA) Systems

Dipak K. Setua, Biswa R. Das, and N. Eswara Prasad
Defence Research and Development Organisation, Kanpur, India
Neeraj Mandlekar and Mangala Joshi
Indian Institute of Technology, New Delhi, India

CONTENTS

DOI: 10.1201/9780429432996-4

4.1 Introduction

Coated textiles consist of at least two layers. One is a base fabric which is continuously coated on either one or both sides of its surfaces with a polymeric material which generates either a combination or synergistic properties of the constituent materials. A polymer or an elastomer/rubber, usually in a viscous form either by dissolving them in a solvent and mixed with other additives, is applied directly onto the fabric surface using a variety of coating techniques. The coated fabric is finally cured or vulcanized (in the case of elastomer/rubber) in an oven. A polymer-coated fabric is an engineered composite material, where the fabrics provide the necessary tear and breaking strength, elongation, and dimensional stability. The polymeric coating confers some new functional properties to the product, e.g., improvement of impermeability to liquids or gases, resistance to degradation on exposure to the outdoor environment under ultraviolet radiation (UV) from sunlight, etc. [1]. The coating also offers general improvement of physico-mechanical properties e.g., flexibility and fatigue, good toughness, and lower abrasion loss [2]. Therefore, due care must be taken in the selection of basic polymers and fabrics, and specialty additives for the development of an effective coating to meet the ultimate functional properties of the finished fabrics.

Coated fabrics are used extensively in defence, transportation, health care, architecture, and sports product. Furthermore, impermeable coated textiles find a very large application in aerospace, inflatable equipment for defence, which includes aerostats and airships for passive air defence and surveillance. Other examples of inflatable textile-coated inflatables are life rafts, and life boats, air bags, life jackets, and gas holding balloons.

Most of the low-altitude aerostat and high-altitude airship hull materials are coated fabrics produced by a coating process or the combination of coating and lamination process in which one or several layers of materials are deposited on the surface of a textile substrate [3]. The majority of coating materials are polymer compounds having gas retention and weathering resistant properties. The success of the development of the polymer-coated textiles for the LTA application depends also on the coating technology or lamination technique used, process monitoring to adjust quality standards as well as safety and automation. Hence, this chapter explores the polymer-based textile coatings developed for improving the gas barrier and weather resistance property of LTA envelope material. The lamination fabric for developing the LTA envelope is discussed in a separate chapter, thus, it is not included here.

4.2 History and Technical Aspects of the LTA System

A historical overview of the significant materials used for the construction of LTA vehicles has been reported by Vadala, Air Vehicle Technology Department of the US Navy [4]. The design and development of a 2.5 meter diameter spherical helium tethered aerostat capable of deployment for long periods of time and modelling of nonlinear stresses under varied wind flow conditions by static finite element analysis have been described by Miller and Nahon [5]. The use of high strength, lightweight, and flexible composite made with a liquid crystal polymer (LCP)-based woven fabric core and a polyvinylidene fluoride (PVDF) film duly secured with polyurethane (PU) adhesive for low gas permeability have also been patented [6, 7]. An overview of technologies on the development of several aerial delivery systems, e.g., parachutes, aerostat, and aircraft arrester barriers, has been thoroughly reviewed by Gupta [8]. The application of ballooning technology of high-altitude airships (HAAs) and the relative merits of different hull materials composed of either polyethylene, polyester, laminated polyester, or biaxial oriented nylon film in association with different adhesives and varied joint strengths vis-à-vis temperatures have been reported by Raven Industries, USA [9]. Lockheed Martin Corp., USA [10] has patented on the use of metalized (silver, aluminium, etc.) flexible coating of thickness 1200 Å on a bias-ply layer (which is previously coated with a polyimide barrier film and bonded by PU adhesive to a single-ply yarn for construction of hulls of LTA vehicles.

Envelope fabric should have a high strength-to-weight ratio in order to reduce the size of the aerostat for particular payload capacity and hence increasing its efficiency. The coated fabrics should be resistant to environmental degradation by agents like UV, temperature, moisture regains, and hydrolysis. During operation in the stratosphere – which extends from 17 to 50 km above the earth's surface, and with 90% of the earth's ozone concentrated in this region at about −60°C, the coating may become brittle and eventually may lead to a catastrophic failure of the system. In addition, low gas permeability, higher tear and joint strength, low creep, and good abrasion resistance of the coated fabric are required for an enhanced service life of the envelope material. Therefore, a single fabric alone cannot meet all these requirements, and a composite structure composed of several layers of fabrics with varied composition or use of polymer membranes in between becomes necessary, as shown in Figure 4.1. The construction gives rise to an optimum balance between strength, endurance, flight performance, and deformation properties. Role of the warp and weft fibres, their coupled

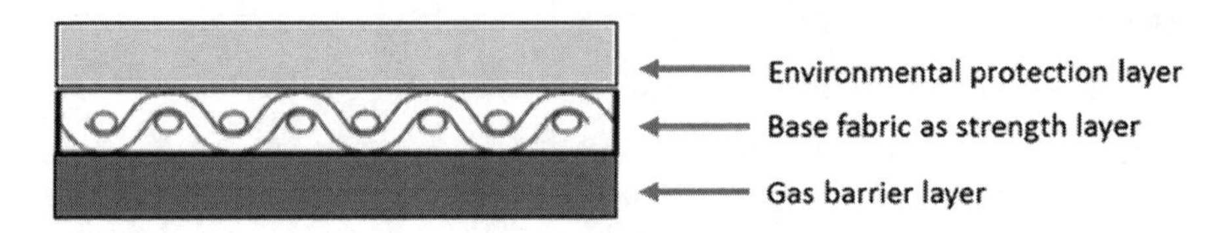

FIGURE 4.1
Different layers of typical multi-layered textile coated structure for the LTA system.

interactions, nonlinearity and viscoelasticity, and constitutive equation for micro-mechanical modelling according to the invariant theory of a typical envelope have been recently proposed by Meng and Lv [11]. An overview on the challenges on material properties i.e., use of strong yarn (fibre) suitable for the stratospheric environment, polymers and films used for coated fabrics with a requisite gas barrier/creep/fatigue/moisture and UV resistance, the effect of different sealing/seaming techniques for fabrication of structurally efficient joints, etc. have been reported by Zhai and Euler [12], and also by research scientists working at a defence organization in India [13, 14].

Environmental protection layer: The outer shell layer of the strength layer of a hull requires materials with very good weathering resistance, low-temperature flexibility, and better thermal reflectivity/emissive properties. PU-coated textiles provide high tear and abrasion resistance, low-temperature flexibility, good weatherability, UV and ozone resistance. Thermoplastic polyurethane (TPU) can be coated and adhesively bonded. Other candidates for LTA applications are plasticized polyvinyl chloride (PVC), poly-tetra-fluoro-ethylene (PTFE, Teflon) and polyvinyl fluoride (PVF, Tedlar) films, low-density polyethylene (LDPE)/polyester (Mylar), PVDF/vinyl chloride copolymers (PVDC, Saran), etc. [15].

Gas barrier layer: The inner shell in the woven fabric of the strength layer should be coated or sealed completely to achieve the gas barrier property. This can be accomplished either by coating the fabric with a suitable polymeric material or by laminating thin polymeric film with the strength layer. This layer serves as a gas barrier to retain helium gas inside the hull for a longer duration. In addition, the polymeric material should have excellent low-temperature flexibility, as well as good bondability or sealability with its adjusted layers, which increases the service life of the aerostat/airship. Generally, a polymer having good helium gas barrier properties such as neoprene, thermoplastic polyurethane (TPU), polyvinylidene chloride (PVDC), polyester (Mylar), and ethylene-vinyl alcohol copolymer (EVOH) are used in this layer [12, 15].

4.3 Construction of the LTA Envelope

The aerostat and the airship are aerodynamic bodies based on LTA technology as shown in Figure 4.2. Aerostats are kept near stationary using tethers, while airships are steerable. Typically, aerostats and airships are made based on a "balloon-within-a-balloon" concept where the envelope of the inner balloon is called the "ballonet" and the outer envelope is called the "hull" [16]. The ballonet is frequently filled with air during inflation or releases air during deflation, for maintaining the proper altitude as per requirement.

Any aerostat or airship hull material should have the following properties for providing a long service life [12]:

1. Good weather resistance especially against intense radiations (UV, ozone, etc.), temperature, and rain/moisture.
2. High barrier properties against helium gas to reduce the loss of filled gas and also to maximize the station time.
3. Flexibility at a wide range of temperatures, especially at a lower temperature, to avoid brittleness of the material and loss of the gas barrier property.
4. High tensile strength as it determines the maximum possible size of the aerostat.
5. Low weight and high strength-to-weight ratio to increase the payload capacity.
6. High tensile and tear strength to prevent catastrophic failure, which is very common.
7. Good abrasion resistance or wear resistance for better handling.
8. Low creep to avoid the change of aerodynamic shape with time and surrounding conditions.
9. Good bondability or sealability for obtaining good joint strength, avoiding leakage of gases.

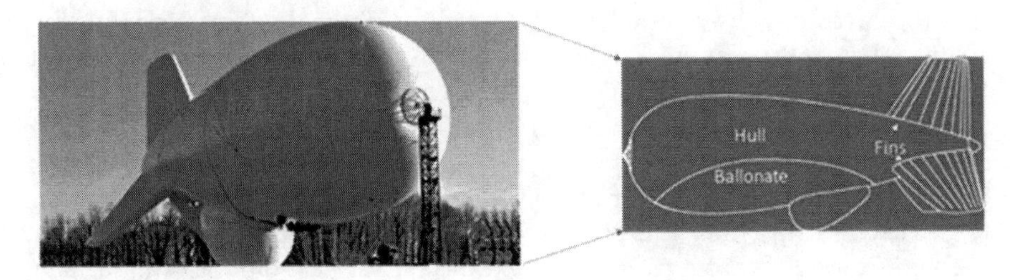

FIGURE 4.2
Digital photo of aerostat and its design as a balloon within a balloon.

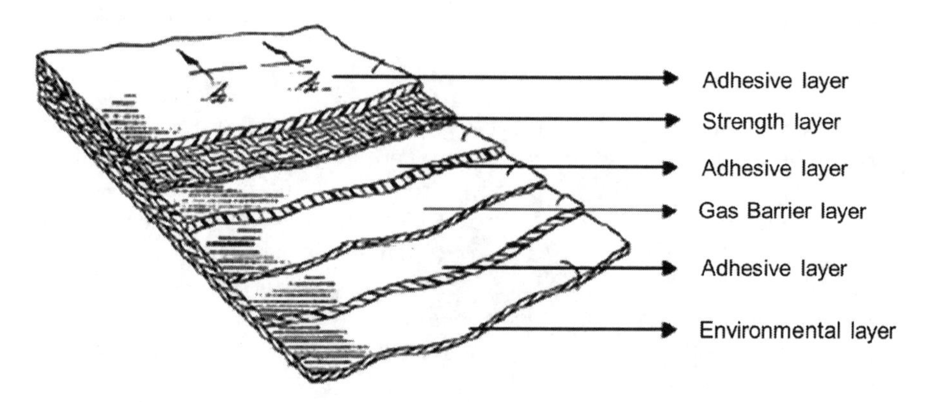

FIGURE 4.3
A multi-layered structure of a typical LTA envelope hull material [14].

In fact, any single fabric or film could not fulfil all these requirements. Therefore, to meet all the desirable properties there is a need for the envelope of the aerostat to be made of multi-layer laminated or coated fabrics. The typical construction of the hull material is shown in Figure 4.3. In a multi-layered structure, the individual layer performs a specific task, and in combination, the structure can fulfil all the requirements.

4.4 Textile Coating Techniques

The coating is a process in which a polymeric layer is applied directly to one or both surfaces of the fabric. The polymer coating must adhere to the textile and a blade or similar opening controls the thickness of the viscous polymer. The coated fabric is heated and the polymer is cured (i.e., polymerized). Required coating thickness is built up by applying successive coating layers, layer on layer. Interlayer adhesion must, therefore, be high [17]. Finally, a thin top layer may be applied for aesthetic or technical enhancement of the coating. Depending upon the end-use requirements, heavy-duty technical textile coatings may be applied at a high weight, while other end-uses for high-technology apparel may require coating with very low thickness. The chemical formulation of the coating, coating thickness and weight, number of layers, the form of the technical textile, and the nature of any pre-treatment (e.g., the desizing of fabric before coating) are of great importance [18].

There are several coating techniques are known, such as solution coating, melt coating, and transfer coating. Solution coating on textile/fabric can be done by various methods such as knife coating (knife-over-roll or floating knife), roll

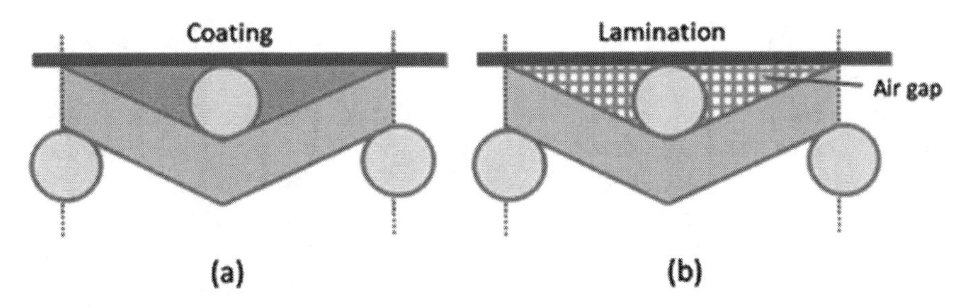

FIGURE 4.4
Schematic of coating and lamination process [19].

coating (direct-roll, kiss-roll, and gravure-roll coating), rotary screen coating, etc. Among these different coating techniques, the floating knife technique is preferred for applying tie-coat on the fabric, while the knife-over-roll coating technique is used for giving more uniform successive coatings on the tie-coat. Melt coating is mainly of two types: extrusion coating and powder coating. The hot-melt extrusion-based Zimmer process is very popular for making coated fabric for inflatables [18].

4.4.1 Comparison of Coating and Lamination Processing techniques

Both coating and lamination processes are used in the construction of LTA envelope material [19]. Lamination is the process of joining two or more layers into one structure where generally at least one layer is a textile material, bonded closely together by means of an added adhesive, or by the adhesive properties of one or more of the component layers. In coated and laminated textiles, the polymeric layer interacts differently with the textile fabric. Figure 4.4(a) shows how a coating formulation covers the surface of the fabric which facilitates coating formulation to penetrate the fabric structure, filling the air pockets and bridging the interstices. However, in the case of a laminate (Figure 4.4(b)), the polymeric film sits on the fabric surface and the fabric retains most of its air pockets, showing fewer points of contact. Therefore, in the coating process, better interaction is generally obtained between polymer and fabric compared to lamination [20].

4.5 Types of Coating Process

There are several processes for the application of the coating to the textile material depending upon the requirement of the end product. Nevertheless, some of the most popular coating techniques used in the LTA envelope material development are discussed below.

4.5.1 Solution Coating

Solution coating is the simplest coating procedure and known as a direct coating method, generally accomplished by knife coating techniques. There are two such techniques: the 'floating knife', or knife over air; and 'roll coating', or knife over roll technique. The tie-coat is made by a floating knife technique, which provides better penetration into the substrate and hence improves adhesion of the coating to the substrate, as shown in Figure 4.5(a). Another technique, knife over roll, gives a more uniform coat, and hence it is used for the subsequent coats where the fabric is stretched flat to form an even uniform surface and transported under a stationary doctor blade [21]. As the fibre moves forward, it is scraped by the knife and the polymer resin compound is spread evenly over the surface as per Figure 4.5(b). The amount of polymer applied, the 'add-on', depends on the concentration of polymer in solution – this is the so-called 'solids content'. In general, fairly tightened woven fabrics capable of being pulled flat and uniform can be coated by the direct coating method [22].

4.5.1.1 Hot Melt Extrusion Coating (Zimmer Coating)

The knife coating technique has a tendency to develop pinholes and makes potential sources of leakage. However, the hot-melt extrusion coating method provides a film-like coating and hence provides a better gas retention property. This method is used for thermoplastic polymers such as polyurethane, polyolefins, and PVC, which are applied by feeding granules into the nip between moving heated rollers as depicted in Figure 4.6. The popular design in use is Zimmer machine, which has two melt rollers [22]. The latter is a smaller version of calendars but differs in the need for a fabric (or paper or a film) as a substrate on which to deposit the film as it is produced. This method can apply resin (in the form of films) on fabric at a faster rate than that can be achieved by transfer coating or direct coating. This process is used to produce lightweight coverings or tarpaulins [23].

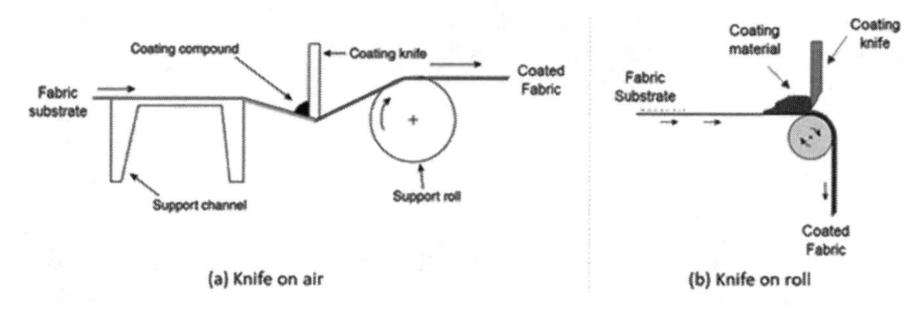

FIGURE 4.5
Solution knife coating, (a) knife on air, (b) knife on roll.

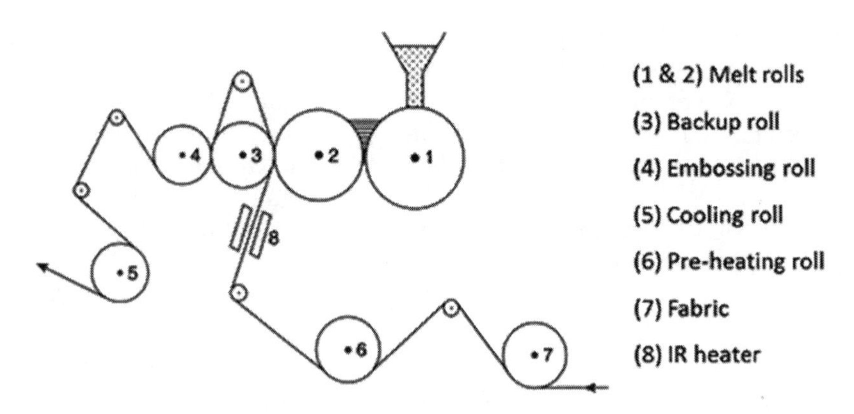

FIGURE 4.6
Schematic of the Zimmer coating process.

Various other coating methods are used in the textile industry and some time for coating the fabric surfaces for the LTA applications. Other methods are discussed next:

Kiss roll/Gravure coating: In this method, the coating material is picked up by the roll and is transferred to the textile substrate as it passes through the nip between the Gravure roll and the back-up rubber roll. The arrangement is suitable for very light coating (< 0.02 g/m^2). The kiss roller can be rotated either in the same running direction of the substrate or in the opposite direction.

Dip coating: This method is more popular for finishing the textile surfaces. In this process, the substrate fabric is immersed in a tank and passed through squeeze rolls. The excess material is then squeezed out by passing through nip rolls calibrated for the exact add-on.

Rotary screen coating: It is used for printing of fabric with add-on up to ~500 g/m^2, which is controlled by the coating material's viscosity. The number of perforations (mesh number) made on one or more nickel cylindrical screens serves to push the coating through the perforations of the screen. The coating released in the form of dots merges to form a continuous layer on the fabric surface.

Spray coating: In this technique, a spray gun system creates the provision of passing heated air with a molten polymer adhesive to produce many continuous filaments through a nozzle. The filaments are applied to the moving fabric. Once the substrate is uniformly coated, it is passed between two rolls and pressure is applied to achieve a permanent adhesion.

Calender coating: Calendering is the versatile method of coating of fabrics with polymers. A calender consists of a number of rollers, at least three or sometimes more in various configurations. The compounded polymer dough is fed between the roll nips which rotate the dough and smooth it into films of uniform thickness. The thickness of the film is determined by the gap of separation of the rollers, but there is usually a limit to the thinness of films which may be produced by this method. More numbers of rollers produced the film thickness with more accuracy.

4.6 Coating Formulation for LTA Envelope Material

The formulation of a coating is complicated, and it can contain a wide range of chemicals depending upon the nature of the polymer, the necessary additives for the specific end-use, whether the coating has to be foamed before application, and the type of coating machinery to be used. It is important to realize that coating formulations consist of several additives. Amongst these are UV radiation and heat stabilizers, antioxidants, fillers to improve the mechanical properties, fillers for gas retention, pigments, and other compounds necessary to confer further special properties or to assist with processing [24]. It is necessary to make sure that all the ingredients are compatible with each other.

Thermoplastic polyurethane (TPU) is one of the few synthetic polymers which has found a wide industrial application for the processing and imparting properties for the LTA envelope structure. TPU based coated fabric is compounded with a variety of additives. Some of the important additives are as follows:

Fillers for gas barrier: Various types of nanomaterials have been used with the polymer to enhance the barrier property of a coated fabric. The shape of the nanofillers, such as granular, spherical, and platelets, has a strong effect on gas permeation [25]. In particular, nanofillers having platelet geometry, such as layered silicate (nanoclay) graphene, have a huge potential for improving the gas barrier properties of the polymer when dispersed in polymer matrices. Adak and Joshi have exploited various types of nanofillers to enhance the gas barrier of PU nanocomposite [26, 27]. Joshi, Adak, and Butola [28] reviewed extensively the polyurethane nanocomposite coatings with a high gas barrier property for LTA applications. In such a nanocomposite structure, a very low gas permeation is occurred due to the increase in the *tortuous path* of the permeate gas molecule in the presence of layered-structured nanofillers [29].

UV absorbers: Most of the polymeric materials are sensitive to UV radiation in the wavelength region around 300–360 nm. The individual polymer absorbs and is degraded by UV within a much narrower region. A number of organic and inorganic UV absorbers have been used in polymers. A good UV absorber should be able to transform the absorbed UV into less harmful energy, which will prevent polymer degradation [30]. The most common organic UV absorbers are low molecular weight derivatives of o-hydroxybenzophenone, o-hydroxybenzotriazole, or o-hydroxyphenyl salicylate. For example, Tinuvin 326 [2-tert-butyl-6-(5-chlorobenzotriazol-2-yl)-4-methylphenol], Tinuvin 327 [2,4-di-tert-butyl-6-(5-c hlorobenzotriazol-2-yl)phenol], and Tinuvin 1130 [methyl 3-[3-(benzotriazol-2-yl)-5-tert-butyl-4-hydroxyphenyl] propanoate] have been found effective UV absorber for TPUs [31–33]. Hindered amine light stabilizers (HALS) also act as scavengers for free radicals that may be generated during UV initiated oxidation of polymer compounds [34]. They are typically organic compounds having 2,2,6,6-tetramethylpiperidine ring structure. HALS shows high efficiency at relatively low concentrations [30, 35]. At only 0.1%, HALS provides protection equivalent to typical UV absorbers used at 3% or 4%.

Besides, carbon black which is an excellent UV absorber, inorganic nanofillers and semiconductors such as zinc oxide (ZnO) [36], titanium oxide (TiO_2) [37], cerium oxide (CeO_2) [38] and silica (SiO_2) [39] are important inorganic UV absorbers. Inorganic oxides are non-volatile, non-migratory, light, and thermally and chemically stable; they are added in TPU coating solution at about 3 to 5 wt %. Solid band theory can be used to explain why TiO_2, ZnO, and CeO_2 can absorb UV light by 'band gap exciton theory', converting the harmful UV radiation to harmless infrared radiation [40]. Moreover, these fillers can also reflect UV light if a continuous top coating is applied. Graphene can also act as a UV absorber [41].

Cross-linkers: The primary role of a cross-linker is to form a bond between the polymer and textile surfaces. Cross-linkers are low molecular weight reactive chemicals creating a linkage by evolving a gaseous or water molecule. Generally, isocyanate and amine-based cross-linkers are widely used for thermoplastic polyurethane coating bonding with a fabric surface.

Antioxidant: In recent years, various nano-antioxidants have been developed for diverse applications. Antioxidants may also be covalently attached to PU elastomers to provide stabilization against oxidation. Amide-based antioxidants are more popular for thermoplastic polyurethane.

Processing aids: The role of a lubricant is to facilitate processing and control the processing rate. Mineral oil, silicone oils, vegetable oils, and waxes are common lubricants. The compatibility of lubricants is low, resulting in their exudation at processing conditions

4.7 Types of Fibres and Fabrics Used for Coating for LTA Applications

In the typical LTA hull material, the strength layer is a woven fabric, either single or multi-layered. The most challenging aspect of designing the strength layer is the identification of a structural fibre. The low strength-to-weight ratio of traditional natural fibres is not able to meet the strict mechanical property requirements of high-altitude LTA systems. With the continuous development of synthetic fibres, the textiles used in the airship industry have witnessed many significant breakthroughs, achieving increasingly high strength-to-weight ratios.

At an early stage of aerostat and airship development, polyester and nylon fabrics were used commonly in the strength layer of the hull. The combination of moderate strength (≈10 g/den) and moderate extension (> 10%) of commercial polyester and nylon fibres results in good work of rupture, making them suitable for aerostat envelopes. In particular, high-tenacity polyester fabrics are extensively used for making aerostat envelopes due to the following advantages. In the late 20th century, high-performance fibres (high tenacity: 20 g/den and high modulus: over 300 g/den) became available. Examples of some of these high-performance fibres are Zylon, Spectra, Vectran, Kevlar, and M5 [12]. With the invention of these new high-performance fibres, the performance of airships also increased in terms of longer service life, higher payload, and higher altitude of floating of an airship. Some high-performance fibres, including Vectran and Kevlar, are less strong than Zylon but possess advantageous properties [42]. Some of the properties of high-performance fibres for LTA applications are compared in Figure 4.7.

Various high-strength textile fabrics are available as substrate for coating as the following [12, 42]:

Nylon 6, or 66: Basically, they are the aliphatic polyamide which exhibits very good strength to mass ratio, good elastic recovery, and excellent flex resistance.

Polyester: Polyester fibres are similar to nylon in terms of their various functional features and better UV resistance but poorer elastic recovery than nylon. However, considering the high strength-to-weight ratio, low creep, low moisture regain, and improved hydrolysis resistance, the polyester becomes a very good choice for LTA applications.

Aramid: Essentially, these are aromatic polyamide fibres from DuPont (Kevlar) which possess a very high strength-to-mass ratio along with high thermal resistance. But their resistances to photo-degradation, flexibility

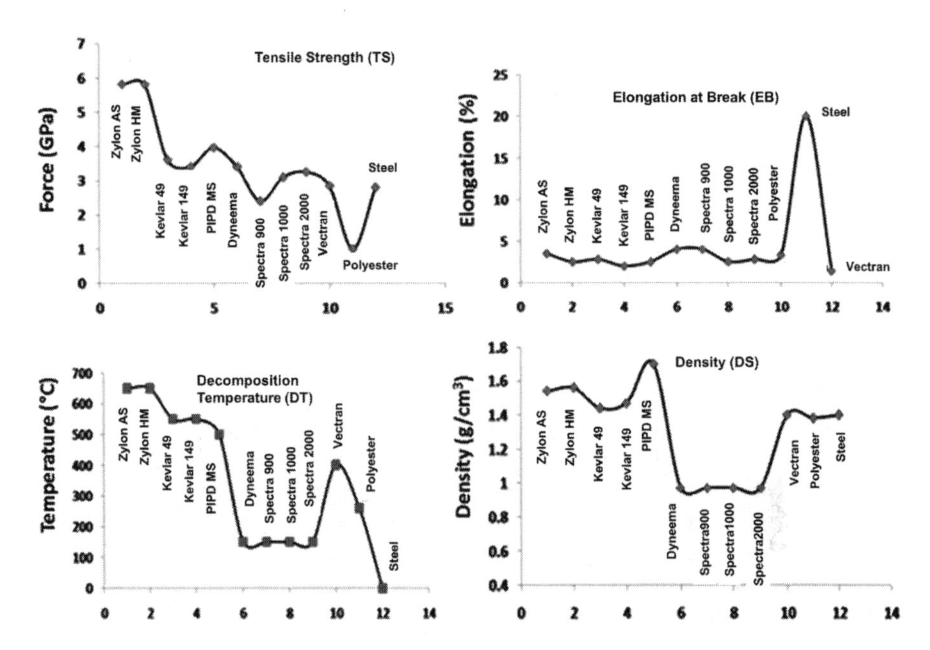

FIGURE 4.7
Properties of some high-performance fibres for the strength layer of the LTA envelope.

and crease retention are poor. Technora, a product from Teijin, is made up of the blend of para- and meta-aramid and found to possess higher chemical resistance, elongation at break, very good abrasion resistance, and lower moisture regain compared to Kevlar [9, 43].

Vectran: It is a modified type of aromatic polyester spun from the LCP. Vectran exhibits exceptional strength-to-mass ratio and modulus as well as displays superior flex property compared to Kevlar.

Zylon: Chemically, it is Poly (p-phenylene-2,6-benzobisoxazole) (PBO) and represents varieties of thermoset LCPs which gives exceptional strength to mass ratio and rigidity and it is 60% stronger than Vectran in specific strength.

Ultra-high-molecular-weight polyethylene (UHMWPE): Dyneema (Toyobo Co., Osaka, Japan) and Spectra (Honeywell Co., NC, USA) are gel-spun and highly oriented strands of synthetic polyethylene known as UHMWPE. They exhibit outstanding toughness and a very high strength-to-mass ratio; e.g., for Dyneema, it is about 40% higher than aramid. They are also resistant to alkali, moisture, UV degradation, and display good abrasion resistance, but the main drawback is poor creep resistance.

M5: Polyhydroquinone-diimidazopyridine (PIPD) is a high-strength synthetic material, currently produced by the Magellan Chemical Co., USA, and is stronger than Kevlar, Technora), Dyneema, and Spectra.

An LTA application requires high-strength, high-modulus (>300 g/den) and high-tenacity (> 20 g/den) fibers [44]. Specific laminated fabrics comprised of at least one woven fabric, with yarns having twist less than six turns per inch, high aggregate strength > 10 g/den, and containing a gas barrier layer have been found to provide very good tensile, tear, and flex fatigue resistance suitable for an airship envelope. In another publication, a low-weight and high-strength triaxial fabric coated with a heat-sealable polymer on one side and a gas-impermeable film to the other side, a total thickness of 150 to 250 micron, weight < 271 GSM, tensile strength > 136 kgf, and tear strength > 11.5 kgf have been reported [45]. Modelling of weave geometry for the airship fabrics have been reviewed by Yousef and Stylios [46]. The use of coating formulation for airship skin material with a barrier layer containing SiO_2 1.5% to 2.5% and particle size of 20 ~ 70 nm has been found to significantly improve the barrier effect of helium (permeability rate of 0.005 ~ 0.02 L/ m^2/24h) [47].

4.8 Types of Polymeric Material Used in LTA Application

A coating or film is necessary to provide weatherability and gas retention characteristics. The material chosen for this layer/layers must also have good shear stiffness, bondability, and thermal reflective/emissive properties. A wide range of polymers is used in coating for the LTA application [48, 49]. The functional properties of the coated fabrics depend on the chemical nature of the polymer, additives used in polymer compounding, and the processing technique adopted. Table 4.1 indicates some major types of polymers and products with coated textiles.

Polyvinyl chloride (PVC): Plasticized polyvinyl chloride (PVC) is a versatile material, commonly used in commercially coated fabrics. It has good low-temperature flexibility, exhibits good weathering (5 years), ozone resistance, and heat sealability, and is inexpensive. Inherent fire resistance and excellent oil and solvent resistance make it a perfect material to be used in the LTA envelope. PVC has been exploited in a number of applications such as tarpaulins, tents, and architectural products, upholstery items like synthetic leather, protective clothing, and aerostat and aircraft arrester barrier nets.

TABLE 4.1

Examples of Polymers Used in Textile Coating and Product Types

Material	Gas Barrier Property	Weatherability	Flex Fatigue	Adhesion to Fabric/Film	Heat Sealability
Polyvinyl chloride (PVC)	Fair	Good	Good	Excellent	Yes
Polyvinylidene chloride (PVDC)	Excellent	Poor	Fair	Fair	Yes
Thermoplastic polyurethane (TPU)	Fair	Good	Excellent	Excellent	Yes
Poly-tetra-fluoro-ethylene (PTFE)	Good	Excellent	Good	Poor	Yes
Butyl rubber (BR)	Excellent	Fair	Excellent	Fair	—
Polychloroprene rubber (CR)	Good	Fair	Fair	Good	Yes
Chlorosulphonated polyethylene rubber (CSM)	Good	Excellent	Fair	Yes	No

Polyvinylidene chloride (PVDC): It is a copolymer of vinylidene chloride/vinyl chloride known as Saran, which is an excellent barrier material and used extensively in packaging applications. However, it is not recommended for use in a composite LTA fabric because of its poor flex life especially at low temperatures. oxygen permeability, transparent and high gloss, heat weldable, unaffected in sunlight, resistant to acid, harder and more brittle than PVC, soluble in the organic solvent, turns yellow or brown if exposed to alkalis, produces pungent odour and evolves toxic HCL on burning upholstery items. A thin-film for food wrapping, it can be blended with acrylics and PVC to improve flame retardancy in coatings.

Thermoplastic polyurethane (TPU): Polyurethane coating on textiles gives a wide range of properties to meet diverse end uses like apparel, artificial leather, fuel and water storage tanks, inflatable rafts, and containment liners. Polyurethane is available in many formulations and possesses an excellent balance of properties. It has outstanding overall toughness, high tensile strength, tear strength, abrasion resistance requiring much less coating weight, low-temperature flexibility, fair gas permeability, good handling properties, crease resistance, and good weatherability and ozone resistance. Thermoplastic polyurethane can be heat sealed, adhesively bonded, and laminated to other substrates.

Poly-tetra-fluoro-ethylene (PTFE): DuPont FEP fluorocarbon film offers excellent optical properties and outstanding weather resistance. Teflon film can be heat-sealed, thermoformed, welded, metalized, and laminated to many other materials. The low permeability to gases, high service life, and low-temperature toughness (service temperature range from −240°C to 205°C) make Teflon film a good candidate for high-altitude LTA applications. PTFE has excellent resistance towards UV exposure, acid, alkali, chemical, solvent, oil, weathering, and flame but its high cost and anti-sticking properties limit its application LTA envelope hull material.

Butyl rubber (BR): Butyl rubber offers excellent impermeability to gases, better flexing, good abrasion, shock-absorption, electrical, heat, oxidation, ozone, water, and acid and chemical resistance, but its solvent and fire-resistance are poor. They can be blended with NR and chemically modified to chlorinated IIR NBC clothing e.g., boots, gloves, impermeable suit, inflatable tank, rain jackets, air cushion bed, pneumatic spring, acid lining material, life vests, etc.

Polychloroprene rubber (CR): Its features include high tensile strength, good adhesion, resistance to oxidation, ozone, fats and oils, flex cracking, and inherent flame retardant, and usefulness as an adhesive. It stiffens at low temperature, and use of plasticizer can improve flexibility and lower resilience. Coated textile for polyester and nylon fabrics are used for a flexible water tank for potable water, inflatable items with self-extinguishing property, underwater suit, etc.

Chlorosulphonated polyethylene rubber (CSM): It offers excellent resistance to flame, ozone, corrosive chemicals, electricity, water, oil, and colour stability upon light exposure, but it is very expensive. It is used in hot conveyor belts, roof sheeting and pond liners, acid-resistant tank linings/hoses, static seals, and membranes.

4.9 Rubber-Based Textile Coating for LTA Applications

A considerable variety of elastomer coatings have been used to provide a good gas barrier in the construction of fabrics for the hull and ballonets. In actual use during the 1930s, the permeability of the gel latex coating was 2.0 L/m^2/24 hours. In the next decade, natural rubber was used to provide a good gas barrier for many years. The natural rubber was used between cotton plies and as a coating for the interior and exterior of the fabric [4]. Advancements in materials led to the replacement of natural rubber with neoprene. Figure 4.8 shows the construction of fabric used for the

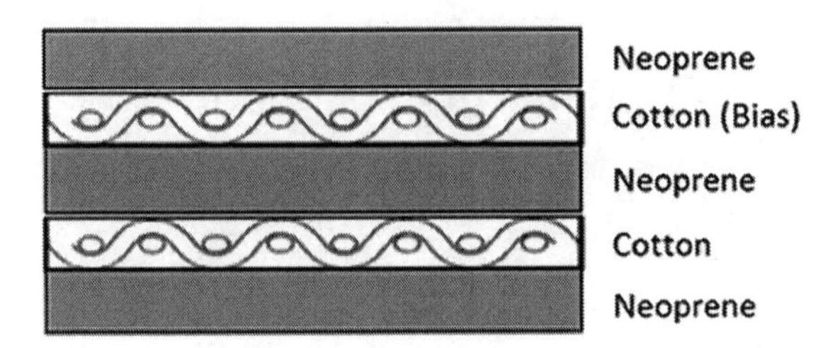

FIGURE 4.8
Construction of airship hull material based on neoprene during the 1950s [4].

early barrage balloons and airships hulls of the 1950s using neoprene as the gas barrier. The permeability was between 2.5 and 3.0 L/m^2/24 hours [4].

In the 1960s the cotton was replaced by polyester fabric. The smooth surface of continuous polyester yarns produce a flatter fabric and less coating was needed on polyester fabric for permeability and weathering resistance. In order to improve the weather resistance of this polyester fabric, the outer surfaces were coated with chlorosulphonated polyethylene (CSM) known commercially as Hypalon. This coating has good ozone resistance and provides good weather resistance. In addition, the outer layer was coated with aluminized layer to provide strong protection against environmental factors. This laminated fabric of Dacron had a slightly higher cylinder bursting strength and half the permeability, and it weighed approximately 102 GSM less than the comparable cotton laminated fabric. The weight reductions on larger airships were greater, as high-tenacity yarns were used in both plies. Studies were made to determine the minimum coating weight for permeability and weathering properties.

The permeability of the coated and laminated fabric can be affected by the shearing stresses applied during service. Some fabrics developed for airship envelopes have shown a considerable increase in permeability when tested under tension. When the tension is removed, the fabric rapidly returns to a low permeability state. The permeability of fabrics subjected to this cyclic test showed a good correlation with the permeability of similarly constructed fabrics removed from airships. The evaluation procedures performed on the Dacron laminated fabrics were based on conditions occurring during the airship service. In service, however, these various conditions usually occur simultaneously and are not duplicated by various combinations of laboratory tests. Thus, exposure in service was necessary for a complete evaluation. Actual records indicated that the loss of helium was considerably less in the Dacron envelopes than that experienced with the cotton-laminated airship fabric. Dacron-laminated fabrics were used to make the last few airships made for the Navy. Even today, Goodyear makes 200,000 cubic-foot-volume airships of Dacron and neoprene [50].

4.10 Preparation of a Rubber-Coated Fabric: A Case Study

Two methods were used for the preparation of rubber-coated fabrics as the following [51]:

(a) *Dry method*: The dry method uses rubber compounded in a mixing mill or Banbury. A three-roll rubber calender was used to apply the dry rubber to the textile substrate. The rolls of the calender can adjust the correct thickness and combine it with the textile fabric.

(b) *Solution or cement method*: Rubber compound is dissolved in toluene or methyl ethyl ketone (MEK) and spread upon a textile substrate. Only a small amount of coating could be applied by the solution method, and heavier coatings were applied by the dry method. The rubber compound material in sheets was cut into small pieces and dissolved in a churn, to 35% solids. The application of rubber to the textile was done by impregnating or by a spreading machine with a knife over a rubber roller with ovens attached, as shown in Figure 4.9. The fabric is pulled through the gap between the knife and the rubber roll, and a three-zone oven. The temperatures of the ovens were varied, depending upon the type of solvents and the rubber, into different zones of 150°F, 200°F, and 275°F.

(c) *Curing*: For elastomer coated fabric, vulcanization is necessary to give the proper physical properties and is generally accomplished by festooning in a dry heat chamber under specific conditions of time and temperature. Coated fabrics are also vulcanized in roll form. The material is dusted with mica or talc and wrapped around metal

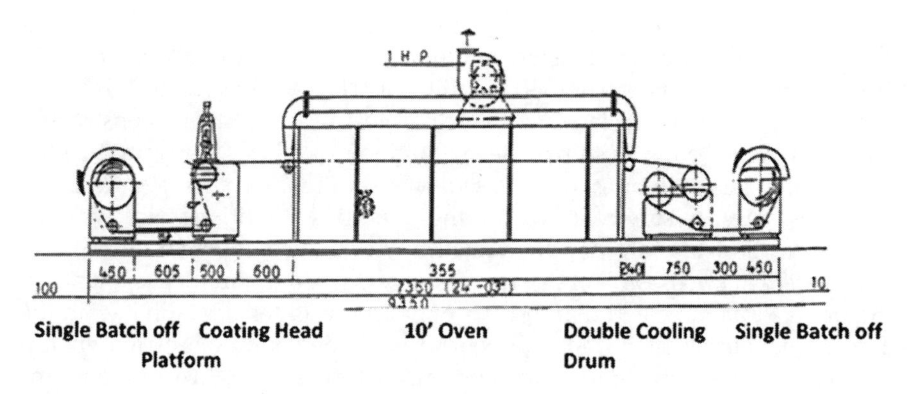

FIGURE 4.9
Direct coating system manufactured by Sanjay Industrial Engineers Pvt. Ltd., Gujarat, India.

drums. Water-resistant covers are put over the metal drums and the total assembly can be rolled into a closed chamber steam vulcanizer.

The rubber coated textile, thus prepared, was used in extreme cold climate (ECW) clothing and for items for NBC applications. Strategic applications of this coated fabric necessitate flame retardancy, oil and water repellency, and sufficient air permeability. Substrates were of wide choice, e.g., cotton, nylon, polyester, fiberglass, asbestos, and blends of these fabrics. The list of elastomers used includes nitrile rubber (NR), styrene-butadiene rubber (SBR), nitrile butadiene rubber (NBR), chloroprene Rubber (CR), chlorosulphonated rubber (CSM), butyl rubber (BR), and Silicone rubber [51]. In the case of woven fabrics, they were first subjected to a boiling-off process that removes all the finishes and then treated on a contact hot drum at 163–190°C in order to minimize the shrinkage of fabric at processing temperatures.

4.11 Recent Advances in Polymer Coated Textiles

4.11.1 Electrospinning Technique for Preparation Nanofiber Web Coating

The electrospinning process is used to produce polymeric nanofibers by electrically charging a suspended droplet of polymer melt or solution ejected by a spinneret. The method uses a high electric field ~70–80 kV and nanofibers of diameter varying from 10 μm to 1000 nm can be produced. The schematic diagram of the electrospinning set up is shown in Figure 4.10. The polymeric solution is held at the tip of a capillary tube by surface tension, and the applied electric field charges the polymer solution which causes a

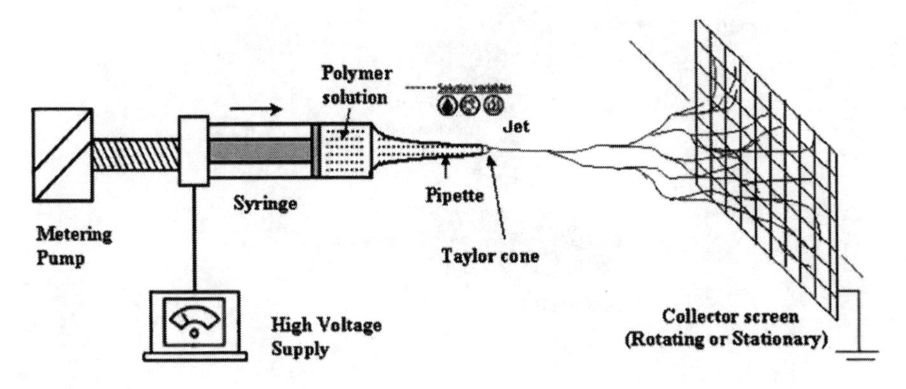

FIGURE 4.10
Schematic diagram of the electrospinning apparatus.

force directly opposing the surface tension and forms a conical shaped liquid known as a Taylor cone. The increase of the electric potential overcomes the surface tension effect and results in the emission of the charged jet of polymer which eventually thins down in air by solvent evaporation and transforms into randomly oriented nanofibers. The nanofibers are collected on a metallic collector/drum. The experimental set-up provides an opportunity to control the thickness and composition of the nanofibers, together with the generation of porosity for their huge surface area. Other techniques for the production of polymer nanofibers include melt blowing, solvent casting, and multi-component processes involving extrusion techniques [52]. Although these methods have significantly higher productivity than electrospinning, the latter gives more control over the formation of finer fibres. Hence electrospinning is favoured for the nanostructuring of materials as fine fibres. The latest trends in electrospinning include coaxial spinning, which produces fibres with different core and shell morphology [53]. Figure 4.11 shows the schematic diagram of a coaxial electrospinning setup for the in-situ generation of a nano-fibre-reinforced polymer composite. Various other modifications are also available: (i) a multi-layering technique in which different polymer solutions are deposited layer by layer; and (ii) multi-jet electrospinning of different polymers to form composite structures or hybrid nanofibers [54]. Very recently the concept of reactive electrospinning has emerged, whereby

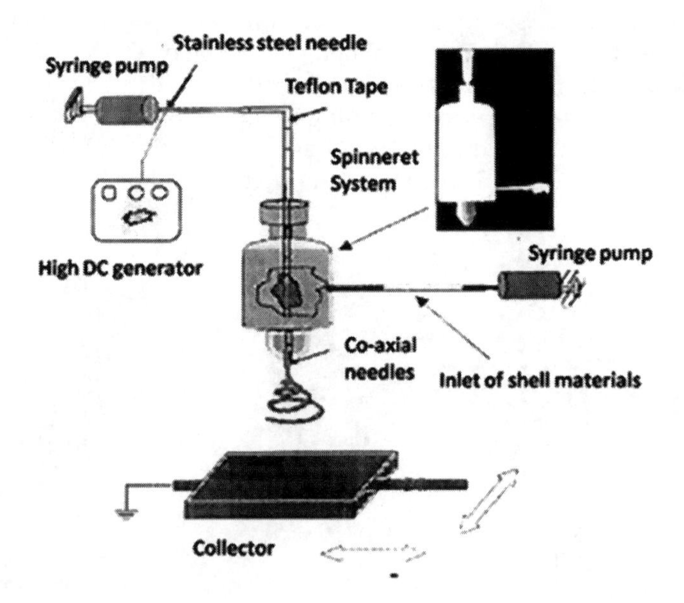

FIGURE 4.11
Schematic diagram of a coaxial electrospinning setup for in-situ generation of a nano-fibre-reinforced polymer composite.

cross-linking of polymeric fibres is done during spinning. This technique avoids an additional step of post-spinning cross-linking for water-based solvent-polymer systems [55]. The latest development in 'needle-free' electrospinning is Elmarco's 'Nanospider (NS) technology', which provides significant scaling-up benefits in terms of production rate and product quality as compared to conventional electrospinning [56]. This NS technology is well recognized by industry leaders as providing excellent web and fibre uniformity, high productivity, and the ability to consistently meet the key performance characteristics [52]. In this technology, droplet formation and jet initiation is a self-organized process, whereby numerous jets are formed simultaneously from droplets. NS technology provides versatility in terms of processing a variety of polymeric nanowebs. The coating of nanowebs on conventional textiles allows the possibility to augment chemical protection, great potential for enabling lighter-weight textiles with multi-functional capabilities such as water repellency, UV protection, and flame retardation [57].

4.12 Smart and Intelligent Fibres and Fabrics

"Smart textile' is derived from the conventional textiles coated with smart or intelligent polymeric materials by various fibre-forming processes e.g., melt/wet/electro spinning, etc. In the past two decades, clothing industries have evolved smart textiles where more functions, e.g., protection for adverse climatic conditions, thermo-regulatory, safety, fashionable, or other conveniences, can be accommodated. However, these endeavours have largely benefited from the use of new polymers and are capable to combine them into fibres and fabrics. Broadly five different types of functions can be envisaged in a smart textile: (i) sensor, (ii) data processing, (iii) actuation, (iv) communication, and (v) data storage. All these functions have their designated role, although not all of them may be present at the same time in each of the items. The functions may be due to the intrinsic property of the materials or realized while constructing a smart structure. The polymers can change the surface property by altering the light-scattering coefficient. Incorporating the fibre sensors in the cloth will enable the measurement of temperature, stress-strain properties, sensing of the presence of gases, toxicants, etc. The most important polymeric materials which are at present in use for construction of smart and intelligent textiles are classified as microencapsulated phase change materials (PCM), shape memory polymers, stimuli-responsive polymers, chromic materials, conductive materials, and electronics incorporated textiles [58, 59].

The bicomponent fibre-spinning technique has made it possible to incorporate the PCM materials inside the hollow fibres, where the core consists

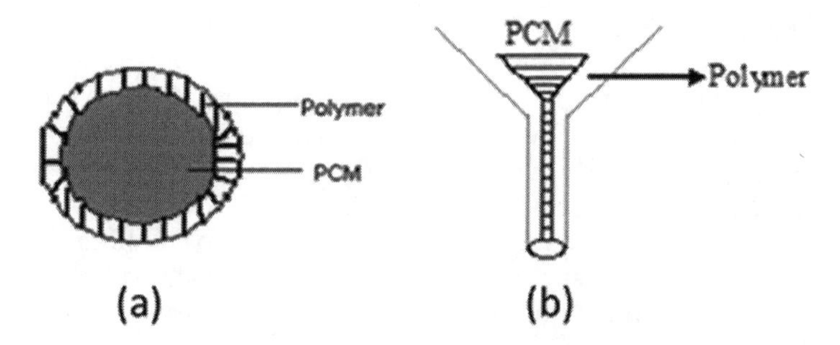

FIGURE 4.12
Polymer encapsulated PCM fibre developed by melt spinning; (a) encapsulated fibre, (b) spinneret assembly.

of a PCM material and the sheath could be of any polymer like polyethylene, polypropylene, nylon, or polyacrylonitrile [60]. By this process, simultaneously, two different polymers/materials through the same orifice can be extruded by dry or melt spinning to combine a core with a top sheath component, as shown in Figure 4.12 [61]. Two sets of electric drive motors, gear boxes, drive shafts, and pumps are required in order to feed by twin reservoirs via capillary tubes using two sets of pressure transducers with a central temperature measurement system. The capillaries from the reservoirs feed into a central mixer head, allowing the core precursor to feed directly to the core spinneret, whilst the sheath capillary divides via a cannula into two capillaries rounded on either side of the core capillary.

Concerning the materials and processes, important parameters are: (i) molecular weight of the polymer and choice of solvent, (ii) core-to-sheath spinneret ratio, (iii) ratio of the wall thickness of the core spinneret to the total diameter of the spinneret orifice, (iv) draw ratio or the reduction in diameter due to extrusion, (v) length of the chamber, and (vi) the direction of gas flow. The fibres can be subsequently cured using similar techniques to those employed to cure monolithic (single component) fibres. The microcapsules of PCMs can also be incorporated in foam which enables a higher amount of microcapsules to be incorporated into these fibres.

Smart photosensitive fibres are used to develop interactive electronic textiles (IETs), fabrics with auto camouflage, etc. The photosensitive polymers can be incorporated in the fabric or transformed into fibres themselves. On excitation with a photon of a characteristic wavelength, the amplification can produce electricity that can be stored and used in electronic devices in the system for military application.

4.12.1 Nanocomposite-Based Textile Coatings

Polymer nanocomposite is the new class of materials with an ultrafine dispersion of nanofillers or nanoparticles in a polymeric matrix, where at least one dimension of nanofillers is smaller than about 10 nm [58]. Due to the nanoscale dimension and very high surface-area-to-volume ratio, the nanomaterials have the potential to enhance the properties of the polymer when dispersed properly in the polymeric matrix. The use of nanocomposite-based processes is growing fast in all fields of science and technology. Nanocomposite coating is an attractive way to impart the new functionality or properties which are not achievable by conventional coating. The textile industry is also experiencing the benefits of nanotechnology in its diverse field of applications. Textile-based nanoproducts starting from nanocomposite fibres, and nanofibres to intelligent high-performance polymeric nanocoatings, are not only getting their way in high-performance advanced applications; nanoparticles are also successfully being used in conventional textiles to impart new functionality and improved performance [58, 59]. With these improved set of properties, they show promising applications in developing advanced textile materials such as nanocomposite fibres, nanofibres and other nanomaterial-incorporated fibres and coated textiles for applications in medical, defence, aerospace and other technical textile applications such as filtration, protective clothing besides a range of smart and intelligent textiles. Polymer nanocomposite-based high-performance coated fabrics are being developed, which might be potential materials for preparing envelopes of futuristic LTA systems. Recently, Prof. Mangala Joshi has edited a book on nanotechnology in textiles, with special attention to polymer nanocomposite-based textile coating for gas barrier and weather resistance properties [62]. Extensive research is being carried out around the world to explore the potential of different nanomaterials to improve specific properties of the polymer. In this regard, novel polyurethane (clay)-based nanocomposites as coatings for inflatables have been explored by Joshi and Adak [19, 63]. The coated fabrics showed improved gas barrier property without affecting the transparency and tear strength.

4.12.2 Polymer Nanocomposite Coating for Gas Barrier Property

The nanomaterials such as nanoclays, nanosilica, nano carbon (e.g., graphene, graphene oxide, carbon nanotube, etc.), nano hybrid material (e.g. POSS), metal oxides (e.g., TiO_2, ZnO_2, and Al_2O_3) have great potential to improve the gas barrier properties of polymers [64–68]. Particularly the nanoplatelet-structured nanofillers have a very high aspect ratio and substantially enhance the impermeability of coated substrate against diffusion of different gases at very low filler loading. These nanomaterials are also called passive fillers, which are effective in improving the gas barrier properties of polymer film without any chemical interaction with a polymer matrix. Actually, with proper exfoliation of these nanoplatelets in the polymer matrix, gas permeability is reduced significantly by increasing the tortuous path length for diffusing the gas molecules through polymer nanocomposite [29, 41]. Except for gas barrier properties, the nanoclay and graphene also have good potential to increase mechanical strength and modulus, thermal stability, etc. [31, 69].

An enormous number of studies have been reported for investigating the effect of polyurethane structure on gas permeability [70–72]. The gas barrier property of polyurethane generally increases with the increase in the ratio of hard and soft segments, T_g and crystallinity of soft segments, and cross-linking density. As already discussed, nanoclay and different derivatives of graphene have great potential to improve the gas barrier property of polymers. A significant number of investigations on the improvement of the gas barrier property of polyurethane by incorporation of nanoclay [63, 67] and graphene [68, 73] have been carried out in academia and in the industrial research community.

A joint study conducted by IIT Delhi and ADRDE, Agra, India, on the PU coated nylon fabric by dispersing nanoclay in the polymer matrix showed significant improvement of hydrogen impermeability of the coated fabric having a coating thickness of 0.25 mm (e.g., 1.38 lit/sq. m/day without clay vs. 0.88 lit/sq. m/day for 3% of nanofiller) [63]. In addition, Joshi et al. developed a polyurethane/clay nanocomposite-based coated fabric with improved hydrogen gas barrier, mechanical, and thermal properties for inflatables. Nylon fabric was coated (coating thickness, 0.25 mm) with a polyurethane/DMF solution containing 3 wt % clay, and about 36% reduction in hydrogen gas permeability was observed [26].

Recently, the same research group from IIT Delhi explored the potential of graphene for improvement of weather resistance and helium gas barrier properties of thermoplastic polyurethane (PU) film [27]. PU/functionalized-graphene nanocomposite (PGN) films were produced by solution masterbatching and subsequent melt mixing, followed by compression moulding. The helium gas barrier of nanocomposite films improved gradually with increasing graphene concentration, showing about a 30% reduction in gas-permeability at 3 wt % graphene loading. Besides, owing to the excellent UV absorption capability of graphene, the weather resistance of nanocomposite films improved significantly, resulting in a lower photo-oxidative and carbonyl index. The best weather resistance in terms of retention in tensile and gas barrier properties was obtained with 2–3 wt % of graphene [27].

4.12.3 Polymer Nanocomposite Coating for Weather Resistance Property

It has been established that nanosized TiO_2, SiO_2, CeO_2, and ZnO particles are more efficient at absorbing and scattering UV radiation than conventional-sized particles and thus were better able to block UV radiation as they have a much larger surface area-to-volume ratio. A lot of efforts have been made on the application of UV blocking treatment to fabrics using nanocomposite coating based on the aforementioned nanoparticles [36–39]. The coating of these nanomaterials protects against harmful ultraviolet (UV) radiation.

UV radiation is one of the major causes of the degradation of the LTA envelope material. Solid band theory can be used to explain why semiconductors such as ZnO and TiO_2 absorb UV radiation [40, 74]. ZnO has a wide band gap (3.37 eV) and a large excitation binding energy of 60 meV. Therefore, it can absorb light that matches or exceeds the band gap energy and which lies in the UV range of the solar spectrum (< 400 nm). TiO_2 is also effective for UV absorption because it presents a similar band gap (3.23 eV for rutile and 3.57 eV for anatase). TiO_2 (particularly, anatase) has a higher refractive index than most other semiconductors, which can increase the contribution to stabilization from UV scattering mechanisms [75]. Moreover, these nanomaterials can also reflect UV light if a continuous top coating is applied. Graphene can also act as a UV absorber, resulting in the improvement of the weather resistance property of polymer [76]. Therefore, a good synergistic weather resistance property can be obtained by choosing a suitable combination of inorganic nano-additives and some organic UV additives [77]. Use of TPU nanocomposite with three potential additives (e.g. UV stabilizer, nanoclay, and graphene) and coated on a woven polyester (PET) fabric was developed for aerostat envelope for enhanced weathering and helium barrier properties [78].

Chatterjee et al. [41] used a polyester fabric coated with a series of TPU-based nanocomposite varied with the concentration of nanoclay (Cloisite 30B), graphene, and organic UV stabilizers (a mixture of 40% UV absorber, 40% hindered amine light stabilizers [HALS], and 20% antioxidant) for aerostat hulls. There was a significant reduction in the 'loss of gas barrier property' after 100 h exposure to accelerated weathering, and an optimum result was obtained for the coating containing 3.03 wt % graphene, 1.36 wt % nano-clay, and 0.61 wt % organic UV stabilizer. Y. Wang et al. [32] and H. Wang et al. [33] developed several UV- and ozone-resistant polyurethane-based coatings utilizing the synergistic effect of nano UV additives ($ZnO/CeO_2/TiO_2$) and organic UV additives (UV absorbers and antioxidants). Saadat-Monfared et al. [38] observed that addition of nano-CeO_2 in PU coatings could offer a gradual increase in the UV absorbency with a significant reduction of 'change in colour' due to accelerated weathering and claimed these coatings and films have the potential for the outer layer of the aerostat/airship envelope.

4.13 Polymer Coated Textiles for Aerostat Systems of DRDO, India

ADRDE, Agra, and DMSRDE have successfully developed various types of coated fabrics for application in different defence systems viz., balloon barrage system for passive air defence, aerostat for surveillance, emergency landing/flotation system of helicopters on the sea, flotation system for space/reentry vehicles, shelter/tent/mobile hospitals, etc. In initial stages, the polymers used are mostly PU, PVC, CR, and IIR and their blends. The list of fabrics includes cotton, nylon, polyester, Kevlar, and Vectran. Subsequently, there were attempts to prepare nylon fabric coated with PVC on both sides for LTA envelope fabrication. A team of researchers has reviewed the state-of-the-art materials (fibres and polymers) and technologies for the flexible envelope material for inflatable systems used in the defence area [13, 14]. PU has immerged as an excellent choice of polymer for the fabrication structure of aerostat envelopes. PU possesses an excellent balance of properties, such as toughness, excellent low-temperature flexibility, good tear strength, good abrasion resistance, fairly good gas permeability, and weatherability, that makes PU a preferred material for coating of textiles for aerostat/airship envelopes [19, 28]. Balraj Gupta highlighted the establishment of ADRDE, Agra, India, and the design and technologies developed to strengthen India's aerial delivery system and its future plan [8]. ADRDE DRDO, AGRA has developed the first indigenous aerostat system "Nakshatra" of 160 m^3 volume mounted with an electro-optical (EO) sensor that could lift a payload up to 300 metres above sea level. Subsequently, there was the development of a medium-sized aerostat system of 2000 m^3 using PU-coated nylon fabrics, a payload capacity of 300 kg and flying altitude of 1000 m under the project named 'Akashdeep', a medium-size, helium-filled Aerostat carrying long-range electro-optic (LREO) and other types of payloads in airborne surveillance and intelligence Systems [14].

4.14 Failure Properties of the Coated Textiles

The properties of a coated fabric depend on: (i) type of polymer and additives used in coating, (ii) the nature of the textile substrate; woven, knitted, or nonwoven constructions, and (iii) the coating method employed for the manufacture of the fabric. PVC-coated fabric has virtually no bending stiffness which prevents excessive deflection and flutter. Some useful measures adopted for these textiles for envelopes were to improve the weather resistance property with a top coating of PVDF film. However, adhesion of PVDF with PVC was poor for sealability, and in comparison with Tedlar/polyester-laminated fabric, the PU-coated fabric shows inferior gas impermeability and strength in lower mass. PVC coated fabric is always manufactured by plain woven fabrics which commonly encountered tearing and puncture during their service life. Zhong, Pan, and Lukas [79] used the Ising model combined with the Monte Carlo simulation to study the phenomenon of single tongue tear failure for the coated fabric. Maekawa et al. [80] established the relationship between tear strength and actual tear propagation characteristics of an airship envelope material which is layered based on Zylon fabrics. Mayo et al. [81] investigated the quasi-static and dynamic puncture behaviour of TPU impregnated aramid fabric and showed an increased cut resistance and reduced windowing compared to neat fabric. Wilson-Fahmy, Narejo, and Koerner [82] provided a theoretical approach to design the inclusion of geomembrane protection materials with high puncture resistance. Pal et al. [83] described the degradation behaviour of polyurethane (PU) coated nylon fabric and woven webbings made of nylon and polyester fibres and interpreted that the loss in gas impermeability, effect of UV radiation and addition of UV absorbers, moisture, and work of rupture are significant for outdoor ageing as well as accelerated ageing under xenon arc exposure. The literature on ageing and environmental degradation of envelope material is not well documented. A detailed discussion on the state-of-the-art on this topic is covered in a separate chapter in this book.

4.15 Conclusions and Future of Coated Textiles

Coated textiles are one of the most important textile products for the global market with phenomenal growth driven by the automotive and transportation industries, protective clothing, advanced parachutes, LTA, and unmanned aerial vehicles (UAVs), beside commodity and high-volume markets of general-use products based on coated textiles. There is a need to explore advanced additives for UV protection and improvement in the gas barrier property of coated textiles. The growth trend considers investment and manufacturing infrastructure, eco-friendly process and sustainability, the scope of recycling, innovation, and emerging materials, e.g., nanomaterials and energy-saving manufacturing techniques.

There is no universally superior method for coated textiles; the best type can be chosen for a given application. The parameters of consideration are coating weight or thickness requirements and desired accuracy, coating properties, substrate properties, tension limitations, operating range and speed limitations of different coating methods, the requirement of finish and uniformity, and, finally, the cost of manufacture of products using any of the various methods. The trends of failure of envelope hull materials for airships/aerostats suggest that material degradation due to weathering, fall in barrier property, and helium leakage and tearing and joint opening are the major causes of failure. However, the present challenges to the material designer are that no accepted model exists to simulate hull or ballonet servile life in extraordinary environments in high altitude and stratospheric region. Current status shows that TPU-based hull envelope material shows significant tensile and tear losses besides the loss of gas retention capability in less than two years. The problems with ballonet material are the low-temperature flex life, permeability of gas inside the hull, and inferior shelf and service life. Therefore, a need exists for further studies under both natural outdoor weathering, accelerated tests in laboratory apparatus, and simulated service conditions, as well as modelling characteristic of life of the envelope materials.

References

1. Billah, S. M. R. 2019. "Textile Coatings." In *Functional Polymers*, edited by M. Jafar Mazumder, H. Sheardown, and A. Al-Ahmed (eds), 825–882. Springer, Cham. doi: 10.1007/978-3-319-95987-0_30.
2. Sen, A. K. 2008. *Coated Textiles - Principles and Applications*. Edited by A. K. Sen. Second ed. CRC Press Taylor & Francis Group. https://www.textileebook.com/2019/10/coated-textiles-principles-and.html.
3. Shahid Ul-Islam, and B. S. Butola. 2018. Advanced Textile Engineering Materials. *Advanced Textile Engineering Materials*. Wiley. doi: 10.1002/9781119488101.
4. Vadala, E. T. 1977. "Assessment of Materials for Application to Modern Lighter-Than-Air (LTA) Vehicals." https://apps.dtic.mil/dtic/tr/fulltext/u2/a081364.pdf.
5. Miller, J. I., and M. Nahon. 2007. "Analysis and Design of Robust Helium Aerostats." *Journal of Aircraft* 44 (5): 1447–1458. doi: 10.2514/1.25627.
6. Howland, C., W. Bebber, and W. Gregory. 2006. High Strength Lightweight Composite Fabric with Low Gas Permeability - Google Patents. US20060084336A1, issued 2006. https://patents.google.com/patent/US20060084336A1/en?q=High+Strength+Lightweight+Composite+Fabric+Low+Gas+Permeability&oq=High+Strength+Lightweight+Composite+Fabric+with+Low+Gas+Permeability.
7. Kelly, D. J., and C. K. Lavan. 2005. Flexible Material for Lighter-than-Air Vehicles. US6979479B2, issued 2005. https://patents.google.com/patent/US6979479B2/en?oq=US+Patent+6979479B2.
8. Gupta, B. 2010. "Aerial Delivery Systems and Technologies." *Defence Science Journal* 60 (2): 124–136. doi: 10.14429/dsj.60.326.
9. Smith, M. S., and E. L. Rainwater. 2003. "Applications of Scientific Ballooning Technology to High Altitude Airships." In *AIAA's 3rd Annual Aviation Technology, Integration, and Operations (ATIO) Forum*. Denver, Colorado. doi: 10.2514/6.2003-6711.
10. Liggett, P. E., D. L. Carter, A. L. Dunne, D. H. Darjee, G. W. Placko, J. I. Mascolino, and L. J. McEowen. 2013. "Metallized Flexible Laminate Material for Lighter-than-Air Vehicles." US8524621B2, issued 2013. https://patents.google.com/patent/US8524621B2/en?oq=US8%2C524%2C621+B2.
11. Meng, J., and M. Lv. 2017. "The Constitutive Relation of a Fabric Membrane Composite for a Stratospheric Airship Envelope Based on Invariant Theory." *Computers, Materials and Continua* 53 (2): 73–89.

12. Zhai, H., and A. Euler. 2005. "*Material Challenges for Lighter-than-Air Systems in High Altitude Applications." AIAA 5th ATIO and the AIAA 16th Lighter-than-Air Systems Technology Conference and Balloon Systems Conference* 3 (September): 1756–1767.
13. Raza, W., G. Singh, S. B. Kumar, and V. B. Thakare. 2016. "Challenges in Design & Development of Envelope." *International Journal of Textile and Fashion Technology (IJTFT)* 6 (2): 27–40.
14. Dasaradhan, B., B. R. Das, M. K. Sinh, K. Kumar, B. Kishore, and N. E. Pra. 2018. "A Brief Review of Technology and Materials for Aerostat Application." *Asian Journal of Textile* 8 (1): 1–12. doi: 10.3923/ajt.2018.1.12.
15. Islam, S., and P. Bradley. 2012. "Materials." In *Airship Technology*, edited by G. A. Khoury, 2nd ed., 113–144. New York: Cambridge University Press. doi: 10.1108/eb035732.
16. Kanoria, A. A., and R. S. Pant. 2012. "Winged Aerostat Systems for Better Station Keeping for Aerial Surveillance." *Advanced Material Research* 433–440: 6871–6879. https://www.scientific.net/AMR.433-440.%206871.
17. Gulrajani, M. L., and D. Gupta. 2011. "Emerging Techniques for Functional Finishing of Textiles." *Indian Journal of Fibre & Textile Research* 36, 388–397. http://nopr.niscair.res.in/bitstream/123456789/13233/1/IJFTR%2036(4)%20388-397.pdf
18. Meirowitz, R. E. 2016. "Coating Processes and Techniques for Smart Textiles." In *Active Coatings for Smart Textiles*, 159–177. Elsevier Inc. doi: 10.1016/B978-0-08-100263-6.00008-3.
19. Adak, B., and M. Joshi. 2018. "Coated or Laminated Textiles for Aerostat and Stratospheric Airship." In *Advanced Textile Engineering Materials*, 257–287. Hoboken, NJ, USA: John Wiley & Sons, Inc. doi: 10.1002/9781119488101.ch7.
20. Singha, K. 2012. "A Review on Coating & Lamination in Textiles: Processes and Applications." *American Journal of Polymer Science* 2 (3). Scientific and Academic Publishing: 39–49. doi: 10.5923/j.ajps.20120203.04.
21. Farboodmanesh, S., J. Chen, Z. Tao, J. Mead, and H. Zhang. 2010. "Base Fabrics and Their Interaction in Coated Fabrics." In *Smart Textile Coatings and Laminates: A Volume in Woodhead Publishing Series in Textiles*, 42–94. Elsevier Ltd. doi: 10.1533/9781845697785.1.42.
22. Shim, E. 2010. "Coating and Laminating Processes and Techniques for Textiles." In *Smart Textile Coatings and Laminates: A Volume in Woodhead Publishing Series in Textiles*, 10–41. Elsevier Ltd. doi: 10.1533/9781845697785.1.10.
23. Ritter, A. 2016. "Smart Coatings for Textiles in Architecture." *In Active Coatings for Smart Textiles*, 429–453. doi: 10.1016/B978-0-08-100263-6.00018-6.
24. ParvinzadehGashti, M., E. Pakdel, and F. Alimohammadi. 2016. "Nanotechnology-Based Coating Techniques for Smart Textiles." In *Active Coatings for Smart Textiles*, 243–268. Elsevier Inc. doi: 10.1016/B978-0-08-100263-6.00011-3.
25. Wolf, C., H. Angellier-Coussy, N. Gontard, F. Doghieri, and V. Guillard. 2018. "How the Shape of Fillers Affects the Barrier Properties of Polymer/Non-Porous Particles Nanocomposites: A Review." *Journal of Membrane Science*. doi: 10.1016/j.memsci.2018.03.085.
26. Adak, B., M. Joshi, and B. S. Butola. 2018. "Polyurethane/Clay Nanocomposites with Improved Helium Gas Barrier and Mechanical Properties: Direct versus Master-Batch Melt Mixing Route." *Journal of Applied Polymer Science* 135 (27). 46422. doi: 10.1002/app.46422.

27. Adak, B., M. Joshi, and B. S. Butola. 2019. "Polyurethane/Functionalized-Graphene Nanocomposite Films with Enhanced Weather Resistance and Gas Barrier Properties." *Composites Part B: Engineering* 176 (November). 107303. doi: 10.1016/j.compositesb.2019.107303.
28. Joshi, M., B. Adak, and B. S. Butola. 2018. "Polyurethane Nanocomposite Based Gas Barrier Films, Membranes and Coatings: A Review on Synthesis, Characterization and Potential Applications." *Progress in Materials Science*. Elsevier Ltd. doi: 10.1016/j.pmatsci.2018.05.001.
29. Adak, B., B. S. Butola, and M. Joshi. 2018. "Effect of Organoclay-Type and Clay-Polyurethane Interaction Chemistry for Tuning the Morphology, Gas Barrier and Mechanical Properties of Clay/Polyurethane Nanocomposites." *Applied Clay Science* 161 (September). Elsevier Ltd: 343–353. doi: 10.1016/j.clay.2018.04.030.
30. Hawkins, W. L. 1984. "Stabilization Against Degradation by Radiation." In *Polymer Degradation and Stabilization*, edited by H. J. Harwood, 74–90. Springer. doi: 10.1007/978-3-642-69376-2_5.
31. Osawa, Z., K. Nagashima, H. Ohshima, and E.-L. Cheu. 1979. "Study of the Degradation of Polyurethanes. VI. The Effect of Various Additives on the Photodegradation of Polyurethanes." *Journal of Polymer Science: Polymer Letters Edition* 17 (7). Wiley: 409–413. doi: 10.1002/pol.1979.130170703.
32. Wang, Y., H. Wang, X. Li, D. Liu, Y. Jiang, and Z. Sun. 2013. "O 3/UV Synergistic Aging of Polyester Polyurethane Film Modified by Composite UV Absorber." *Journal of Nanomaterials* 2013. Hindawi Publishing Corporation. doi: 10.1155/2013/169405.
33. Wang, H., Y. Wang, D. Liu, Z. Sun, and H. Wang. 2014. "Effects of Additives on Weather-Resistance Properties of Polyurethane Films Exposed to Ultraviolet Radiation and Ozone Atmosphere." doi: 10.1155/2014/487343.
34. Klemchuk, P. P., M. E. Gande, and E. Cordola. 1990. "Hindered Amine Mechanisms: Part III-Investigations Using Isotopic Labelling." *Polymer Degradation and Stability* 27 (1). Elsevier: 65–74. doi: 10.1016/0141-3910(90)90097-Q.
35. Saha, S., D. Kocaefe, Y. Boluk, and A. Pichette. 2011. "Enhancing Exterior Durability of Jack Pine by Photo-Stabilization of Acrylic Polyurethane Coating Using Bark Extract: Part 1-Effect of UV on Color Change and ATR-FTIR Analysis." *Progress in Organic Coatings* 70 (4), 376–382. https://doi.org/10.1016/j.porgcoat.2010.09.034
36. Ammala, A., A. J. Hill, P. Meakin, S. J. Pas, and T. W. Turney. 2002. "Degradation Studies of Polyolefins Incorporating Transparent Nanoparticulate Zinc Oxide UV Stabilizers." *Journal of Nanoparticle Research* 4 (1–2). 167–174. doi: 10.1023/A:1020121700825.
37. Allen, N. S., M. Edge, A. Ortega, C. M. Liauw, J. Stratton, and R. B. McIntyre. 2002. "Behaviour of Nanoparticle (Ultrafine) Titanium Dioxide Pigments and Stabilisers on the Photooxidative Stability of Water Based Acrylic and Isocyanate Based Acrylic Coatings." *Polymer Degradation and Stability* 78 (3). 467–478. doi: 10.1016/S0141-3910(02)00189-1.
38. Saadat-Monfared, A., M. Mohseni, and M. Hashemi Tabatabaei. 2012. "Polyurethane Nanocomposite Films Containing Nano-Cerium Oxide as UV Absorber. Part 1. Static and Dynamic Light Scattering, Small Angle Neutron Scattering and Optical Studies." *Colloids and Surfaces A: Physicochemical*

and Engineering Aspects 408 (408). Elsevier B.V.: 64–70. doi: 10.1016/j.colsurfa.2012.05.027.
39. Jalili, M. M., and S. Moradian. 2009. "Deterministic Performance Parameters for an Automotive Polyurethane Clearcoat Loaded with Hydrophilic or Hydrophobic Nano-Silica." *Progress in Organic Coatings* 66 (4): 359–366. doi: 10.1016/j.porgcoat.2009.07.011.
40. Yang, H., S. Zhu, and N. Pan. 2004. "Studying the Mechanisms of Titanium Dioxide as Ultraviolet-Blocking Additive for Films and Fabrics by an Improved Scheme." *Journal of Applied Polymer Science* 92 (5). John Wiley & Sons, Ltd: 3201–3210. doi: 10.1002/app.20327.
41. Chatterjee, U., B. S. Butola, and M. Joshi. 2016. "Optimal Designing of Polyurethane-Based Nanocomposite System for Aerostat Envelope." *Journal of Applied Polymer Science* 133 (24): 1–9. doi: 10.1002/app.43529.
42. Li, A., R. Vallabh, P. D. Bradford, and A. F. M. Seyam. 2019. "Textile Laminates for High-Altitude Airship Hull Materials – A Review." *Journal of Textile and Apparel, Technology and Management* 11 (1).
43. "Aramid Fibers." 2000, January. Pergamon, 199–229. doi: 10.1016/B0-08-042993-9/00044-9.
44. McDaniels, K., R. J. Downs, H. Meldner, C. Beach, and C. Adams. 2009. "High Strength-to-Weight Ratio Non-Woven Technical Fabrics for Aerospace Applications." In *AIAA Balloon Systems Conference*. Reston, Virigina: American Institute of Aeronautics and Astronautics. doi: 10.2514/6.2009-2802.
45. Inflatable lighter-than-air article composed of a coated triaxial weave construction - Google Patents. 2020. US3974989A. Accessed June 4. https://patents.google.com/patent/US3974989.
46. Yousef, M. I., and G. K. Stylios. 2015. "Legacy of the Zeppelins: Defining Fabrics as Engineering Materials." *Journal of the Textile Institute* 106 (5): 480–489. doi: 10.1080/00405000.2014.926606.
47. Airship Sheath Material and Preparation Method Thereof. Google Patents. 2020. CN102416739A-. Accessed June 4. https://patents.google.com/patent/CN102416739A/en.
48. Akovali, G. (Ed.) 2012. *Advances in Polymer Coated Textiles*. Smithers Rapra Technology Ltd. http://www.polymer-books.com.
49. Wilson, J. 2011. "Fibres, Yarns and Fabrics: Fundamental Principles for the Textile Designer." In *Textile Design*, 3–30. Elsevier. doi: 10.1533/9780857092564.1.3.
50. Goodyear Aerospace Corporation. 1977. "*Feasibility Study of Modern Airships, Phase 11 - Executive Summary*." Akron, Ohio. https://ntrs.nasa.gov/search.jsp?R=19780003055.
51. Setua, D. K., A. K. Pandey, K. K. Debnath, and G. N. Mathur. 1999. "Flame Retardant and Impermeable Rubber Coated Fabric - Technische Informationsbibliothek (TIB)." *Kautschuk Gummi Kunststoffe*. https://www.tib.eu/en/search/id/BLSE:RN065673137/Flame-Retardant-and-Impermeable-Rubber-Coated-Fabric?noCHash=bee3b9d0e62855e4264bcab92cc0815c.
52. Das, B. R. 2010. "Electrospun Nanofibrous Scaffolds for Tissue Engineering." *Asian Journal of Textile* 19 (10): 61–65.
53. Zhang, Y., Z. M. Huang, X. Xu, C. T. Lim, and S. Ramakrishna. 2004. "Preparation of Core-Shell Structured PCL-r-Gelatin Bi-Component Nanofibers by Coaxial

Electrospinning." *Chemistry of Materials* 16 (18). American Chemical Society: 3406–3409. doi: 10.1021/cm049580f.

54. Murugan, R., and S. Ramakrishna. 2007. "*Design Strategies of Tissue Engineering Scaffolds with Controlled Fiber Orientation." Tissue Engineering*. Tissue Eng. doi: 10.1089/ten.2006.0078.
55. Kim, S. H., S. H. Kim, S. Nair, and E. Moore. 2005. "Reactive Electrospinning of Cross-Linked Poly(2-Hydroxyethyl Methacrylate) Nanofibers and Elastic Properties of Individual Hydrogel Nanofibers in Aqueous Solutions." *Macromolecules* 38 (9): 3719–3723. doi: 10.1021/ma050308g.
56. Huang, Z. M., Y. Z. Zhang, M. Kotaki, and S. Ramakrishna. 2003. "A Review on Polymer Nanofibers by Electrospinning and Their Applications in Nanocomposites." *Composites Science and Technology* 63 (15): 2223–2253. doi: 10.1016/S0266-3538(03)00178-7.
57. Sinha, M. K., B. R. Das, K. Kumar, B. Kishore, and N. Eswara Prasad. 2017. "Development of Ultraviolet (UV) Radiation Protective Fabric Using Combined Electrospinning and Electrospraying Technique." *Journal of The Institution of Engineers (India): Series E* 98 (1). Springer India: 17–24. doi: 10.1007/s40034-017-0094-z.
58. Joshi, M., and A. Bhattacharyya. 2011. "Nanotechnology - A New Route to High-Performance Functional Textiles." *Textile Progress* 43 (3): 155–233. doi: 10.1080/00405167.2011.570027.
59 Joshi, M., and B. Adak. 2019. "Advances in Nanotechnology Based Functional, Smart and Intelligent Textiles: A Review." *In Comprehensive Nanoscience and Nanotechnology*, 1–5: 253–290. doi: 10.1016/B978-0-12-803581-8.10471-0.
60. Iqbal, Kashif, and Danmei Sun. 2014. "Development of Thermo-Regulating Polypropylene Fibre Containing Microencapsulated Phase Change Materials." *Renewable Energy* 71 (November): 473–479. doi: 10.1016/j.renene.2014.05.063.
61. Tran, N. H. A., M. Kirsten, and C. Cherif. 2019. "*New Fibers from PCM Using the Conventional Melt Spinning Process*." In *AIP Conference Proceedings*, 2055–350:060002. American Institute of Physics Inc. doi: 10.1063/1.5084834.
62. Joshi, M. (Ed.). 2020. *Nanotechnology in Textiles Advances and Developments in Polymer Nanocomposites*. Jenny Stanford Publishing.
63. Joshi, M., K. Banerjee, R. Prasanth, and V. Thakare. 2006. "Polymer-Clay Nanocomposite Based Coatings for Enhanced Gas Barrier Property." *Indian Journal of Fiber and Textile Research* 31: 202–214.
64. Xiang, C., P. J. Cox, A. Kukovecz, B. Genorio, D. P. Hashim, Z. Yan, Z. Peng, et al. 2013. "Functionalized Low Defect Graphene Nanoribbons and Polyurethane Composite Film for Improved Gas Barrier and Mechanical Performances." *ACS Nano* 7 (11). American Chemical Society: 10380–10386. doi: 10.1021/nn404843n.
65. Stratigaki, M., G. Choudalakis, and A. D. Gotsis. 2014. "Gas Transport Properties in Waterborne Polymer Nanocomposite Coatings Containing Organomodified Clays." *Journal of Coatings Technology and Research* 11 (6): 899–911. doi: 10.1007/s11998-014-9594-7.
66. Madhavan, K., D. Gnanasekaran, and B. S. R. Reddy. 2011. "Poly(Dimethylsiloxane-Urethane) Membranes: Effect of Linear Siloxane Chain and Caged Silsesquioxane on Gas Transport Properties." *Journal of Polymer Research* 18 (6): 1851–1861. doi: 10.1007/s10965-011-9592-8.

67. Choudalakis, G., and A. D. Gotsis. 2009. "Permeability of Polymer/Clay Nanocomposites: A Review." *European Polymer Journal.* doi: 10.1016/j.eurpolymj.2009.01.027.
68. Cui, Y., S. I. Kundalwal, and S. Kumar. 2016. "Gas Barrier Performance of Graphene/Polymer Nanocomposites." *Carbon.* doi: 10.1016/j.carbon.2015.11.018.
69. Pattanayak, A., and S. C. Jana. 2005. "Thermoplastic Polyurethane Nanocomposites of Reactive Silicate Clays: Effects of Soft Segments on Properties." *Polymer* 46 (14): 5183–5193. doi: 10.1016/j.polymer.2005.04.035.
70. Chen-Yang, Y. W., Y. K. Lee, Y. T. Chen, and J. C. Wu. 2007. "High Improvement in the Properties of Exfoliated PU/Clay Nanocomposites by the Alternative Swelling Process." *Polymer* 48 (10): 2969–2979. doi: 10.1016/j.polymer.2007.03.024.
71. Huang, R., P. Chari, J.-K. Tseng, G. Zhang, M. Cox, and J. M. Maia. 2015. "Microconfinement Effect on Gas Barrier and Mechanical Properties of Multilayer Rigid/Soft Thermoplastic Polyurethane Films." *Journal of Applied Polymer Science* 132 (18). John Wiley & Sons Inc.: n/a-n/a. doi: 10.1002/app.41849.
72. Hsieh, K. H., C. C. Tsai, and S. M. Tseng. 1990. "Vapor and Gas Permeability of Polyurethane Membranes. Part I. Structure-Property Relationship." *Journal of Membrane Science* 49 (3). Elsevier: 341–350. doi: 10.1016/S0376-7388(00)80647-X.
73. Berry, V. 2013. "Impermeability of Graphene and Its Applications." *Carbon.* doi: 10.1016/j.carbon.2013.05.052.
74. Janotti, A., and C. G. Van de Walle. 2009. "Fundamentals of Zinc Oxide as a Semiconductor." *Reports on Progress in Physics* 72 (12): 126501. doi: 10.1088/0034-4885/72/12/126501.
75. Chen, X. D., Z. Wang, Z. F. Liao, Y. L. Mai, and M. Q. Zhang. 2007. "Roles of Anatase and Rutile TiO_2 Nanoparticles in Photooxidation of Polyurethane." *Polymer Testing* 26 (2): 202–208. doi: 10.1016/j.polymertesting.2006.10.002.
76. Nuraje, N., S. I. Khan, H. Misak, R. Asmatulu, H. Jafari, and J. I. Velasco. 2013. "The Addition of Graphene to Polymer Coatings for Improved Weathering." *ISRN Polymer Science 2013.* Hindawi Publishing Corporation. doi: 10.1155/2013/514617.
77. Mahltig, B., H. Böttcher, K. Rauch, U. Dieckmann, R. Nitsche, and T. Fritz. 2005. "Optimized UV Protecting Coatings by Combination of Organic and Inorganic UV Absorbers." *Thin Solid Films* 485 (1–2). Elsevier: 108–114. doi: 10.1016/j.tsf.2005.03.056.
78. Chatterjee, U., S. Patra, B. S. Butola, and M. Joshi. 2017. "A Systematic Approach on Service Life Prediction of a Model Aerostat Envelope." *Polymer Testing* 60 (July): 18–29. doi: 10.1016/j.polymertesting.2016.10.004.
79. Zhong, W., N. Pan, and D. Lukas. 2004. "Stochastic Modelling of Tear Behaviour of Coated Fabrics." *Modelling and Simulation in Materials Science and Engineering* 12 (2). IOP Publishing: 293. doi: 10.1088/0965-0393/12/2/010.
80. Maekawa, S., K. Shibasaki, T. Kurose, T. Maeda, Y. Sasaki, and T. Yoshino. 2008. "Tear Propagation of a High-Performance Airship Envelope Material." *Journal of Aircraft* 45 (5): 1546–1553. doi: 10.2514/1.32264.

81. Mayo, J. B., E. D. Wetzel, M. V. Hosur, and S. Jeelani. 2009. "Stab and Puncture Characterization of Thermoplastic-Impregnated Aramid Fabrics." *International Journal of Impact Engineering* 36 (9). Pergamon: 1095–1105. doi: 10.1016/j.ijimpeng.2009.03.006.
82. Wilson-Fahmy, R.F., D. Narejo, and R.M. Koerner. 1996. "Puncture Protection of Geomembranes Part I: Theory." *Geosynthetics International* 3 (5). Industrial Fabrics Association International: 605–628. doi: 10.1680/gein.3.0077.
83. Pal, S. K., V. B. Thakare, G. Singh, and M. K. Verma. 2011. "Effect of Outdoor Exposure and Accelerated Ageing on Textile Materials Used in Aerostat and Aircraft Arrester Barrier Nets." *Indian Journal of Fibre & Textile Research*. Vol. 36.

5

Laminated Textiles for the Envelope of LTA Vehicles

Mangala Joshi
Indian Institute of Technology, New Delhi, India
Bapan Adak
Kusumgar Corporates Pvt Ltd, Gujarat, India
Siddhanth Varshney and Rishabh Tiwari
Indian Institute of Technology, New Delhi, India

CONTENTS

DOI: 10.1201/9780429432996-5

5.1 Introduction

Nowadays, different non-rigid lighter-than-air (LTA) systems such as aerostats and airships are getting renewed interest for many applications related to defence, weather forecasting, traffic management, monitoring air pollution in cities, telecommunication relay, etc. The airship is essentially a free-flying, low-speed LTA vehicle, and an aerostat is a tethered system. Both of these aircraft are lifted and fly by using some buoyant gas such as helium or hydrogen. Moreover, these LTA systems work in a very harsh atmosphere where they have the exposure of UV radiation, ozone, rain, humidity, huge temperature and pressure variation. Therefore, multi-layered coated or laminated structures are used to fulfil all the requirements where a specific layer of film/fabric/foil/coating provides specific functionality. The history of material developments in this area shows that the laminated structures always received greater importance than coated structures. However, there are many challenges in developing laminated envelope materials for LTA systems, and research is continuing for better design and property enhancement, with many other added functionalities to the LTA systems.

This chapter will discuss the typical laminated structure for an LTA system, their advantage and disadvantage over coated structures, different laminated structures developed by various research organizations of different countries and the materials used in these laminates.

5.2 A Typical Laminated Structure for LTA Systems

A typical laminated structure of an LTA system such as an aerostat and airship contain four main layers (Figure 5.1), which are: (i) the base layer or strength layer, (ii) the weather-resistant or protective layer, (iii) the gas barrier layer and (iv) the adhesive layer.

5.2.1 Base Layer or Strength Layer

The base layer, which is also called the 'strength layer', provides strength and stability to the laminated structure. Generally, woven fabric of different designs (mainly, plain, ripstop, twill and leno) are used in the base layer where the yarn may be conventional fibre-based (high tenacity polyester or nylon) or high-performance fibre-based (Vectran, Kevlar, M5, Spectra, Dyneema, etc.) [1]. One study also reported the use of nonwoven material in the base layer, for better flexibility, higher specific strength and better long-term fatigue and creep resistance [2]. Sometimes base fabric is treated with plasma or e-beam or corona to make it chemically active for better adhesion with an adjoining layer.

5.2.2 Weather-Resistant Layer

A weather-resistant layer provides protection to the whole laminated structure. Generally, polymeric films which have a high weather resistance property, such as Tedlar and Teflon, are used in a protective layer. Sometimes, the top surface of the protective layers is aluminized to provide an UV-reflective surface. As most of these protective films (Tedlar or Teflon) are chemically inert, they are generally treated with plasma or corona for improving adhesion property with other layers. Recently, different UV-shielding nanomaterial and organic UV additive-based weather-resistant thermoplastic polyurethane nanocomposite films have been found to be a potential material for the protective layer [3, 4].

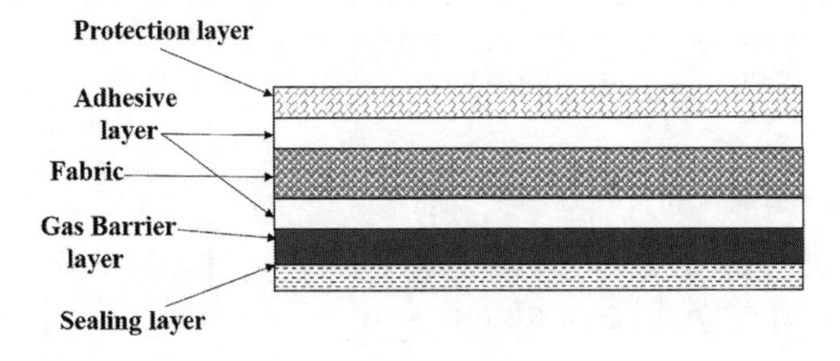

FIGURE 5.1
Typical laminated structure used in the envelope of LTA aircrafts.

5.2.3 Gas Barrier Layer

A gas barrier layer restricts the passage of gas from the inner side to the outer side of LTA system. Generally, biaxially oriented polyester (Mylar), polyvinylidene chloride (PVDC), and ethylene vinyl alcohol copolymer (EVOH) are used in this layer [1, 5]. Currently, in our research group, we are working on the synthesis of gas barrier polymer nanocomposite films. In this context, nanoclay- or graphene-incorporated polyurethane nanocomposite films can be potential materials for the gas barrier layer of the aerostat/airship envelope [6, 7], which will be discussed in detail in Chapter 6.

5.2.4 Adhesive Layer

An adhesive layer joins two successive layers in a laminate. In the context of laminated material for LTA system, the main properties which are needed for this layer are: (i) good compatibility with both the layers or substrates, (ii) good adhesion property, (iii) low temperature flexibility and (iv) weather resistance property for property retention for the long run. A wide variety of adhesives are available in the market, like thermoplastic polyurethane (TPU), reactive polyurethane (PUR), thermoplastic polyolefin (TPO) and ethylene-vinyl acetate (EVA). However, as flexibility is a big concern for an LTA system, TPU-based adhesives are preferred for this particular application. Sometimes different additives such as cross-linker, rheology modifier (plasticizer or thickener) or nanomaterials are incorporated to the adhesive to improve its specific property. This layer may not be required for laminating the substrates having good adhesion property and can be joined by applying heat.

5.2.5 Sealing Layer

The two ends of the multi-layered laminated material need to be joined or sealed properly to provide a particular shape of an LTA system. Therefore, the inner layer, i.e. the barrier layer, also should have a good sealing property. If not, then, in addition to these four main layers, an additional layer called a 'sealing layer' is incorporated at the bottom of the 'gas barrier' layer. Generally, a sealable thermoplastic polymer like polyurethane is used in this layer for proper sealing by thermal or RF sealing technique.

5.3 Advantage of Lamination over Coating for Making Envelope for LTA Systems

There are many advantages of lamination/laminated substrate over coating/coated substrate, which are summarized in this section.

Firstly, in industry, solution coating is more popular where different toxic solvents are used, which causes environmental pollution. However, in lamination, no toxic solvent is used except when solvent-based adhesives are used in rare cases. Therefore, lamination is more environmentally friendly in comparison to coating.

Secondly, due to use of excessive solvents, the coating process requires high energy to evaporate the solvents during drying. In this regard, the energy requirement is much less in the case of lamination compared to coating.

Thirdly, generally, in case of lamination, penetration of adhesive inside the fabric structure or spacing between warp and weft threads is much less than the penetration of coating formulations, which results in less reduction in tear strength of the laminated fabric compared to coated fabric.

Fourthly, the air bubble formation in coating formulation diminishes the surface appearance as well as the final properties of the coated fabric. No such issue is there with lamination, as the surface finish of laminated fabric depends on the film surface.

5.4 Challenges in Making Laminated Structures

The main issues of lamination technique are: (i) lesser covering and more air-pockets generation, (ii) delamination tendency and (iii) higher stiffness, in comparison to coating. These points are discussed here in detail.

If we consider the schemes in Figure 5.2, it can be observed that in the case of coating, the gaps between the interstices of warp and weft yarns are filled with coating formulation, resulting in less air-pocket generation. On the contrary, lamination has a greater tendency towards air-pocket generation as the interstices are not properly filled. Therefore, in case of coating, a better interaction is possible between polymer and fabric, in comparison to laminated structure. However, it may result in better tear strength in the case of a laminated structure in comparison to the coated structures.

The main disadvantage of laminated structures is delamination tendency because of the use of different polymeric materials in different layers. It also results in deterioration in overall mechanical performance of the laminates. However, the inter-laminar adhesion strength of laminates can be improved significantly by using a suitable adhesive that has compatibility or reactivity to both substrates. For an aerostat or airship laminate, flexibility at low temperature is one of the main requirements and, hence, thermoplastic polyurethane-based adhesive is preferred. Another option for increasing

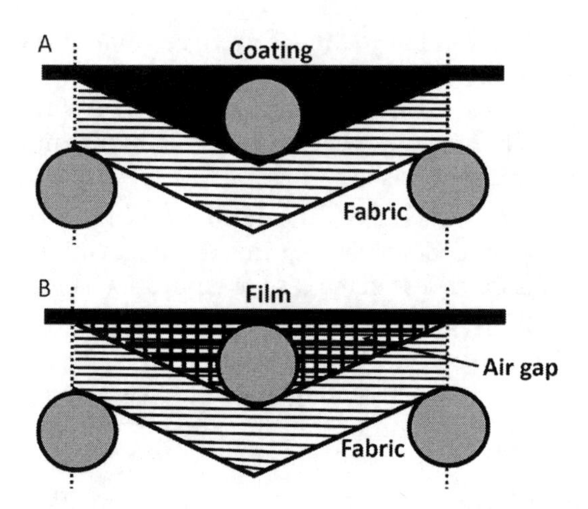

FIGURE 5.2
Schematic of cross-section of (a) coated and (b) laminated textiles.

inter-layer adhesion strength in laminate is surface-etching or surface-functionalization of the fabric or film surface by corona or plasma treatment. It is more applicable to the fabric or film having an inert surface or no functional groups, like polyester fabric and Tedlar or Teflon film. For example, virgin Tedlar film has a surface tension of about 30 dynes/cm, which is lower than the surface tension of most commonly used adhesive which is about 40 dynes/cm and thus results in inadequate adhesion strength. By treating PET fabric and Tedlar film by corona, the surface energy increases to about 44–48 dynes/cm, resulting in a significant improvement in adhesion strength [8].

Another disadvantage with laminated structures is their issue with flexibility. In comparison to coated structures, laminates are generally stiffer. However, sometimes coated fabrics also become very stiff because of higher penetration of the coating formulation inside the fabric structure.

5.5 Development of Laminated Structures for LTA Systems: Case Studies

In the last few decades, many countries, such as the USA, Japan, Korea, UK, Russia and India, have been designing and developing different LTA structures and have used different permutations and combinations in term of materials for achieving the desired goal. The contribution of each country

TABLE 5.1

Materials Used by Different Countries in Different Layers of Laminated Envelope for LTA Systems

Layer	Material	Country
Strength Layer	Zylon	Japan
	Vectran	Korea, USA, UK, Japan
	High-strength nylon	India
	High-strength polyester	India
	Aramid	USA
Protective Layer	Tedlar	USA, Japan, Korea, India
	Teflon	USA
	Kapton	USA, UK
Gas Barrier Layer	Mylar	USA, Japan, UK, India
	EVOH	Japan
Adhesive Layer	Polyurethane	USA, Japan, Korea, UK, India

and their material usage in laminated LTA aircraft envelope are summarized in Table 5.1, and the developments in this area by different countries are discussed in the next section.

5.5.1 LTA Aircraft Envelope Material Development by India

India has been working on aerostat structures for a long time and has seen many successful test short flights over the last two decades. Organizations such as TIFR (Tata Institute of Fundamental Research), IIT Mumbai (Indian Institute of Technology, Mumbai), ADRDE (Aerial Delivery Research and Development) and IIT Delhi (Indian Institute of Technology, Delhi) have all been working towards perfecting the aerostat design and also bringing novelty to the materials used in the work. Facilities provided by institutions like CIPET Ahmedabad and Industry partners like M/s Kusumgar Corporates, Mumbai (India), and M/s Entremonde Polycoaters, Nashik (India), have been a strong support in research and development.

5.5.1.1 Aerial Delivery Research and Development Establishment, ADRDE Agra

ADRDE Agra has been working on the indigenous design and development of aerostats and airships for a long time. Conventionally, polymer-coated nylon fabrics were being used by them, but in the past few years they have undertaken a project to develop laminated fabric for LTA systems because of several advantages over conventional polyurethane-coated nylon fabrics, such as higher expected life, better mass efficiency, better weather resistance and better gas barrier property.

In an ongoing project, which was initiated in 2006, they have made some progress in development of a 'laminated fabric', targeting an inflatable platform filled with a payload capacity of 300 kg at an altitude of 1 km. The fabric is targeted to achieve breaking strength of 250 Kgf/5 cm (both in warp and weft directions), hydrogen permeability of 1.0 cc/m^2/day at 25 cm water column.

5.5.1.2 National Balloon Facility (Hyderabad)

The National Balloon Facility (NBF) was founded by Tata Institute of Fundamental Research (TIFR) Mumbai in 1971. This facility combines R&D, production and launch facilities all under one roof. All balloon components are made indigenously, proving to be a complete solution in scientific ballooning. They have successfully launched low-range aerostats more than 35 times in the past few decades and continuously do marvellous work in the field. They could scale up to more than 780,000 m^3 of volume using an LLDPE film (Antrix). They have also developed balloon load tapes which have a breaking strength of around 182 kg and introduced S-band telemetry with a commendable timer cut off unit in the flight hardware [9].

5.5.1.3 IIT Mumbai (Department of Aerospace Engineering)

In 2001, a national R&D project, "Program on Airship Design and Development" (PADD) was initiated at IIT Mumbai for development of airships [9]. They had designed and developed two remote-controlled airships: (i) the PADD mini (payload capacity of 3.5 Kg), and (ii) the PADD micro (payload capacity of 1.0 Kg) in 2003. This project was sponsored by Technology Information Forecasting and Assessment Council (TIFAC) [10].

In 2004, a lighter-than-air (LTA) systems laboratory was set up in the Aerospace Engineering Department of IIT Bombay, in which many R&D and technology development projects related to design, analysis, fabrication, and testing of LTA systems have been carried out. Several ongoing and completed projects at IIT Mumbai are listed in Table 5.2 [10].

5.5.1.4 IIT Delhi (Department of Textile and Fibre Engineering)

Our research group at IIT Delhi has been associated with the research and development of polymer nanocomposite-based coated and laminated textiles for inflatables, which is sponsored by ADRDE Agra, since 2004. However, till now, most of the developments in this area are mainly focused on development of novel polyurethane nanocomposite coating using different nanomaterials such as nanoclay and graphene for improving the gas barrier property and UV additives for improving weather resistance properties of the coated structure, which will be discussed in Chapter 6 with more details.

TABLE 5.2

Ongoing and Completed Projects under PADD

Sr. No	Project Titles	Commencement Date	Status
1	Sizing and optimization of high-altitude airship (HAA) platforms	2008 onwards	Ongoing
2	Conceptual study of unmanned aerial platforms for real-time monitoring of avalanche parameters	2008 onwards	Ongoing
3	Design of an aerostat for aerial observation at SASE, Manali, India	2008–2009	Completed
4	Pre-feasibility study of powered hot air balloon for Multipoint River Ferry	2008–2009	Completed
5	Long distance wireless communication for rural connectivity, using tethered aerostat	2006–2007	Completed
6	Studies on optimum shapes of envelopes for aerostats and airships	2004–2006	Completed
7	Development of dynamic neural models and their hardware fabrication for the nonlinear control of airships	2004–2006	Completed
8	Feasibility study of a stratospheric airship as high-altitude platform for high-integrity pseudolite based precision navigation system	2004–2005	Completed

5.5.2 LTA Aircraft Envelope Material Development by the USA

5.5.2.1 Cubic Tech Corporation

Cube Tech Corporation is developing materials to meet the challenges of current and future aerospace applications. Material properties of these can be tailored for strength and modulus in multiple arbitrary directions and can be produced in seamless two-dimensional flat or three-dimensional complex curved structures exceeding 40 ft in width and exceeding 100 ft in length. Materials can be joined with seams that are stronger than the base laminate material and capable of carrying structural loads for extended periods without failure, slippage or creep. They had developed three different types of composites, i.e., lightweight, medium-weight and heavyweight [2].

UHMWPE-based lightweight-oriented multidirectional composite laminates may offer much better performance than silicon-coated nylon or polyester fabric. In comparison to silicon-coated nylon fabric of similar weight and thickness, the composite laminates developed by Cube Tech is 80%

stronger, has ten times higher modulus and four times higher tear strength [2]. These lightweight laminated structures can be used in applications like parachutes, balloons and parafoils.

Vectran- and Aramid-based medium-weight flexible composite laminates, developed by Cube Tech, may be used for other LTA systems. In addition to low gas permeability, these composite laminates have high mechanical properties and conversion efficiencies, excellent low temperature performance and pressure retention. On the other hand, PBO- and UHMWPE-based lightweight flexible composite laminates may be used in tension structures, heavy-lift airships and flexible pressure vessels having high strength, linear stress-strain behaviour, low elongation, tear/damage resistance and very high conversion efficiencies (> 80%).

5.5.2.2 Defense Advanced Research Projects Agency (DARPA), USA

With a goal of developing a stratospheric airship-based sensor that can detect both air and ground targets for years, DARPA started their program Integrated Sensor Is Structure (ISIS). DARPA has claimed that their program ISIS will develop a low-areal-density (90.6 g/m^2) advanced hull material with matrix glass transition temperature ~ −101°C. They have achieved a fibre strength-to-weight ratio of 1274 kN-m/kg, with a strength retention capacity greater than 85% up to 22 years [11]. As per their claim, the developed airship has a payload capacity greater than 30% of the system mass while the conventional has only 2–3%.

Walrus HULA (Hybrid Ultra Large Aircraft) project was another DARPA-funded experiment to create an airship capable of transporting up to 1000 tons across international distances. In distinct contrast to earlier generation cigar-shaped airships, the Walrus HULA would be a heavier-than-air vehicle and would generate lift through a combination of aerodynamics, thrust vectoring, and gas buoyancy generation. DARPA said the advances in envelope and hull materials, buoyancy and lift control, drag reduction and propulsion combined to make this concept feasible. The WALRUS could potentially expand and speed the strategic airlift capability of the United States substantially with simultaneous reduction of costs. However, the project was cancelled in 2010 [12].

Recently, DARPA has initiated a program "Adaptable Lighter Than Air" (ALTA) to develop and demonstrate a high-altitude LTA vehicle capable of wind-borne navigation over extended ranges. The balloons can fly at altitudes of more than 75,000 feet, while they do not have independent propulsion. The ALTA vehicle is designed to navigate by changing altitude and thus taking advantage of different wind profiles aloft. A state-of-the-art Winds Aloft Sensor (WAS) is also being developed on the program, which is intended to provide real-time stratospheric wind measurements [13].

5.5.2.3 National Aeronautics and Space Administration (NASA) and Jet Propulsion Lab

The Jet Propulsion Lab imposed an effort to develop a new balloon for Venus, jointly with the team of engineers from the NASA-Wallops Flight Facility (WFF) and ILC Dover. The developed balloon was 5.5 m in diameter and was designed to carry a 45 kg payload at an altitude of 55 km. The balloon material was a 180 g/m^2 multi-component laminate (Figure 5.3). It was a single-shell material in which each element of the laminate serves a different purpose: the Teflon outer layer protects from sulfuric acid aerosols, the metallization on the Teflon minimizes solar heating via a second-surface mirror effect (low solar absorption to IR emissivity [a/e] ratio), the metalized Mylar film minimizes helium permeation, the Vectran fabric gives high strength to tolerate steady-state super-pressure and transient deployment forces, and the urethane coating (inside) facilitates fabrication of interior gore-to-gore seams. This construction provides the required balloon functional characteristics of low gas permeability, sulfuric acid resistance and high strength for super-pressure operation. The design burst at a super-pressure of 39,200 Pa which was predicted to be 3.3 times the worst case value expected during flight at the highest solar irradiance in the mission profile [5].

The High-Altitude Venus Operational Concept (HAVOC) study of NASA has developed an evolutionary exploration plan for Venus that can meet scientific objectives for planetary science while offering another destination for human exploration beyond Earth. The team found the FEP-Teflon shows promise as a material for protecting solar panels and other elements of the airship from the ambient sulfuric acid [14].

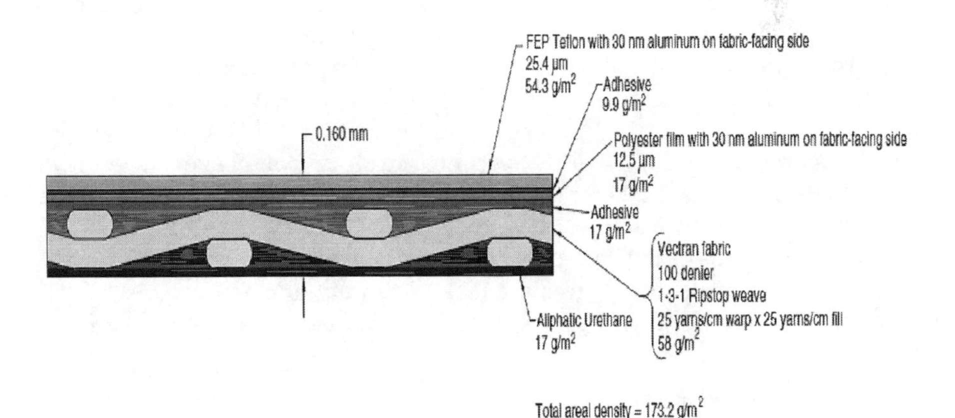

FIGURE 5.3
Venus balloon laminate developed by the Jet Propulsion Lab jointly with other US agencies [5]. Reprinted with permission from Elsevier.

FIGURE 5.4
NASA's pumpkin balloon [15].

NASA's Jet Propulsion Laboratory has developed high-altitude super-pressure balloons with 'meshed pumpkin' design (Figure 5.4) which are capable of functioning as long as 100 days. The weight of the Pumpkin Balloon is much less than a spherical balloon of equal payload capacity [15].

NASA's pumpkin balloon (as shown in Figure 5.4), which has a total mass of 2800 kg, is made of a complex composite structure (areal density 62 g/m^2) consisting of polyester fabric, polyethylene film, polyester film and adhesive. The balloon is capable of carrying a payload of 1600 kg at an altitude of 33 km. One corresponding meshed-pumpkin design calls for reinforcement of the membrane with a 1-inch-by-1-inch mesh of polybenzoxazole scrim fibre (25 denier). With this reinforcement, the complex composite membrane could be replaced by a simple polyethylene film of thickness about 12.7 μm), reducing the mass of the balloon to $<$ 400 kg. The mesh is capable of providing a strength of 400 N/m, giving a factor of safety of 5, with respect to the required strength for a pumpkin balloon with a bulge radius of 8 m [15].

5.5.2.4 Lockheed Martin Corporation

In a US patent (2008), Lockheed Martin Corporation has revealed two flexible laminated structures for LTA vehicles as shown in Figure 5.5. In one structure, a polyimide (Kapton) film and a PVDF film were laminated on a fabric made of a liquid crystalline polymer fiber (Vectran) using a polyurethane-based

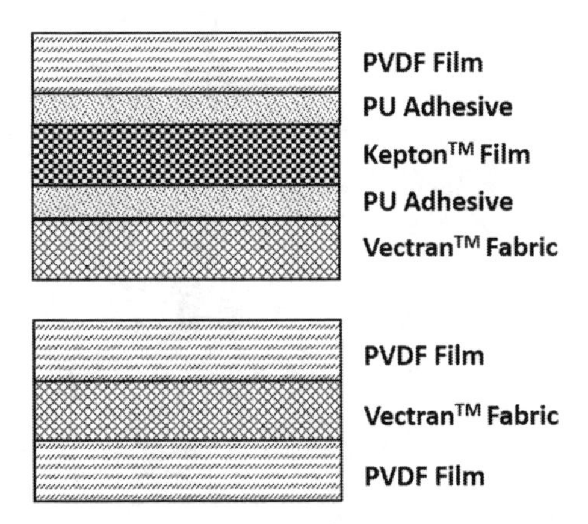

FIGURE 5.5
Schematic structure of the laminated fabric developed by Lockheed Martin Corporation [16].

adhesive (Figure 5.5a). The Kapton film acted as a gas barrier layer, while the top PVDF layer acted as a protective layer due to its excellent resistance against ozone and ultraviolet light. In another structure, a Vectran fabric was sandwiched between two layers of PVDF film by applying heat and pressure using nip roll set-up. This embodiment provided almost all the benefits and attributes of previous structure with additional advantage of – (i) elimination of polyurethane adhesive, (ii) less weight and (iii) much easier manufacturing. The Vectran fabric was advantageous in term of high strength with light weight and also excellent creep resistance and flex fatigue resistance [16].

Lockheed Martin Corporation has also developed and filed two US patents on metalized, flexible and power-generating (piezoelectric and pyroelectric) laminate for airship [17, 18].

Figure 5.6 shows the schematic of a metalized flexible laminated structure developed by Lockheed Martin Corporation. The commercially available monofilament yarns which can be used in layer 1 are polyamides, polyesters, aramids, liquid crystal polymers, carbon, polybenzoxazole and ultra-high molecular weight polyethylene (UHMWPE). In certain embodiments, a high-tenacity yarn such as carbon, or those designated as M5 (Magellan Systems/DuPont), Vectran, Zylon, Dyneema and Spectra may be employed. Layer 3 represents a second yarn monofilament yarn layer. Layer 2 and 4 is an adhesive layer. Film layer 5 may include any kind of polymeric film. In one or more embodiments, film layer 5 includes a high modulus film, such as polyamide, liquid crystal polymer, polyethylene teraphthalate (PET), etc., for functioning as a gas barrier for retaining helium or the like. Metal coating layer 6 is adhered to the outer surface of polymeric film layer 5. Suitable metals include highly reflective metals such as silver and aluminium. Reflectance enhancing layer 7 may include a polymer film such as 3M photonic filter films, or dielectric materials such as titanium dioxide, silicon

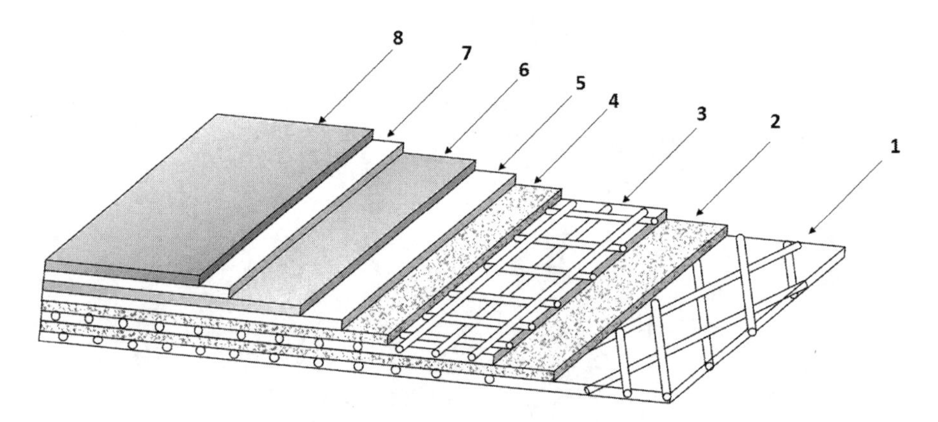

FIGURE 5.6
Metalized flexible laminated structure developed by Lockheed Martin Corporation [18].

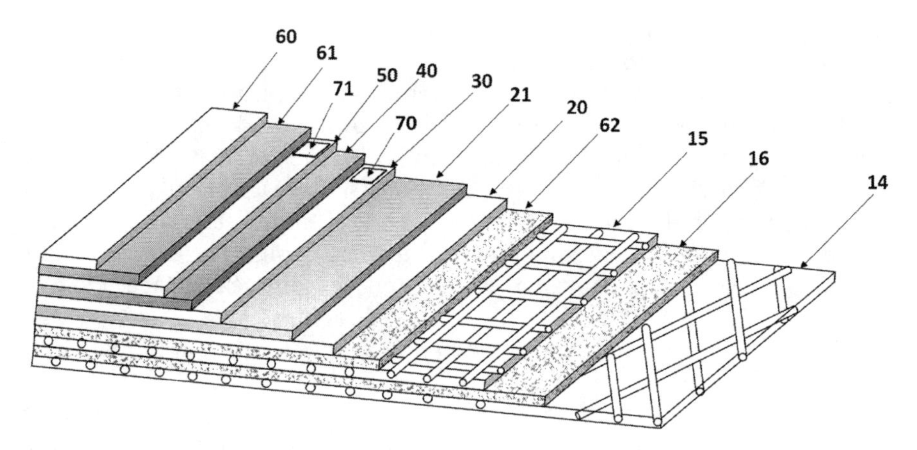

FIGURE 5.7
Power-generating laminated structure developed by Lockheed Martin Corporation [17].

dioxide or hafnium dioxide. Clear film cover layer 8 may include any film that is resistant to ozone and ultraviolet radiation. Useful films also include corrosion protector films. Examples of suitable films include polyvinylidene fluoride [18].

Figure 5.7 represents a multi-layered power-generating laminated structure, where the straight-ply fabric 14 provides an inner surface bonded upon the bias-ply fabric 15 with the adhesive layer 16. Layer 15 is bonded via an adhesive layer 62 with a barrier layer 20. The barrier layer 20 may comprise polyamide material, such as Kapton, which serves to provide an impervious or nearly impervious barrier to escape the air and lifting gas maintained within the inner volume of the airship. Layer 20 is bonded via an adhesive layer 21 with an inner metal layer 30. A piezoelectric layer 40, formed of piezoelectric polymer, such as polyvinylidenefluoride (PVDF) film is disposed upon the inner metal layer 30. The inner and outer metal layers 30 and 50 may be laminated or otherwise bonded to the piezoelectric layer 40 via various processes. The electrical power may then be supplied to a power processing system that is coupled to the metal layers 30 and 50 via respective electrical terminals 70 and 71, so as to provide power to the various electrical components maintained aboard the airship. The outer metal layer 50 is covered and protected by cover layer 60, which is bonded thereto by an adhesive layer 61. In particular, cover layer 60 may be formed of polyvinylidene fluoride (PVDF), and it serves as a dielectric to electrically isolate the outer metal layer 50 from the external environment [17].

5.5.2.5 TCOM

Currently, TCOM is one of the leading manufacturers of aerostat systems. They design and manufacture aerostats for intelligent and persistent surveillance solutions, covering three different classes – tactical, operational and strategic aerostat systems [19]. The specification of different types of TCOM aerostat systems are summarized in Table 5.3.

5.5.2.6 California Institute of Technology

California Institute of Technology, in association with Lamart Corporation and Near Space Corporation, worked for a stratospheric aerobot for exploration of 'Titan'. The material is a laminate of Mylar film glued onto a polyester fabric. This material has an areal density of 94 g/m^2, a tensile strength of 9100 N/m at 298 K, and a tensile strength of 16,400 N/m at 77 K [20, 21]. They have compared the properties of the laminate with many other materials, and the Mylar film plus fabric laminates demonstrated significantly better performance on a cyclic loading performance compared to the other materials [21].

5.5.2.7 Google X

Google X (presently known as X, which operates as a subsidiary of Alphabet Inc.) initiated a research and development project, called 'Project Loon'. The main objective of this project is to provide internet access to rural and remote areas poorly served by existing provisions, and to improve communication during natural disasters to affected regions [22]. The company uses high-altitude stratospheric balloons at an altitude of 18 km to 25 km to create an aerial wireless network with up to 4G LTE speeds [23].

The balloons are constructed from materials like metalized Mylar, BoPET, or a highly flexible latex or rubber material like chloroprene [24]. There is no

TABLE 5.3

Specifications of Different Types of Aerostat Systems Manufactured by TCOM

Specification	12M Tactical	17M Tactical	22M Operational	28M Operational	71M Strategic	74M Strategic
Payload Weight	27 kg/60 lb.	90 kg/200 lb.	190 kg/425 lb.	385 kg/850 lb.	1600 kg/3500 lb.	3200 kg/7000 lb.
Nominal Altitude	300 m/1000 ft.	300 m/1000 ft.	900 m/3000 ft.	1500 m/5000 ft.	4600 m/15,000 ft.	3000 m/10,000 ft.
Available Payload Power	500 w	1 kVA	2 kVA	3 kVA	23.5 kVA	70 kVA
Flight Duration	7 days	7 days	14 days	14 days	30 days	30 days
Wind Speeds	Operational, 40 kts/ survival, 55 kts	Operational, 40 kts/survival, 55 kts	Operational, 50 kts/survival, 70 kts	Operational, 50 kts/survival, 70 kts	Operational, 70 kts/survival, 90 kts	Operational, 70 kts/survival, 100 kts

doubt that Project Loon is an asset of Google, by which the world is benefitting so much in terms of easy availability of information, naturally education, health and medical awareness, etc.

5.5.3 LTA Aircraft Envelope Material Development by Russia

RosAeroSystems International Ltd. (Moscow, Russia) has been working on development of the high-altitude airship (HAA) 'Berkut' that is a unique combination of lighter-than-air and space technologies, a cost-effective alternative to geostationary satellites. This project will open a new era in communication and earth observation. Berkut ET, Berkut ML and Berkut HL are three versions designed to work at different latitudes, such as equatorial and tropical countries, middle geographical latitudes and high geographical latitudes. All three versions have the same performance and same basic technical approach [25].

5.5.4 LTA Aircraft Envelope Material Development by Japan

5.5.4.1 Japan Aerospace Exploration Agency (JAXA)

The Japan Aerospace Exploration Agency (JAXA) has developed high strength and lightweight envelope materials with reduced weight in collaboration with Kawasaki Heavy Industries Inc. (Kakamigahara, Gifu, Japan) and Taiyo Kogyo Corporation (Hirakata, Osaka, Japan). In the early phase of the development, high strength was the top priority, and Z4040T-AB (laminated structure) had been developed. The base fabric of Z4040T-AB was Zylon (PBO, Toyobo) and the density was 203 g/m^2. The tensile strength was 1310 N/cm [26]. This material suits the operational airship which was planned to have an overall length of 250 m [27].

They have developed another lightweight laminated structure 'Z2929T-AB', the layer compositions of which are shown in Figure 5.8. A combination of Tedlar with an aluminium layer was used in the protective layer, while

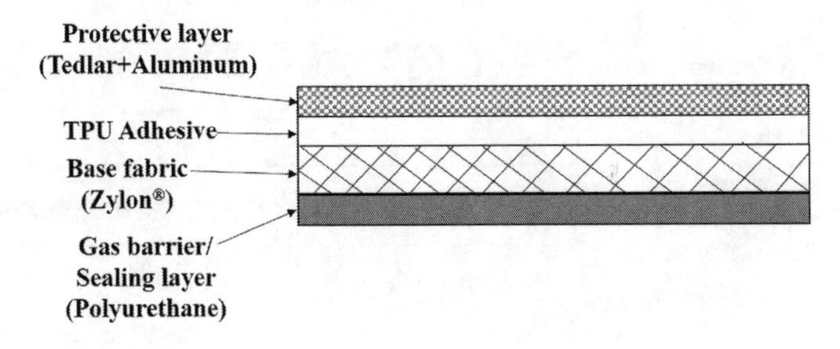

FIGURE 5.8
Envelope material developed by JAXA [27].

polyurethane was used in adhesion layer to bind the base fabric (Zylon) and protective layer as well as in internal barrier layer or sealing layer. The material density was very low at 157 g/m^2 with a tensile strength of about 997 N/cm. The performance of the fabric was analysed by measuring change in joint strength and tensile strength after exposure to simulated UV weathering (Xenon light, 180 W/m^2, for 100 h) and accelerated ozone environment (Ozone, 50 ± 5 ppm which is five times higher than that at the altitude 20 km, for 24 h) [27].

5.5.4.2 National Aerospace Laboratories (NAL), Japan

National Aerospace Laboratories (NAL) has developed laminated envelopes with high-specific strength, as shown in Figure 5.9. In essence, this envelope consists of a strong load-bearing material bonded to an inner gas-tight membrane and a weatherproofed outer layer that will not degrade in sun or ultraviolet radiation. In their Stratospheric Platform (SPF) project, more than 30 different envelope materials were manufactured experimentally [9, 28]. In the load-bearing fabric, polyarylate fiber (Vectran), Kevlar, PBO (p-phenylene benzobisoxazole) fibre (Zylon) and nylon were used. In the protective layer, PVF (Polyvinylfluoride) film (Tedlar) and Tedlar with aluminium deposit was used while the adhesive layer was polyurethane-based. The Tedlar film is well known for excellent resistance to weathering, outstanding mechanical properties and inertness towards a wide variety of chemicals, solvents and staining agents. In the gas barrier layer, ethylene vinyl alcohol (EVOH), Mylar and aluminium stuck EVOH were utilized. Finally, they came up with a sample which consists of 250 denier Zylon fibre of weight 110 g/m^2 with polyurethane and EVOH applied, while the final weight was in the range of 160–175 g/m^2. The laminated structure showed the body strength as well as joint strength (overlap joint type) higher than 1000 N/cm in a temperature range from −80°C to 20°C [28].

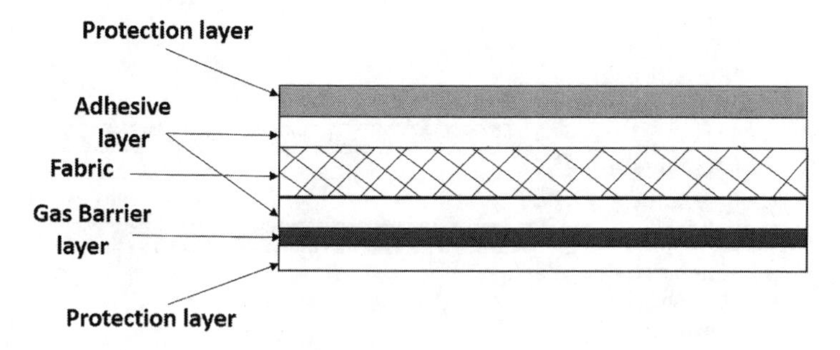

FIGURE 5.9
Typical layers in the laminated structure developed by NAL.

5.5.5 LTA Aircraft Envelope Material Development by the European Union

5.5.5.1 Lindstrand Technologies Limited (United Kingdom)

Lindstrand Technologies Limited over many years has been working on an LTA portfolio that includes a range of aerostats, manned and unmanned airships and gas balloons. They developed the HALE airship which has a potential of working at an altitude of 21 km and for a period of three to five years. The non-rigid airship skin consists of four primary layers, named: (i) environmental shield (Tedlar), (ii) helium barrier (PVDC and EVOH films), (iii) primary structure (Vectran and PBO) and (iv) bonding layer (polyurethane). The final material was of 295 GSM having an ultimate tensile strength of 730 kg/5 cm and a helium permeability of 0.003 L/m^2/24 h.

Lindstrand Technologies has designed an aerostat based on the concept of a solar-powered aerodynamic super-pressure airship. Fuel cells are used for highly efficient energy storage, i.e. for night-time power supply and an intelligent propulsion system which provides a reliable thrust also in the case of heavy winds [29].

5.5.5.2 Airborne Industries Ltd (United Kingdom)

Airborne Industries is one of the largest manufacturers of technical military inflatables which are used for training, communication, surveillance and defence. They are manufacturing aerostat of all three types – tactical (sizes 9–15 m), operational (sizes 19–22 m) and strategic (sizes 23–30 m). These systems are most commonly used for persistent ground surveillance due to the high working altitudes, and they are able to carry multiple payloads [30].

5.5.5.3 CNIM Air Space and Thales Alenia Space

CNIM Air Space (formerly Airstar Aerospace) designs and manufactures tethered aerostats, airships and stratospheric balloons and leads tailor-made projects for its customers. Thales Alenia Space is one of the largest satellite manufacturer based in Europe (French-Italian). CNIM Air Space is a partner of the Stratobus program, led by Thales Alenia Space. Stratobus is an airship, halfway between a drone and a satellite (Figure 5.10), which can be considered as high altitude pseudo satellite (HAPS), operating from an altitude of 20 km [31]. Stratobus is powered with solar energy and has clean energies on board; hence, it has a low carbon footprint.

As a multidisciplinary platform, Stratobus has a potential of completing complementary missions to those accomplished by a telecommunication or an Earth observation satellite, covering security, environmental monitoring (forest fires, beach erosion, pollution, etc), surveillance of borders or high-value sites and telecommunication (4G and 5G). The envelope of Stratobus is

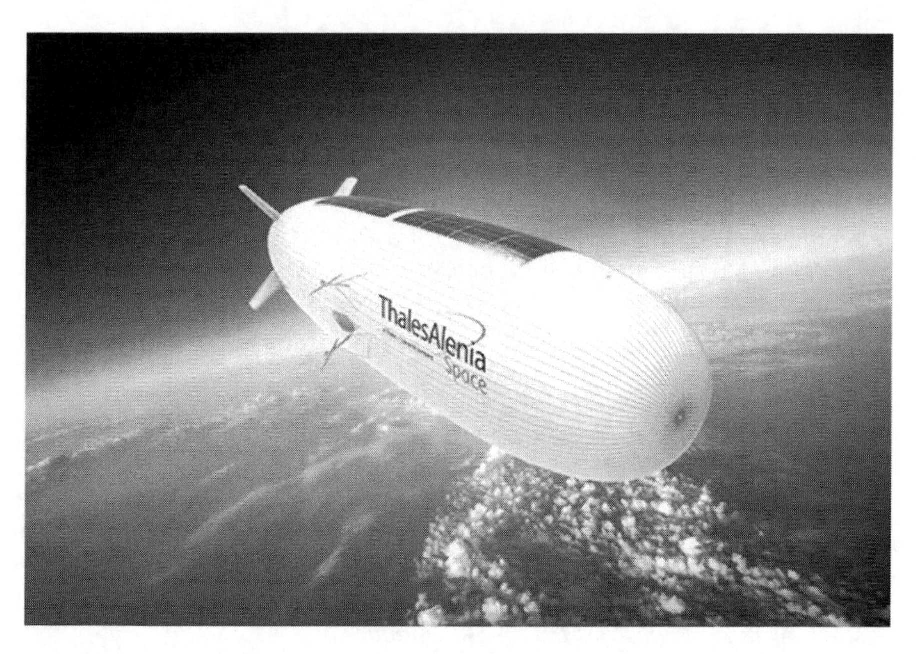

FIGURE 5.10
Photograph of Stratobus [31].

made of laminated textiles which are prepared by assembling various films and hermetic fabrics.

5.5.6 LTA Aircraft Envelope Material Development by Korea

The Korea Aerospace Research Institute (KARI) designed a laminate which consisted of three layers: a helium barrier, a load carrier, and a thermal bonding layer (Figure 5.11). The load carrier was composed of high-performance Vectran HT fibre (200/40 Denier/number of fibres) in the form of plain-woven fabric having threads/cm of 22 in both directions (warp and fill). The plain-woven fabric had an areal density 106 kg/m^2, thickness 0.19 mm and tensile strength 90 Kgf/cm. Tedlar (PVF) film of 38 μm thickness was selected for both the helium barrier and the UV radiation protection layer. It was laminated to the load carrier fabric by a polyurethane adhesive matrix. The supplier claims that clear Tedlar film with UV absorber additives initially blocks more than 99% of the UV light over the energy wavelength range of 290–350 nm, which makes the further reduction of total areal density possible [32]. A 25 μm thermoplastic polyurethane (TPU) film (Dongsung Chemical, Korea) was laminated at the inner side for thermal bonding. For conventional airship, TPU coating is commonly used as a thermal bonding layer and the coated fabric is easier to fabricate since it is very flexible compared to

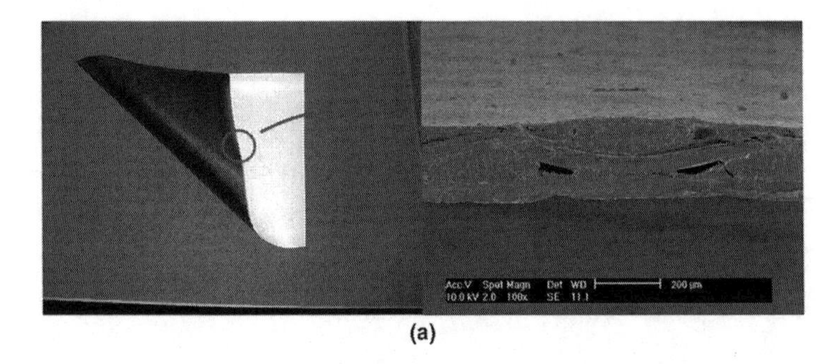

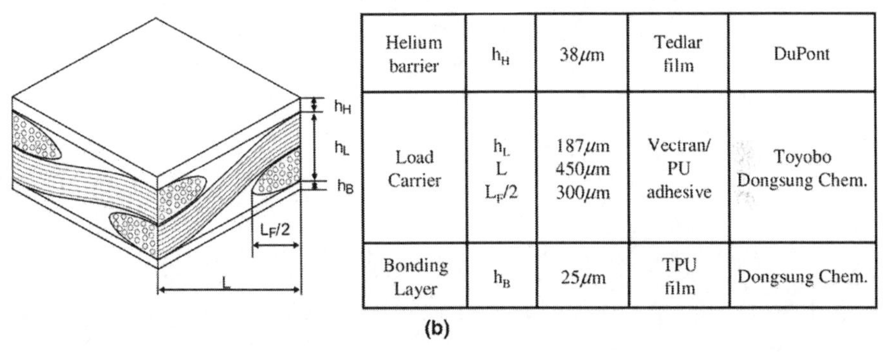

Helium barrier	h_H	38μm	Tedlar film	DuPont
Load Carrier	h_L L $L_F/2$	187μm 450μm 300μm	Vectran/ PU adhesive	Toyobo Dongsung Chem.
Bonding Layer	h_B	25μm	TPU film	Dongsung Chem.

FIGURE 5.11
Microstructure of film-fabric laminate developed by KARI: (a) SEM image of the cross-section and (b) schematic of CAD modelling and different layers [32].

laminated fabric. However, laminated TPU film can serve as another helium barrier inside the envelope. In particular, when the external Tedlar layer is damaged, the internal TPU layer also can block helium leakage, which gives a huge advantage to envelope manufacturers. They added carbon black to the TPU film since the inside black colour was good for inspection after fabrication [32, 33].

5.6 Conclusion

From the start of material development for the envelope of LTA systems, laminated structures always received greater preference over coated structures. Still, there are some technological challenges with the laminated structures, which mainly are delamination tendency and stiffness. Not only these, but recent research in this field is also focusing on many other parameters such as incorporation of airborne moving target indicator, wind-borne navigation over extended ranges, increase in payload capacity, power generation, real-time stratospheric wind measurements and radar risk reduction. Many research institutes, agencies and defence organizations are working in this field to develop and commercialize laminated envelopes for LTA systems. Still, there are many deficiencies existing in the conventional laminates as described in this chapter, and therefore extensive research work is going on all over the world for developing better laminated structures for different LTA systems. Recently, with the advancement in the field of nanomaterials and nanocomposites, researchers are focusing on development of polymer nanocomposite-based films and laminates, which will be discussed in the next chapter.

References

1. H. Zhai, A. Euler, *Material challenges for lighter-than-air systems in high altitude applications*, in: *AIAA 5th ATIO and 16th Light. Sys Tech. Balloon Syst. Conf.*, American Institute of Aeronautics and Astronautics, Reston, Virginia, 2005. doi:10.2514/6.2005-7488.
2. K. McDaniels, R.J. Downs, H. Meldner, C. Beach, C. Adams, *High strength-to-weight ratio non-woven technical fabrics for aerospace applications*, in: *AIAA Balloon Syst. Conf*, 2009. doi:10.2514/6.2009-2802.
3. B. Adak, B.S. Butola, M. Joshi, Calcination of UV shielding nanopowder and its effect on weather resistance property of polyurethane nanocomposite films, *J. Mater. Sci.* 2019. doi:10.1007/s10853-019-03739-7.
4. A. Ahmed, B. Adak, T. Bansala, S. Mukhopadhyay, Green solvent processed cellulose/graphene oxide nanocomposite films with superior mechanical, thermal, and ultraviolet shielding properties, *ACS Appl. Mater. Interfaces* 2020. doi:10.1021/acsami.9b19686.
5. J.L. Hall, D. Fairbrother, T. Frederickson, V. V. Kerzhanovich, M. Said, C. Sandy, J. Ware, C. Willey, A.H. Yavrouian, Prototype design and testing of a Venus long duration, high altitude balloon, *Adv. Sp. Res.* 2008. doi:10.1016/j.asr.2007.03.017.
6. B. Adak, M. Joshi, B.S. Butola, Polyurethane/functionalized-graphene nanocomposite films with enhanced weather resistance and gas barrier properties, *Compos. Part B* 2019, 107303. doi:10.1016/j.compositesb.2019.107303.
7. B. Adak, B.S. Butola, M. Joshi, Effect of organoclay-type and clay-polyurethane interaction chemistry for tuning the morphology, gas barrier and mechanical properties of clay/polyurethane nanocomposites, *Appl. Clay Sci.* 161, 2018, 343–353. doi:10.1016/j.clay.2018.04.030.
8. W. Raza, G. Singh, S.B. Kumar, V.B. Thakare, Challenges in design and development of envelope materials for inflatable systems, *Int. J. Text. Fash. Technol.* 6(2), 2016, 2319.
9. K. Komatsu, Pressure proof test for envelope, *NAL Newsl.* 4(2), 2001.
10. Program on Airship Design and Development, n.d. https://www.aero.iitb.ac.in/~ltasys/index.htm

11. Integrated Sensor is Structure (ISIS) (Archived), *Def. Adv. Res. Proj. Agency* n.d. https://www.darpa.mil/program/integrated-sensor-is-structure (accessed January 30, 2020).
12. Walrus HULA, n.d. https://en.wikipedia.org/wiki/Walrus_HULA (accessed January 30, 2020).
13. A.M.G. Walan, Adaptable lighter than air (ALTA), *Def. Adv. Res. Proj. Agency* n.d. https://www.darpa.mil/program/adaptable-lighter-than-air
14. C.A. Jones, D.C. Arney, G.Z. Bassett, J.R. Clark, A.I. Hennig, J.C. Snyder, High altitude venus operational concept (havoc): proofs of concept, *Am. Inst. Aeronaut. Astronaut.* n.d., 1–12. https://ntrs.nasa.gov/archive/nasa/casi.ntrs.nasa.gov/20160006580.pdf
15. Meshed-pumpkin super-pressure balloon design, NASA's Jet Propuls. Lab., Pasadena, CA, 2003. https://www.techbriefs.com/component/content/article/tb/pub/techbriefs/materials/1197
16. C.K. Lavan, D.J. Kelly, Flexible laminate material for lighter-than-air vehicles, U.S. Patent No. 7354636, 2008.
17. P.E. Liggett, Piezoelectric and pyroelectric power-generating laminate for an airship envelope, U.S. Patent 7878453, 2011.
18. P.E. Liggett, H.D. Dennis, L. Carter Anthony, L. Dunne Dhiraj, W.P. Gerald, J.M. James, I. Mascolino Lowell, Metallized flexible laminate material for lighter-than-air vehicles, U.S. Patent 8524621, 2013.
19. Aerostat systems, TCOM. n.d. https://tcomlp.com/ (accessed January 30, 2020).
20. J.L. Hall, V. V. Kerzhanovich, A.H. Yavrouian, J.A. Jones, C. V. White, B.A. Dudik, G.A. Plett, J. Mennella, A. Elfes, An aerobot for global in situ exploration of Titan, *Adv. Sp. Res.* 2006. doi:10.1016/j.asr.2004.11.033.
21. J.L. Hall, J.A. Jones, V. V. Kerzhanovich, T. Lachenmeier, P. Mahr, M. Pauken, G.A. Plett, L. Smith, M.L. Van Luvender, A.H. Yavrouian, Experimental results for Titan aerobot thermo-mechanical subsystem development, *Adv. Sp. Res.* 2008. doi:10.1016/j.asr.2007.02.060.
22. Loon LLC, Project Loon, 2019. https://en.wikipedia.org/wiki/Project_Loon (accessed January 30, 2020).
23. Loon, n.d. https://loon.com/
24. Project Loon, n.d. https://xedknowledge.com/Coverstory_Demo.aspx?id=qJZ85UM6v6RmN6IFBF+t1Q==.
25. Lifter News Sonderausgabe, CargoLifter, Berlin, April 2010. https://d-nb.info/1011256452/34 (accessed October 3, 2021).
26. S. Maekawa, T. Maeda, Y. Sasaki, T. Kitada, Development of advanced lightweight envelope materials for stratospheric platform airship, in: *43rd Aircr. Symp. Aeronaut. Sp. Sci. Japan*, 2005, pp. 120–124.
27. S. Maekawa, K. Shibasaki, T. Kurose, T. Maeda, Y. Sasaki, T. Yoshino, Tear propagation of a high-performance airship envelope material, *J. Aircr.* 2008. doi:10.2514/1.32264.
28. K. Komatsu, M.A. Sano, Y. Kakuta, Development of high-specific-strength envelope materials, AIAA's 3rd Annu. *Aviat. Technol. Integr. Oper. Forum*, 2003, 1–7. doi:10.2322/jjsass.51.158.
29. High Altitude Long Endurance (HALE), Lindstrand Technol. n.d. https://www.lindstrandtech.com/high-altitude-long-endurance-stratospheric-airships/.

30. Unmanned System Technology, Airborne industries, https://www.unmannedsystemstechnology.com/company/airborne-industries-ltd/
31. Thales Alenia Space, Thales Alenia Space and Thales sign concept study contract with French defense procurement agency for a Stratobus type platform, https://www.thalesgroup.com/en/worldwide/space/press-release/thales-alenia-space-and-thales-sign-concept-study-contract-french
32. W. Kang, Y. Suh, K. Woo, I. Lee, Mechanical property characterization of film-fabric laminate for stratospheric airship envelope, *Compos. Struct.* 2006. doi:10.1016/j.compstruct.2006.04.060.
33. Y.G. Lee, D.M. Kim, C.H. Yeom, Development of Korean high altitude platform systems, *Int. J. Wirel. Inf. Networks*, 2006. doi:10.1007/s10776-005-0018-6.

6

Polyurethane Nanocomposite-Based Advanced Materials for Aerostat/Airship Envelopes

Mangala Joshi
Indian Institute of Technology, New Delhi, India
Bapan Adak
Kusumgar Corporates Pvt Ltd, Gujarat, India
Upashana Chatterjee
Indian Institute of Technology, New Delhi, India

CONTENTS

DOI: 10.1201/9780429432996-6

6.1 Introduction

A polymer nanocomposite (PNC) consists of nanomaterials (which may be of different shape and geometries such as particles, platelets, rods, spheroids, etc.) that act as a reinforcement and are dispersed in a polymer or copolymer matrix. They exhibit unique physicochemical properties in the PNC that cannot be obtained with individual components acting alone. One of the most extensively used polymers is polyurethane, and its nanocomposite is called polyurethane nanocomposite (PUNC). PU and PUNCs are used in many forms such as fibre, film and membrane, coating, foam, adhesive and many other solid forms which can be prepared by moulding or 3D printing techniques [1–3].

PU is a unique block copolymer with versatile properties because of its unique structure and morphology containing blocks of hard and soft segments. Some special properties of thermoplastic polyurethane are excellent flexibility, very high extensibility and excellent elastic recovery, good chemical resistance, excellent adhesion property, good abrasion resistance etc and most importantly these properties can be tuned by changing the PU chemistry [4]. On the basis of these versatile properties, PU and PUNC based films and coatings getting significant interest to cover wide range of applications ranging from automotive, defence, packaging, biomedical, sensing and actuation, electromagnetic absorption, protection (weather resistance, thermal resistance, gas barrier, etc.), energy storage and many more [2, 4, 5]. Especially for an aerostat or airship envelope where coated and laminated textiles are used, PU is a potential material [6]. However, its gas barrier property and weather resistance property are not very satisfactory for using directly in aerostat/airship envelope material. Therefore, these properties need to be improved for such high-end outdoor applications where both the weather resistance and gas barrier property are very important. By virtue of nanotechnology, researchers have synthesized different nanomaterials such as nano-ZnO, nano-TiO_2, nano-CeO_2, etc., which can improve the weather resistance property of PU [7–11]. While there are different-layered structure nanomaterials, mainly nanoclay and graphene have huge potential in improving gas barrier properties of PU [12–14]. Beside these popular nanomaterials, researchers are developing or working with many other new nanomaterials that have potential for improving weather resistance and gas barrier properties of PU [15–18].

This chapter will highlight the structure-morphology of PU, different routes for synthesis of PUNCs, the potential of PUNCs for development of weather-resistant and gas barrier films as well as coatings, and a few case studies on PUNC-based coated/laminated textiles for aerostat and airship envelopes.

6.2 Polyurethane and Its Versatility

6.2.1 Fundamental Structure of PU

In 1937, at I.G. Ferbenindustri, Germany, Otto Bayer and co-workers discovered PUs in response to the competition arising from Carothers's work on polyamides at E.I. Dupont. Although industrial-scale production of PU had started in 1940, market growth was seriously impacted after World War II [19]. The basic three raw materials required for PU production are diisocyanates, diols, and chain extenders. PUs are classified depending on the different raw materials used in their production and the structure of the segments formed in the main chain. The extensive applicability of PU coatings is due to the versatility in selecting monomeric materials from a vast list of diisocyanates, diols, and chain extenders. The basic monomeric materials and the chemistry of the urethane structure [20, 21] of PU are presented schematically in Figure 6.1.

6.2.2 Structure Property Relationship

6.2.2.1 PU Structures Based on Different Diisocyanates

Diisocyanates used in PU production can be categorized into two types: aromatic diisocyanate and aliphatic diisocyanate. The most important difference between aromatic and aliphatic PU is their resistance to sunlight. On exposure to solar UV radiation, aromatic PU suffers auto oxidation degradation with change in colour and loss of physical properties. The aromatic ring is auto oxidized under such conditions to a chromophore, for example quinone

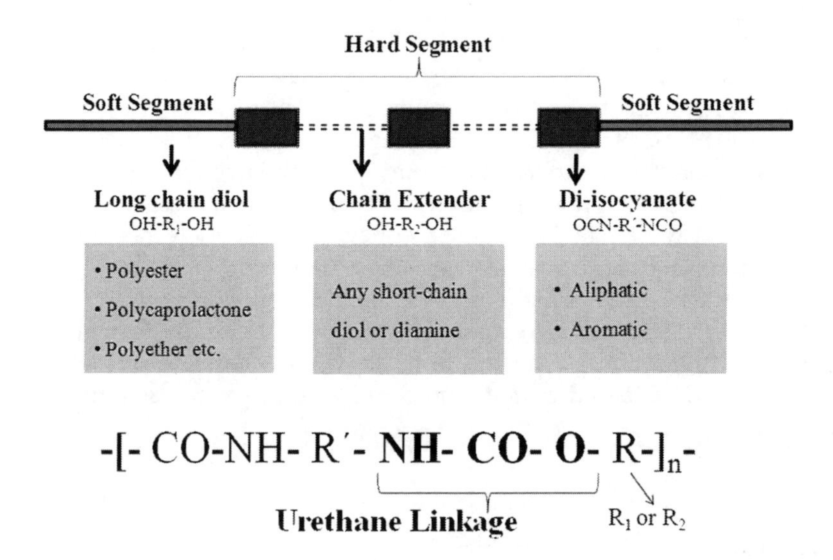

FIGURE 6.1
Basic chemistry of PU.

imide structure, which is yellow in colour. Aliphatic diisocyanate cannot experience such chemical changes, yielding PU with a better stability to weathering. On the contrary, in the absence of oxygen, aromatic PU has better thermal resistance than aliphatic PU. This point should be remembered, as the urethane group is considered to be the least thermal-resistant group in the PU structure. But it should also be noted that resistance to thermooxidation is mainly influenced by the type of diol, rather than by the choice of aliphatic or aromatic diisocyanate.

6.2.2.2 PU Structures Based on Different Diols

There are three main types of diols which are commercially used for the production of PU. These are polyester, polycaprolactone, and polyether. As the different types of diols may provide PU with different characteristics, a basic comparison of their properties is given in this section.

(a) Hydrolysis Resistance

Theoretically, ether groups have much better hydrolysis resistance than ester groups. In case of an ester group, a free acid is liberated as a result of the hydrolysis, which ultimately makes the reaction autocatalytic. Thus, polyester-based PU shows very poor hydrolysis resistance. However, in the case of polycaprolactone PU, hydrolysis resistance is better than normal polyester PU. This is due to the fact that the polyester has only four $-CH_2$ groups between two ester bonds, whereas polycaprolactone has five. The stability of the ester bond increases with the length of the hydrocarbon chain. Moreover, the PU also becomes more hydrophobic, absorbs less water, and thus becomes less prone to hydrolysis. In a nutshell, the hydrolysis resistance would be:

Polyether PU > Polycaprolactone PU > Polyester PU

(b) Microbial Stability

The higher hydrophobicity of a polymer results in better stability against microbial attack. Polyether PU is the most hydrophobic, and thus polyester and polycaprolactone PUs have much lower microbial resistance than Polyether PU. To sum it up, the resistance to microbial attack is in the following order:

Polyether PU > Polycaprolactone PU > Polyester PU

(c) Thermooxidative Stability

Thermooxidative reaction usually initiates with attack on the hydrocarbon chains to produce free radicals followed by several reactions in succession until chain scission occurs. Thus, the thermooxidative stability depends on the labile hydrogen atoms adjoining the hydrocarbon chains. In ethers, the hydrogen bonded to the carbon adjacent to the oxygen is predominantly sensitive to oxidation and readily forms peroxides. In contrast, ester bonds are the most resistant to oxidation. In other words, the order of thermooxidative stability in PUs is:

Polycaprolactone PU = Polyester PU > Polyether PU

(d) Segmental Structure

PU has a two-phased segmental morphology. Hard segments have a considerably higher polarity than the soft segments. The diol with less polarity causes more phase separation in PUs. Thus, using diols of the same molecular weight, polyether PU is found to be more phase separated than polycaprolactone PU which is, again, more phase separated than polyester PU. This is the main reason behind the lowest Tg and better flexibility for polyether PU, while polyester PU provides the highest Tg and the poorest flexibility. So, the order of the phase separation in PUs is:

Polyether PU > Polycaprolactone PU > Polyester PU

(e) Oil and grease resistance

In contrast to hydrolysis resistance, polyester PU has the best oil and grease resistance due to its higher polarity, while the polyether PU has the worst resistance to oil and grease. In other words, oil and grease resistance in PU is in the order of:

$$\text{Polyester PU} > \text{Polycaprolactone PU} > \text{Polyether PU}$$

A summary of PU based on different diols is given in Table 6.1. The properties of the overall PU depend on the specific composition, lengths, and molecular weights of the two segments as well as the sequence of length distribution, anomalous linkages (branching, crosslinking), etc.

6.2.3 PU-Based Coatings

PU-based coatings are generally characterized by their high-performance properties achieved with very thin coatings that do not markedly increase the ultimate product weight. PU can be coated in various forms: as a one-component thermoplastic PU in which chemical reactions have been pre-performed during the production; as a two-component PU with isocyanate cross-linking; as a one-component product that allows dispersion in water which is environmentally friendly; or as a solid product where greater quantities can be coated in each coating passage [22–24]. Six different groups in the PU coating types have been distinguished by the ASTM D16-03 standard. Besides the types discussed in Table 6.2, two-package polyurea and poly(urethane-urea) coatings are also very popular; they are composed of one package that contains amines along with fillers, pigments, and additives, and a second package that contains monomeric multifunctional isocyanates and/ or prepolymeric adducts of diisocyanates [25]. Research has also been done to minimize the volatile organic content of the coatings, and low-dispersion of PU is also a growing technology for coating industries [25]. PU powder coating is a recent

TABLE 6.1

Comparison of Properties of PU Based on Different Diols

Parameters	Polyester-based PU	Polycaprolactone-based PU	Polyether-based PU
Hydrolysis resistance	Very poor	Poor	Excellent
Microbial stability	Poor	Poor	Good
Adhesion strength	Good	Excellent	Poor
Thermooxidative resistance	Good	Good	Poor
Low temperature flexibility	Fair	Good	Excellent
Mechanical properties	Excellent	Excellent	Good
Oil and grease resistance	Excellent	Good	Poor

TABLE 6.2

Various Forms of PU-Based Coatings

Type	Specification	General applications
Type I	One-package pre-reacted; any significant quantity of free isocyanate groups is absent	Wood and floor finishes
Type II	One-package moisture cured; free isocyanate groups are present which react with ambient moisture to form the coating	Sealers for concrete and wood as well as floor and deck finishes
Type III	One-package heat cured; get dried-on curing by thermal release of blocking agents and regeneration of active isocyanate groups which subsequently react with polyols	Coil coating, electrical wire coatings
Type IV	Two-package catalyst; one package containing a prepolymer or adduct having free isocyanate groups capable of combining with a relatively small quantity of catalyst, accelerator, or cross-linking agent such as a monomeric polyol or polyamine of the second package	Not widely used
Type V	Two-package polyol; one package contains a pre-polymer or adduct or other polyisocyanate capable of combining with a substantial quantity of a second package containing a resin having active hydrogen groups with or without the benefit of catalyst	High-performance areas such as automobile refinish, aircraft, industrial-structure maintenance coatings
Type VI	One-package nonreactive lacquer; solution prepared from high-molecular-weight PU with thermoplastic properties	Fabric coating, aerial delivery systems

technology where finely divided powdered PU prepared from polyols, difunctional isocyanates, urethane modified polyesters, and hydroxyl containing polyacrylics is usually cured with melamines or blocked isocyanates [25, 29].

A detailed discussion on the chemistry of different types of PU coatings has been depicted thoroughly by Chattopadhyay and Raju [19], which is beyond the scope of the present discussion.

6.2.4 PU-Based Films

PU films have huge application in preparation of laminated technical textiles. These films are generally used for laminating on the top of a fabric or in between two fabrics, imparting different functionalities. It has an especially huge application in developing waterproof breathable fabrics. In industry, PU films are generally prepared by the following extrusion techniques [30]:

1. Film blowing
2. T-die casting
3. Slit-die casting

In comparison to film blowing, the extrusion casting method (T-die casting and slit die casting) shows much higher productivity with uniform film thickness. However, the PU films with very low thickness can't be produced via the extrusion casting method, which can easily be produced in the blown film technique. We need a special PU grade with required MFI and melt strength for making the film by film-blowing method.

6.3 Polymer Nanocomposite: The Leading-Edge Material

6.3.1 What Is Nanocomposite?

Polymer nanocomposites (PNC) have received much attention in aerospace applications over the past decade. PNC consist of a polymer matrix having nanofillers dispersed into it. Nanofillers may be of different shape (e.g., platelets, fibres, spheroids), where at least one dimension must be in the range of 1–50 nm. Nanofillers provide very high interfacial area for better adhesion to the polymer matrix which ultimately leads to superior property enhancement of the product. A range of nanoreinforcements with different shapes have been used in making polymer nanocomposites. An important parameter for characterizing the effectiveness of reinforcement is the ratio of surface area to volume of reinforcement [31].

Unlike traditional filled polymer systems, nanocomposites require relatively low dispersant loadings to achieve significant property enhancements, which make them a key candidate for aerospace applications. Besides, nanomaterial-based composites can provide various multifunctional properties, such as thermal stability, fire retardancy, electronic properties, field emission, optical properties, improved material durability, high impact resistance, and energy absorption, which are particularly significant for aerospace application [32, 33]. Figure 6.2 describes the evolution of materials for aerospace structures.

6.3.2 Synthesis Routes and Properties of Nanocomposites

6.3.2.1 Synthesis Routes

There are three methods to prepare nanocomposites: solution casting, melt blending, and in-situ polymerization (Figure 6.3). In solvent casting, the polymer and the nanoreinforcement are combined into a solvent and thoroughly mixed (e.g. by ultrasonication), and then the solvent is allowed to evaporate, leaving behind the nanocomposite typically as a thin film. The solvent chosen should completely dissolve the polymer as well as disperse the nanoreinforcement. The solvent used will help in the mobility of the

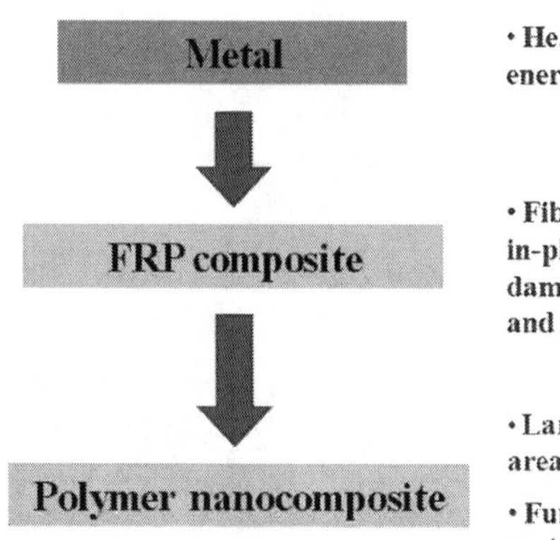

FIGURE 6.2
Evolution of materials for aerospace structures [37].Reprinted with permission from Elsevier.

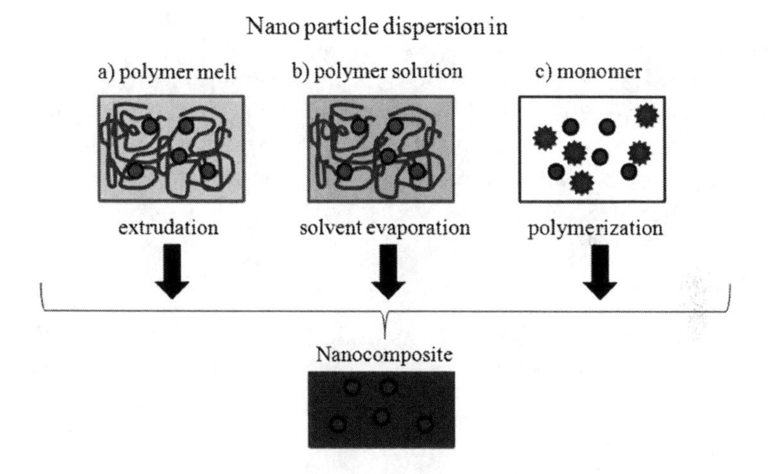

FIGURE 6.3
Schematic demonstration of nanocomposite synthesis [37].Reprinted with permission from Elsevier.

polymer chains, which in turn helps in the intercalation of the polymer chains with the layered nanoreinforcement. In the case of melt blending, extruder or an internal mixer is used. Polymer and nanoreinforcement are added into the extruder and subjected to intensive mixing for some specific time, and then nanocomposite is extruded from the die. In this method, polymer mobility simply comes from thermal energy. In case of in-situ polymerization, initially the monomer and nanoreinforcement are mixed. The monomer is then allowed to intercalate between the silicate layers.

Once the monomer is intercalated, polymerization is started. The polymerization may be due to some surface modification at the silicate surface, or due to any functionalities present which catalyse the reaction [34]. There have been numerous reports on the synthesis of conducting polymer nanocomposites [35, 36].

6.3.2.2 Properties

In today's research, nanotechnology has come out as an efficient and powerful strategy to upgrade the structural as well as functional properties of polymers [37]. Nanoparticles have an extremely high surface-to-volume ratio which dramatically changes their properties when compared with their bulk-sized equivalents. The result is that the composite can be many times improved with respect to the component parts. Some nanocomposite materials have been shown to be 1000 times tougher than the bulk component materials. Numerous studies on polymer nanocomposites have demonstrated that nanofillers can provide various functional properties. Table 6.3 lists some nanomaterials having potential functional properties for aerospace applications.

Nanocomposites currently represent the frontier of high-performance materials. The growing requirement for high-performance military aerospace systems and the growing demands of government-funded defence projects represent an ideal frontrunner for the commercial adoption of nanocomposite materials.

6.3.3 Nanocomposites in Films and Coatings

Polymer nanocomposite-based films and coatings have huge applications in preparation of coated/laminated technical textiles. Different polymeric

TABLE 6.3

Some Nanomaterials and Their Functional Properties Related to Aerospace Applications

Functional Properties	Nanomaterials
Mechanical, scratch resistance	Al_2O_3, SiO_2, ZrO_2
Antimicrobial	CuO, TiO_2, ZnO
Gas barrier	Nanoclays, Graphene
Corrosion	Nanoclays
Electrical conductivity	CNT, Graphene, SnO_2
Fire retardancy	Nanoclays
Heat stability	Nanoclays, CNT, ZrO_2
UV stability	TiO_2, ZnO, $BaSO_4$, CeO_2, Graphene
Impact resistance	SiO_2, TiO_2, $CaSiO_3$, Al_2O_3, CNT, clay

coatings are extensively used for improving water proofness, barrier property, chemical resistance etc. Polymeric coatings are generally applied by two techniques: (i) solution coating, and (ii) hot-melt coating.

Applying nanomaterials to high-performance coatings is a recent concept for newly developed coating systems that has shown better performance than the conventional one [38]. Having exceptionally high surface area-to-volume ratio, nanomaterials give rise to exceptional properties in the new products. A nanoadditive or the filler has at least one dimension in the nanometer scale and thus, owing to very high surface area, the dispersion of only a tiny amount of nanomaterials within the polymer matrix can lead to a tremendous increase in interfacial contacts between the polymer and the filler which ultimately leads to the superior properties of the new product with only a negligible increase in weight. Figure 6.4 shows the schematic diagram of various types of nanoadditives or fillers with nanoscale dimensions that are very popular for a nanocomposite coating application.

Incorporation of nanofillers into the organic coatings might enhance their barrier performance by decreasing the porosity and zigzagging the diffusion path for deleterious species. Thus, the coatings containing nanofillers are expected to have significant barrier properties for corrosion protection and reduce the trend for the coating to blister or delaminate. On the other hand, high hardness could be obtained for metallic coatings by producing the hard nanocrystalline phases within a metallic matrix. Coatings incorporated with nanoadditives cover potential applications areas such as the gas barrier, anticorrosion, antiwear, UV protection, superhydrophobic area, self-cleaning, antifouling/antibacterial area, and electronics [37].

In lab-based research, PNC films are commonly prepared by: (i) hot pressing under compression moulding, or (ii) the solution casting method, followed by evaporation of solvent from the casted formulation or coagulation by using a non-solvent. Whereas, the pilot scale production in lab/industry

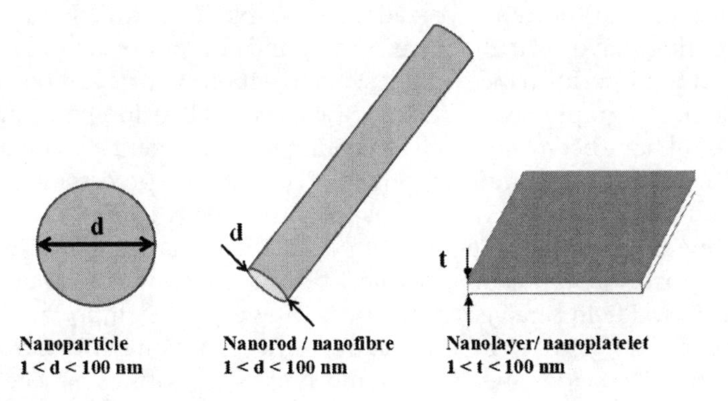

FIGURE 6.4
Scheme of various types of nanofillers or fillers with nanoscale dimensions.

or bulk production in industry of PNC films are generally performed by two process techniques, i.e. extrusion casting or film blowing.

In the film blowing technique as the film is stretched in all directions, if any nanofiller is present in the polymer matrix, the filler as well as the hard segment of PU will be oriented in the polymer matrix in the direction of stretching. Similarly, during uniaxial or biaxial stretching in extrusion casing, the nanofillers and hard segments can be oriented along the direction of stretching, which may result in improvement in the gas barrier property of the polymer [30, 39]. Different types of PUNC films can be produced for developing laminates with various functionalities for various technical applications. For example, PUNC-based weather-resistant and gas barrier films can be potential materials for making aerostat/airship envelope.

6.4 Polyurethane Nanocomposite for an Aerostat/Airship Envelope

6.4.1 PU Nanocomposite-Based Coatings

PU has recently attracted greater interest in functional coating applications. It has outstanding overall toughness, high tensile strength, tear resistance, flexibility, abrasion resistance, fair gas permeability, good handling properties, and ease of processibility [19]. The PU system is used in various advanced coating applications today mainly because of its customized formulation and a wide range of property profiles. The applications of PU-based coatings are widespread, from outdoor construction to automotive areas such as metal, wood, plastic, glass, and textiles. At present, PU is the predominant type of topcoat material used in the aircraft industry and aerial delivery systems like aerostats, high-altitude airships, radomes, etc. Nano-modified developments in PU coatings have led to the current status and new opportunities for the formation of high-performance coatings. Nanocarbon such as carbon nanotube, nanodiamond, graphene and its derivatives, as well as inorganic nanoparticles [40–46] have been employed as coating reinforcements. Incidentally, the type of nanoparticle, nanofiller content, PU structure, processing technique, and conditions may affect the final nanocomposite properties. Consequently, the cutting-edge application areas of polyurethane nanocomposite coatings such as corrosion protection, textiles, biomaterials, and aeronautical and other technical fields are very impressive. However, it is important to understand the challenges of weathering-induced degradation in PU nanocomposite coatings. The wide range of PU nanocomposite coatings may be attained using versatility in the selection of monomeric materials. Moreover, modification of nanofillers and incorporation in appropriate PU matrices has also been adopted as a powerful tool to widen the applicability of nanocomposite coatings in this kind of advanced applications.

Please refer to Chapter 8 for the detailed discussion on the mechanism of photostabilization of PU materials and the recent trends in weather-resistant polyurethane coatings particularly for the aerostat/airship envelope.

6.4.2 PU Nanocomposite-Based Films

Different types of PUNC films with improved weather resistance or gas barrier property are the potential materials for a multi-layered laminate to be used in the envelope of LTA systems such as aerostats and airships.

6.4.2.1 PU/Organic UV Additive-Based Films

Before discussing the current trend of using weather-resistant PUNC films, we should know the conventional approaches for preparing weather resistance PU films. There are several organic UV stabilizers such as antioxidant (primary and secondary), UV absorbers, hindered amine light stabilizers (HALS), quenchers, etc., which have good potential in improving the weather resistance property of polymers like PU.

Many studies have reported the improvement of weather resistance property of PU by incorporating different organic UV additives, mainly in different combinations and even in combination of inorganic/nano UV additives [47, 48]. Chu et al. [47] analysed the efficacy of different UV stabilizers for improving the weather resistance property of an aliphatic PU. Eleven different types of combinations (different UV stabilizers and different concentrations) were used to improve weather resistance of the PU film and were analysed after exposure inside a twin-lamp carbon arc weatherometer (Atlas). The most promising weather-resistant property was obtained in a combination of Tinuvin 328, Tinuvin 3770, ZnO, and an antioxidant.

However, most of the organic UV additives are toxic, and they have a sustainability issue because of their volatile nature. Therefore, nanocomposites are drawing attention for better and sustainable performance in terms of improving the weather resistance property of polymers.

6.4.2.2 PU/UV Shielding Nanomaterial-Based Films

A few nanomaterials which are promisingly used for improving the weather resistance property of polymers are nano-TiO_2, nano-ZnO, and nano-CeO_2. Many research papers [7–11] reported their potential in improving weather resistance properties of PU film.

Gu et al. [49] studied the effect of percentage loading and size of nano-ZnO on UV resistance property of PU/ZnO nanocomposite films. Compared to 1 wt % loading, the 5 wt % loading of ZnO showed a faster rate of degradation.

However, a degradation rate of 5 wt % ZnO loaded PU film was slower than that of neat PU film. On the other hand, with the increasing size of nanoparticles from 20 to 60 nm, the rate of degradation decelerated.

Many literatures reveal that excellent performance or synergy effect could be acquired by using composite stabilizers. Wang et al. [7] studied the effect of UV/Ozone synergy ageing on pure polyester polyurethane film and the modified polyurethane (M-PU) film containing 2.0 wt % inorganic UV absorbers mixture (nano-ZnO/CeO_2 with weight ratio of 3:2) and 0.5 wt % organic UV absorbers mixture (UV-531/UV-327 with weight ratio of 1:1) prepared by the spin-coating technique. The results show that the composite UV absorber provides better protection to the PU system, which distinctly reduces the degradation of PU film. O_3/UV ageing of the films increased with incremental exposure time. Both the photooxidation index (PI) and carbonyl index (CI) of PU and M-PU films increase with increasing exposure time. The PI and CI of M-PU film are much lower than that of PU film after the same time of exposure, indicating their improved weather resistance property.

In another study, Wang et al. [8] prepared modified PU films (PU-ANT and PU-COM) by adding antioxidants-1010 and composite additives (containing nano-ZnO, nano-CeO2, UV-531, UV-327, and the hindered amine-622), respectively, to the polyurethane matrix. The accelerated weathering tests were performed by using self-designed UV/ozone ageing test device to analyse the effects of antioxidants-1010 and composite additives on the UV/ozone resistance properties of polyester PU films, which provide some basis in application of PU to the weathering layer of a high-altitude balloon or airship. The composite stabilizer significantly reduced the UV transmittance of the PU film. The UV transmittances of three films gradually decreased with the increase of the ageing time. Antioxidant-1010 and the composite stabilizer were able to reduce photooxidation index and the carbonyl index of the polyester PU film significantly. The colour difference and the yellowness index of these films gradually increased with the increase in ageing time. In the early stage of ageing, the variations of the colour difference and the yellowness index were not obvious. At ageing time of more than 60 h, the yellowing resistance properties of PU-ANT film were better than those of PU-COM and PU films.

Our group [9] studied the synergistic effect of ZnO nanoparticles and TiO_2 nanoparticles (two different crystal phases, i.e. anatase and rutile) for improving the weather resistance property of PU film. The UV-shielding nanopowder containing ZnO and TiO_2 nanoparticles was calcinated at varying conditions, and the process was optimized to obtain the desired crystal structure. The calcinated UV-shielding nanopowder with the enriched rutile phase of TiO_2 showed good potential in improving weather resistance of PU nanocomposite films in term of retention in tensile properties and surface morphology after exposure under accelerated artificial wreathing conditions.

These PUNC films containing only inorganic/nano UV additives, or in combination of organic and inorganic/nano UV additives, can be potentially used in a weather-resistant layer of multi-layered laminates for an aerostat/airship envelope.

6.4.2.3 PU/Clay Nanocomposites-Based Gas Barrier Films

Recently, PU/clay nanocomposite films are getting enormous attention because of remarkable improvement in properties of the neat polymer such as mechanical, thermal, gas barrier, flame resistance and lots of others. However, focusing on the application of aerostat or airship envelopes, here we will discuss only the gas barrier properties of the PU/clay nanocomposites. Numerous papers [12, 13, 50–56] have reported gas barrier properties of PU/clay nanocomposites.

Qian et al. [51] reported about a 19% reduction in CO_2 permeability through a PU film (MDI and polyether polyol based) with incorporation of CTAB-VMT (modified vermiculite clay). In another study [53], they reported a 40% reduction in CO_2 permeability through the PU film with incorporation of only 3.3 wt % modified vermiculite clay in the PU matrix by master-batch-based mixing. Similarly, in a very recent study, Adak et al. [13] reported that the master-batch preparation by solution mixing and subsequent melt-mixing lead to much better results in comparison to direct melt-mixing in term of exfoliation of clay platelets in the PU matrix and improvement in helium gas barrier properties of PU film. With 3 wt % loading of organomodified montmorillonite clay, about 31% and 39% reduction in helium gas permeabilities were observed for the PU/clay nanocomposite films produced by direct mixing and master-batch based mixing, respectively. In another study [12], we found that organomodified montmorillonite clay (containing hydroxyl group) interacted with PU in a better way than the organomodified-bentonite clay. It resulted in a much better helium gas barrier property and better exfoliation of clay platelets in PU matrix for montmorillonite clay than bentonite clay (Figure 6.5). Interestingly, the helium gas permeability reduced gradually with increasing the concentration of clay to a certain loading (about 3–4 wt %), and after that, gas permeability increased with increasing clay concentration, which might be due to the agglomeration tendency of clay platelets at higher loading (Figure 6.5). A similar trend was also reported by Osman et al. [57], where they compared the oxygen gas barrier property of organomodified montmorillonite (OMMT) clay-reinforced PU and epoxy nanocomposites. The oxygen gas permeability decreased exponentially, with increasing filler loading up to 3 vol % for both nanocomposites, but the gas permeability increased at higher filler loadings (> 3 vol %).

Chang et al. [56] discussed the effect of three different organomodified clays [Cloisite 25A, hexadecylamine-montmorillonite (C16–MMT), dodecyltrimethylammonium-montmorillonite (DTA–MMT)] on oxygen-gas

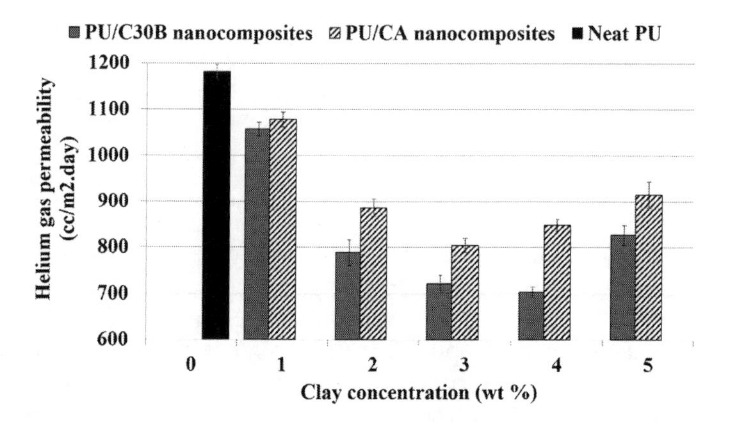

FIGURE 6.5
Helium gas permeability (cc/m^2/day) for neat PU, Cloisite 30B/PU nanocomposite (C30B-CPN) and Claytone APA/PU nanocomposite (CA-CPN) films [12]. Reprinted with permission from Elsevier.

barrier property of PU/clay nanocomposite films. It was reported that the oxygen gas permeability of PU/clay hybrid nanocomposites decreased linearly with increasing clay lading (0–4 wt %), regardless of the type of organoclay, because of (i) increased tortuosity in the path length for gas diffusion, and (ii) increase in rigidity of nanocomposite films. However, the performance of Hexadecylamine-montmorillonite (C16–MMT) was much better than that of the other two organoclays and resulted in about 50% reduction of oxygen gas permeability at 4 wt % loading. In a similar study, Herrera-Alonso et al. [55] reported the extent of improvement in the gas barrier property of PU/clay nanocomposites against both helium and methane gases in this order: PU/Cloisite 30B > PU/Cloisite 10A > PU/Cloisite 20A.

Shamini and Yusoh [50] modified the montmorillonite clay by iron (III) chloride and copper (II) chloride, and the modified clays were utilized in preparation of PU/clay nanocomposite by solution mixing. The nitrogen permeability of PU film decreased by 60% and 50% with incorporation of only 1% modified iron-clay and copper-clay, respectively. Similarly, a remarkable fourfold and threefold decrease in oxygen gas permeability was observed with incorporation of only 1 wt % modified iron-clay and copper-clay in PU matrix, respectively.

Benali et al. [58] explored two strategies for improving gas barrier properties of PU, involving (i) poly(ecaprolactone) (PCL)/organoclay master batch, and (ii) PCL-grafted organoclay hybrids, which were further mixed with an ester-based PU by melt mixing. These were resulted in decrease in the permeability coefficient of about 61% and 41%, with respect to neat PU for PU/PCL-clay master-batch nanocomposite and PCL-grafted organoclay-filled PU nanocomposite, respectively. Maji et al. [52] reported a significant improvement in the helium gas barrier property of hyper-branched PU with incorporation of organomodified clay (Cloisite 30B), as a result of higher retardation to the gas molecule penetration. It resulted in about a 76% reduction in helium gas permeability for third-generation hyperbranched PU with 8 wt % loading of Cloisite 30B. Table 6.4 summarizes the gas permeability value of different PU/clay nanocomposite films, compiled from the existing literature.

6.4.2.4 PU/Graphene Nanocomposite Films/Membranes

Similar to clay, graphene is a layered-structured nanomaterial having strong potential in improving gas barrier properties of PU, as reported in many studies.

Kim et al. [42] investigated thoroughly the role of graphene oxide (GO), thermally reduced graphene oxide (TRG), and isocyanate modified graphene oxide (iGO) in improving the nitrogen gas barrier property of PU/graphene nanocomposite film. The nanocomposites were prepared by all three routes, i.e. melt-mixing, in-situ polymerization and solution mixing (Figure 6.6), where best properties (mechanical and gas barrier) were obtained by the solution-mixing route, due to effective exfoliation of graphene platelets in PU matrix, in comparison to other processes. Meanwhile, a most inferior result was obtained by the melt-mixing route, and TRG showed better exfoliation in PU matrix in comparison to graphite. PU/iGO nanocomposite (with 3 wt % loading of iGO) prepared by solution mixing showed a remarkable 80-fold decrease in N_2 gas permeability because effective exfoliation of 2D graphene platelets having a very high aspect ratio [42].

Kaveh et al. [59] reported about 80% reduction in helium gas permeability with only 1 wt % graphene oxide (GO) content. The functional polar groups that present in GO facilitated the complete exfoliation of GO in the PU matrix by simple sonication, resulting in significant improvement in helium gas barrier property.

Yousefi et al. [60] observed a significant reduction in water vapor permeability of a waterborne PU by incorporating reduced graphene oxide (rGO). In comparison to neat PU, the moisture vapor permeability of PU/rGO nanocomposites reduced by about 76% with 3 wt % rGO loading. It was a consequence of the synergistic effect of a very high aspect ratio and high horizontal alignment of impermeable graphene sheets resulting from strong interaction between rGO and polar groups (urethane and carbonyl groups) present in PU. Another study [3] reported significant improvement in the oxygen barrier property of intumescent flame-retardant polyurethane (IFR-PU) nanocomposites prepared by incorporating rGO, melamine. and microencapsulated ammonium polyphosphate. A good dispersion of rGO in the PU matrix and a strong interaction between them resulted in about 90% reduction in oxygen permeability with incorporation of 2 wt % rGO in PU matrix.

TABLE 6.4

Overview of Gas Barrier Property of PU/Clay Nanocomposite Films

Type of PU	Filler (Clay)	Filler loading	Preparation method	Gas	% Reduction	Ref
Polyether polyol-based aliphatic PU	Cloisite 30B	4 wt %	Master-batch-based melt mixing	He	41	[12]
Polyether polyol-based aliphatic PU	Claytone APA	3 wt %	Master-batch-based melt mixing	He	32	[12]
Polyether polyol and MDI-based PU	CTAB-VMT	5.3 wt %	In-situ polymerization	CO_2	19	[51]
Polyether polyol and MDI-based PU	CTAB-VMT	5.3 wt %	In-situ polymerization	N_2	30	[51]
Polyether polyol and MDI-based PU	CTAB-VMT	3.3 wt %	In-situ polymerization (master-batch mixing)	CO_2	40	[53]
Polyether polyol and MDI-based PU	CTAB-VMT	3.8 wt %	In-situ polymerization (direct mixing)	CO_2	20	[53]
PU	Cloisite 25A	4 wt %	Solution mixing	O_2	34.5	[56]
PU	C_{16}–MMT	4 wt %	Solution mixing	O_2	50	[56]
PU	DTA-MMT	4 wt %	Solution mixing	O_2	15	[56]
Polyester-based PU	MMT-Fe	3 wt %	Solution mixing	O_2	54	[50]
				N_2	62	
Polyester-based PU	MMT-Cu	3 wt %	Solution mixing	O_2	44	[50]
				N_2	52	
Third-generation hyperbranched PU	Cloisite 30B	8 wt %	Solution mixing	He	76	[52]

Note: Cloisite 30B = bis(2-hydroxy-ethyl)methyl tallow ammonium montmorillonite, Claytone APA = benzyl dimethyl stearyl ammonium modified bentonite clay, CTAB-VMT = cetyltrimethylammonium bromide (CTAB) modified vermiculite clay (VMT), Cloisite 25A = N-(hydrogenated tallow)-N,N,N-trimethyl ammonium montmorillonite, C_{16}–MMT = Hexadecylamine montmorillonite, DTA-MMT = dodecyltrimethylammonium montmorillonite, MDI = methylene diphenyl diisocyanate, MMT = montmorillonite clay, MMT-Cu = Na^+ montmorillonite modified by copper (II), MMT-Fe = Na^+ Montmorillonite modified by iron (III).

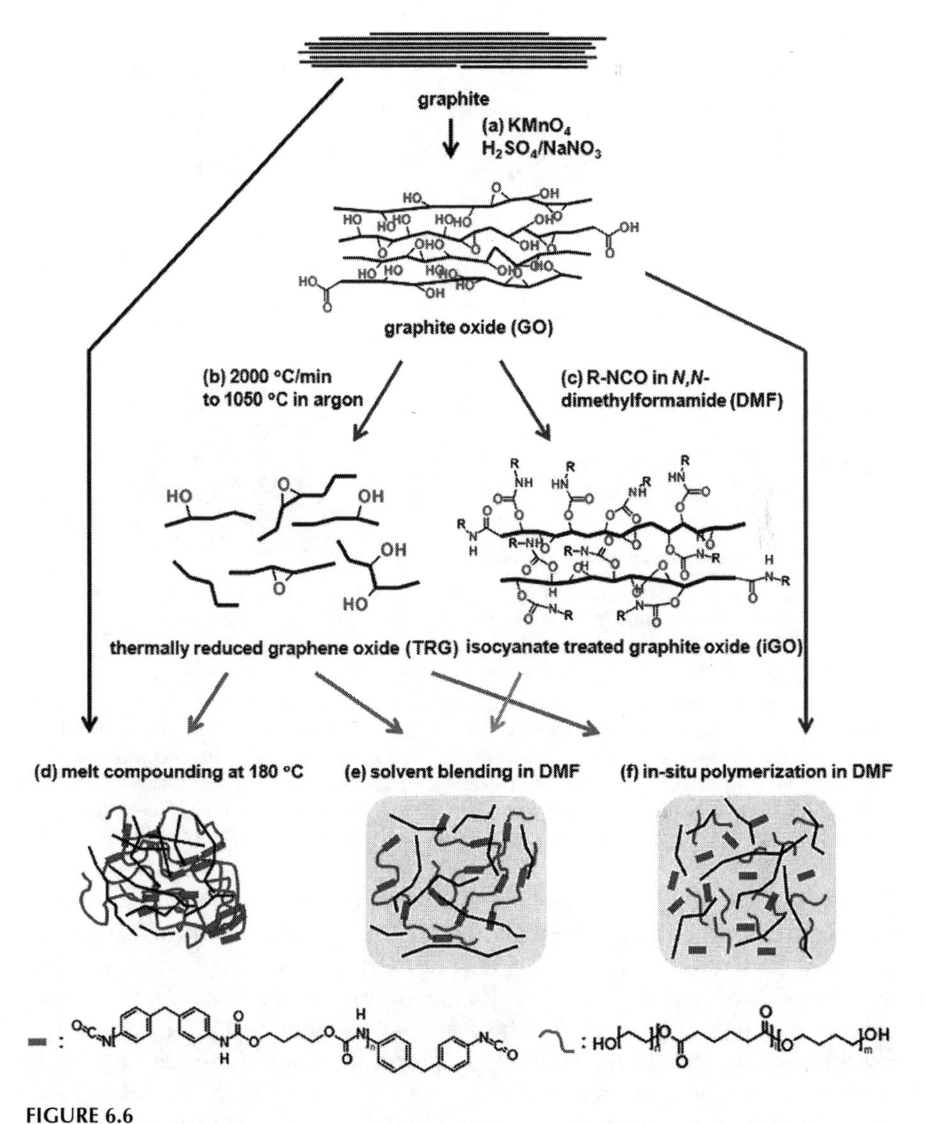

FIGURE 6.6
Schematics of different graphene derivative preparation and different routes for synthesis of PU/GO nanocomposites [42]. Reprinted with permission from ACS.

In a very interesting study, Xiang et al. [61] synthesized a unique, high-gas-barrier PU/graphene nanocomposite film reinforced with hexadecyl-functionalized low-defect graphene nanoribbons (HD-GNRs) by solution casting. The hexadecyl groups on the edges assisted the HD-GNRs to disperse easily in organic solvent, resulting in uniform dispersion in the PU matrix (Figure 6.7). Moreover, the low defects in HD-GNRs structure was highly

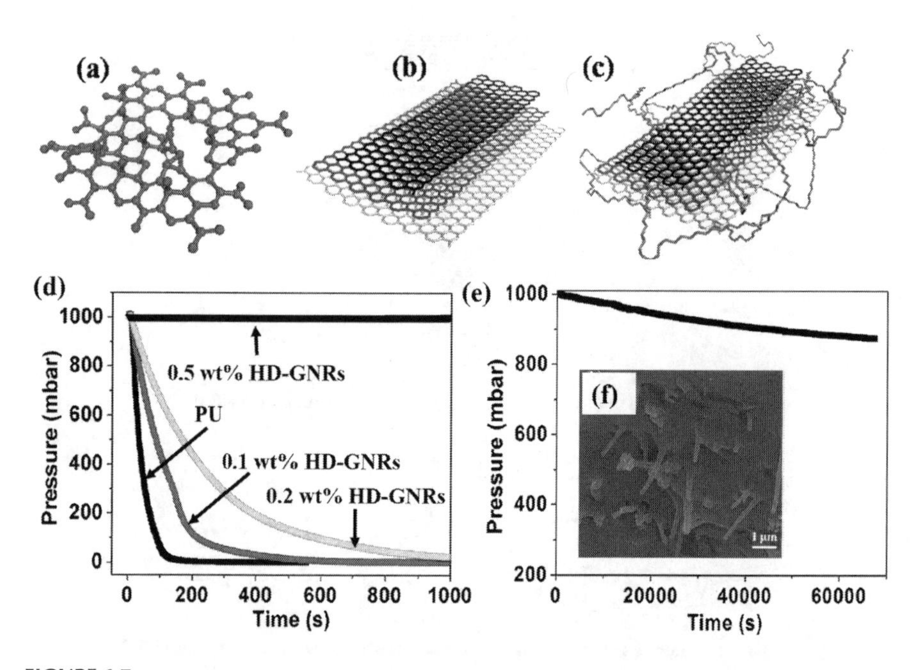

FIGURE 6.7
Schematic representation of chemical structure of (a) graphene oxide (GO), (b) graphene nanoribbons (GNRs), and (c) hexadecyl-functionalized low-defect graphene nanoribbons (HD-GNRs); (d) pressure drop of N_2 gas through PU and PU/HD-GNRs nanocomposite films with time; (e) pressure drop of PU/0.5 wt % HD-GNRs nanocomposite film over a longer time period; (f) SEM of cross-section of nanocomposite film showing dispersion of HD-GNRs in PU matrix [61]. Reprinted with permission from Elsevier.

impermeable to any gas and resulted in remarkable improvement in the gas barrier property with only 0.5 wt % loading of HD-GNRs (Figure 6.7(d–e)).

In a recent study, Adak et al. [14] reported the gas barrier and weather resistance property of PU/functionalized-graphene nanocomposite films prepared by solution master-batching and subsequent melt mixing, followed by compression moulding. Helium gas permeability gradually reduced with increasing graphene content up to 3 wt %. Moreover, with 2–3 wt % loading of graphene, weather resistance of the film increased significantly, which resulted in very good retention in gas barrier and mechanical properties. Table 6.5 summarizes the gas barrier property of different PU/graphene nanocomposite films.

6.4.2.5 PU/POSS Nanocomposite Films/Membranes

Polyhedral oligomeric silsesquioxane (POSS) nanoparticles are organosilicon compounds consisting of rigid silica cores (Si-O-Si) surrounded by functional *organic* groups, and they have huge potential in improving the thermal,

TABLE 6.5

Overview of Gas Barrier Property of Different PU/Graphene Nanocomposite Films

Type of PU	Filler	Filler loading	Processing	Gas	% Reduction	Ref
Polyether polyol-based aliphatic PU	f-Graphene	3 wt %	Master-batch-based melt mixing	He	30	[14]
Polyester-based PU	GO	1 wt %	Solvent mixing	He	78.5	[59]
Polyester-based PU	iGO	1.6 vol %	Solvent mixing	N_2	94-99	[42, 62]
Polyester-based PU	TRG	1.6 vol %	Solvent mixing	N_2	81	[42, 62]
BDO- and MDI-based PU	TRG	1.5 vol %	In-situ polymerization	N_2	71	[42, 62]
Polyester-based PU	TRG	1.6 vol %	Melt intercalation	N_2	52	[42, 62]
BDO- and MDI-based PU	GO	1.5 vol %	In-situ polymerization	N_2	62	[42, 62]
IFR-PU	rGO	2 wt %	Melt mixing	O_2	90.4	[3]
Polyether-based aliphatic PU	HD-GNRs	0.5 wt %	Solution mixing	N_2	99.9%	[61]

Note: f-Graphene = functionalized graphene, GO = graphene oxide, iGO = isocyanate treated graphene oxide, TRG = thermally reduced graphene oxide, rGO = reduced graphene oxide, IFR-PU = intumescent flame retardant polyurethane, HD-GNRs = hexadecyl-functionalized low-defect graphene nanoribbons, MDI= 4,4′ methylene diphenyl diisocyanate, and BDO = 1,4-butanediol.

mechanical, and flame retardancy of polymers. Few studies also divulge the gas barrier property of PU/POSS nanocomposite films, and it depends on the cross-linking density based on reaction chemistry and functional groups available in PU and POSS [63–66].

Madhavan et al. [67] reported the improvement of the gas barrier property PU with incorporation of amine functionalized POSS due to the increase in cross-linking density which resulted in decrease in the solubility and diffusivity of the gases. Moreover, the barrier property of PU/POSS nanocomposites increased with increasing concentration of amine-POSS. The permeability of gases through PU/POSS-amine hybrid nanocomposite film was in this order: $N_2 < O_2 < CO_2$.

Another study [64] reported the gas barrier property of a PDMS-PU based hybrid membrane by incorporating two different types of caged POSS [Octakis(hydridodimethylsiloxy) octasilsesquioxane (POSS-H) and heptacyclopentyl tricycloheptasiloxane triol (CyPOSS)]. The CO_2, O_2, and N_2 gas permeability decreased significantly in the presence of CyPOSS. However, incorporation of a small quantity of POSS-H with CyPOSS resulted in an increase in gas permeability through the membranes, which might be the reduction in compatibility between POSS and PU in the presence of POSS-H.

6.4.2.6 PU/CNT Nanocomposites Films

Some studies reveal that even the carbon nanotube (CNT) has a potential to increase the gas barrier property of polyurethane. Ali et al. [15] reported a 70% reduction in nitrogen gas permeability of a castor oil-based thermoset PU by incorporating only 0.5 wt % multi-walled carbon nanotube (MWCNT). It was a result of a good compatibilization, dispersion, orientation, and lower reaggregation of MWCNT in the PU matrix. However, this type of thermoset nanocomposite film is not useful for developing an aerostat/airship envelope, as this type of film will not be flexible. To the best of our knowledge, no study has been reported on thermoplastic PU/CNT nanocomposite film for improving hydrogen or helium gas permeability as needed for developing an aerostat/airship envelope.

6.5 Case Studies on PU Nanocomposite-Based Coatings and Laminates for a Weather-Resistant and Gas Barrier Aerostat/Airship Envelope

This section provides an overview of the recent research and details of technological and scientific approaches behind the exploration of PU nanocomposites coatings and films in aerostat/airship envelope applications. To that end, a few specific case studies, done by the nanotechnology research group at IIT Delhi, India supervised by Prof. Mangala Joshi, have been illustrated here.

6.5.1 PU Nanocomposite-Based Coatings for Weather-Resistant Aerostat/Airship Envelopes

The study was an attempt to investigate a novel high-performance PU nanocomposite coatings having superior weathering resistance as well as a good impermeability property for textile-based advanced inflatables. The weathering behaviour of two chemically different types of PUs, with and without various classes of conventional UV protective additives, was studied in detail to understand the degradation trend thoroughly. Aliphatic-based PU was found to perform better than the aromatic one. However, organic UV protective additives were found to suffer mainly from physical losses. To overcome these limitations, the study was extended to explore the effect of incorporation of advanced nanomaterials into the coating system. Nanomaterials seem to provide minimum physical loss; at the same time, they can provide additional functional properties, such as resistance to permeability of gases [68]. A combination of conventional UV protective additives as well as advanced nanomaterials such as nanoclay and graphene was investigated for a high-performance, multifunctional nanocomposite coating. An optimal design of PU-based nanocomposite formulation was explored in accordance with a mixture design for three component additives (UV stabilizer, nanoclay, and graphene) [69]. Further, the analysis was extended to explore a novel multi-layered coating where different layers contain a sequence of UV additives and different nanomaterials, for further improvement in performance properties. A series of PU nanocomposite-based coating formulations were synthesized and coated onto high-strength polyester (PET) woven fabric targeting inflatable envelope systems, and the performance properties were analysed in detail. Figure 6.8 shows the morphology of the face (upper) side of the coated fabrics after 300 h of weathering, observed under SEM. This approach of novel multi-layered coating has been filed for patent [39]. Figure 6.9 depicts the tearing behaviour of one particular type (neat aliphatic) of formulated coated fabric. With increasing weathering time, a decreasing trend of tearing strength was found, and the same trend was observed for all the sample prototypes.

The weathering of materials has long been recognized as a natural fact, and the extent of weathering is truly considered to be dependent upon the exposure conditions. However, the natural or outdoor weathering tests of the material take a prolonged time, which is not acceptable in an industrial set-up. Besides, natural conditions have too many uncontrolled and unpredictable variables that make analysis difficult. Thus, in the study, the focus was on artificial accelerated weathering where the exposure conditions are simulated inside the

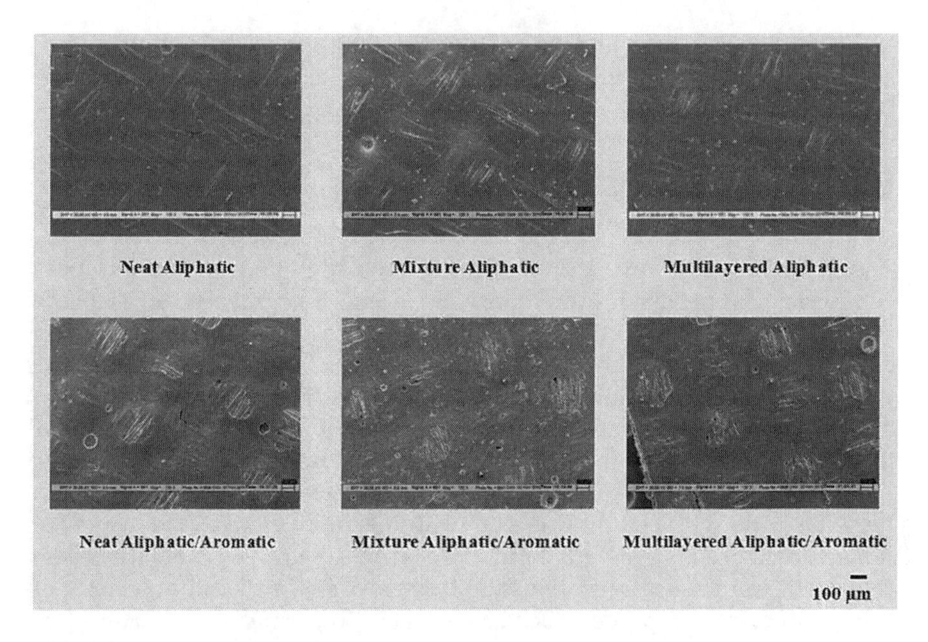

FIGURE 6.8
SEM images of the different formulated coated fabrics after 300 h of artificial accelerated weathering.

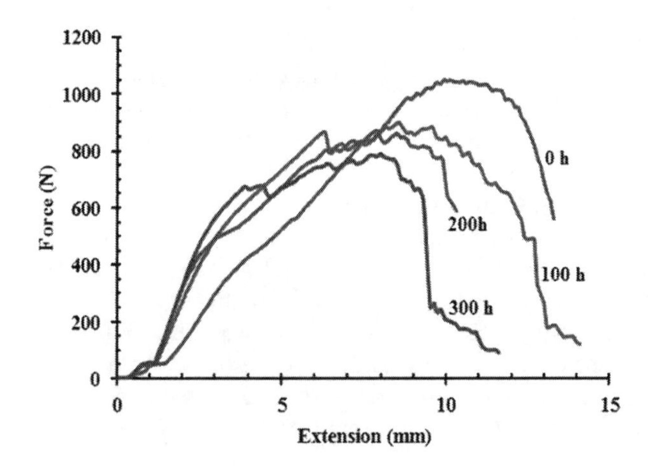

FIGURE 6.9
Tearing behaviour of neat aliphatic coated fabric with artificial accelerated weathering time.

lab at an accelerated level and maintained in a controlled manner. However, too much reliance on accelerated testing can be potentially misleading. In the study, an attempt was also been made to study the degradation behaviour of the coatings under natural weathering conditions, using a specially designed and in-house fabricated outdoor weathering rack (refer to Figure 6.10). Figure

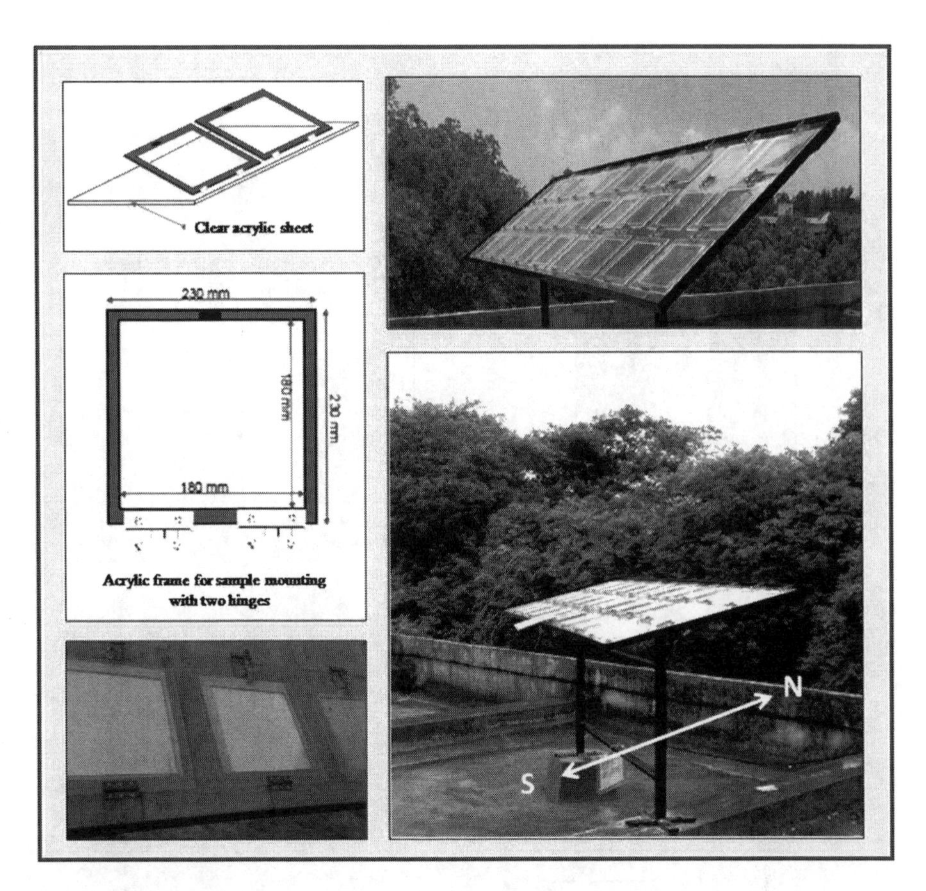

FIGURE 6.10
In-house fabricated outdoor weathering rack at IIT Delhi.

6.11 compares the effect of artificial weathering and natural weathering on the breaking strength of both sides' coated fabrics having several compositions. The degradation behaviour of the samples with exposure time was found to be similar for both cases of exposure conditions, which means that the samples were following the same failure mechanism in both artificial and natural weathering conditions. However, the relative effect of the weathering on all the samples was not found to be exactly the same. It is quite obvious, as artificial weathering techniques do not necessarily mimic natural weathering conditions. Natural weathering always contains some variability. Yet, the ranking within the samples was found to be same for both weathering conditions. The best sample was performing the best, and the worst sample was performing the worst, in the artificial weathering as well as in the natural weathering.

The ultimate aim of a weathering test is to provide a prediction of the performance of the material under the in-service conditions. In the study, a systematic approach (refer to Figure 6.12) of prediction of service life of

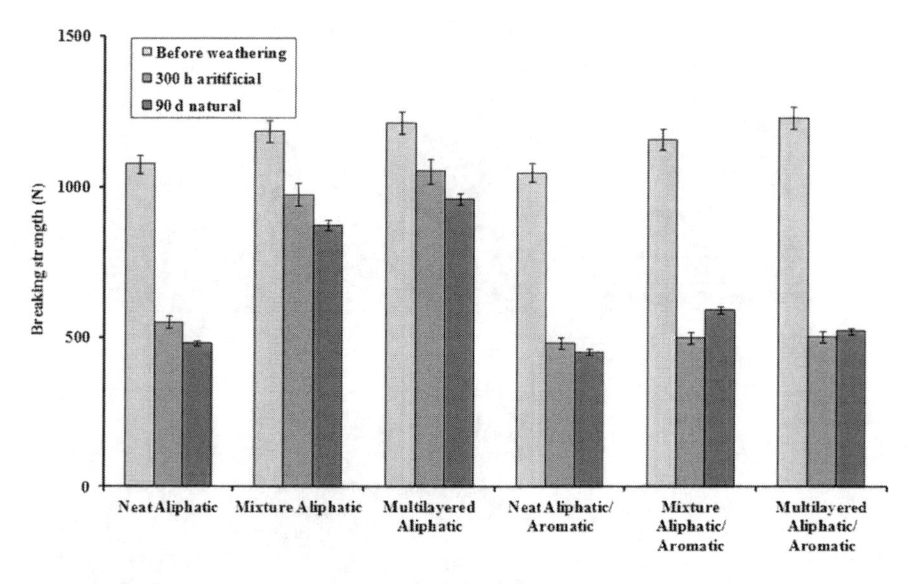

FIGURE 6.11
Comparison of breaking strength of both sides coated fabrics against artificial and natural weathering.

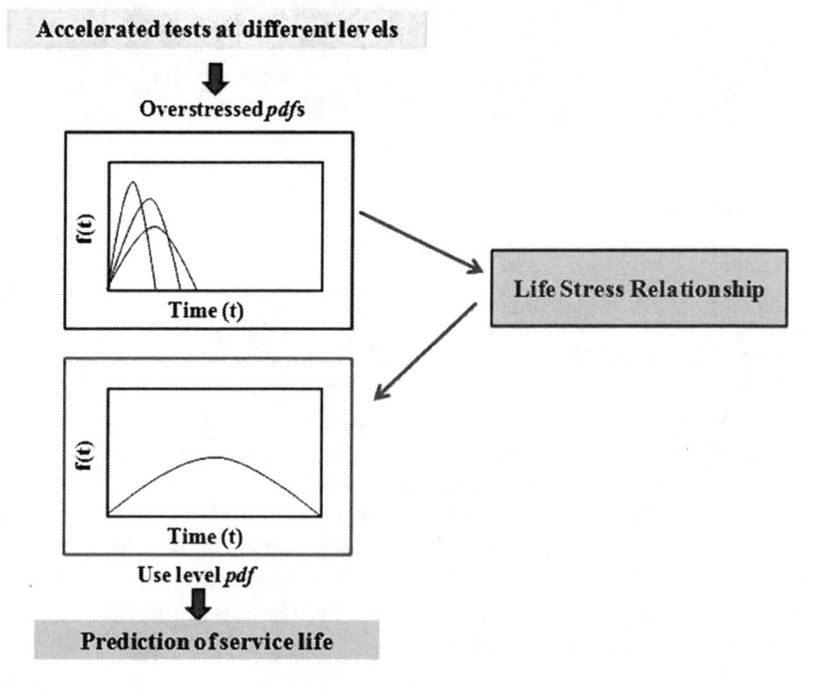

FIGURE 6.12
General scheme of service life prediction model.

the ultimate coated material was also proposed based on a mathematical model [70]. A reliability model with two stress types (UV radiation and temperature) was applied for the estimation of the service life of three different types of PU-coated systems based on the loss of helium gas barrier property. It was found from the study that, in comparison to the neat PU coating, nanocomposite-based coatings have significant effect in enhancing the service lifetime of the product. Validation of the proposed model was also made by performing natural weathering of the materials.

Linearized Weibull cumulative density function (*cdf*) of gas permeability of the three types of coatings was drawn separately using all the failure times of each type of coating under natural weathering (Figure 6.13). It could be noted from the figure that to get the same value of probability of failure $F(t)$, the required time was much higher in the case of Mixture as well as

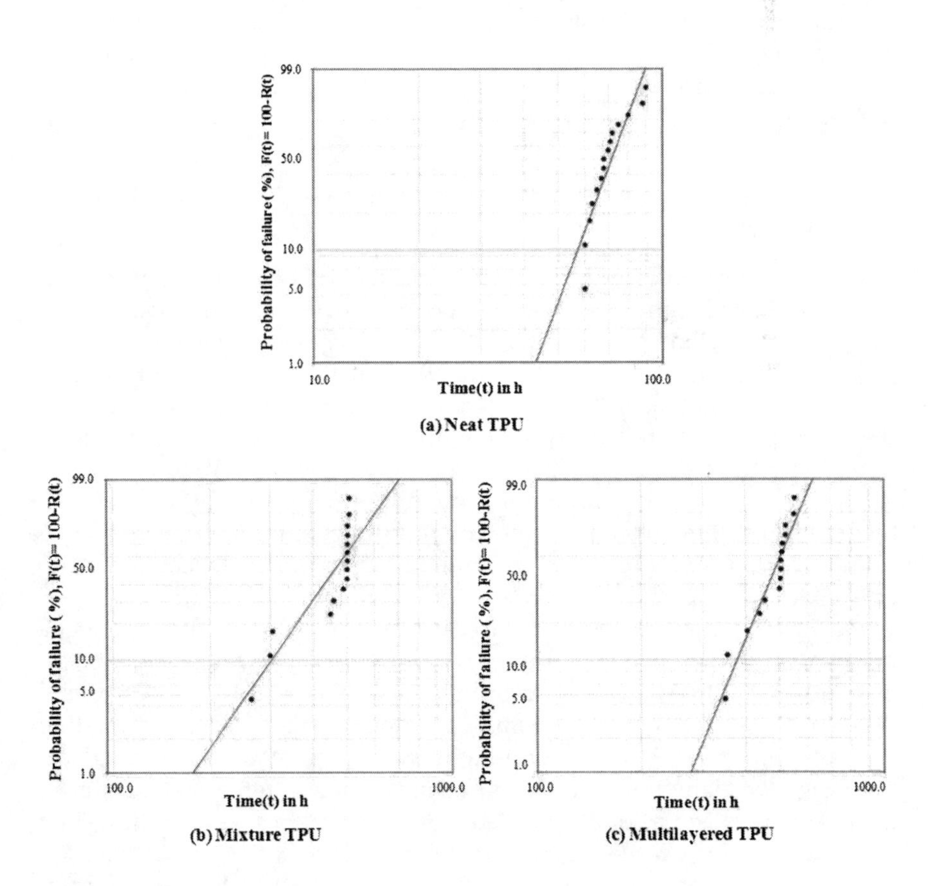

FIGURE 6.13
Cumulative density function for the three types of coating formulations: (a) neat TPU, (b) mixture TPU, and (c) multi-layered TPU; exposed in natural weather.

TABLE 6.6

Comparison of Predicted and Actual Service Life for Different Types of Coatings

Coating formulations →	Neat TPU		Mixture TPU		Multi-layered TPU	
	Mean life (h)	**Warranty 90 % (h)**	**Mean life (h)**	**Warranty 90 % (h)**	**Mean life (h)**	**Warranty 90 % (h)**
Predicted value, accelerated weathering	131	103	531	416	541	497
Actual value, natural weathering	71	83	485	414	521	496

Multi-layered TPU coating formulations in comparison to Neat TPU coating formulation. Table 6.6 compares the actual service life obtained from the natural weathering with that of the predicted values derived from accelerated weathering for different types of coating formulations. The predicted life from the service life prediction model was found in close agreement with the actual natural weathering data.

6.5.2 PU Nanocomposite-Based Films and Laminates for Aerostat/Airship Envelopes

The objective of this study was the synthesis of gas barrier and weather-resistant PUNC films and preparation of textile laminates using these PUNC films. Helium gas barrier PUNC films have been prepared by incorporating different organoclays and functionalized graphene in a PU matrix by solvent-based master-batch preparation, subsequent melt mixing, and then film preparation [12–14]. Different weather-resistant nanocomposite films have been developed by incorporating organic UV additives, UV shielding nanomaterials, and graphene individually and in different combinations, which resulted in significant improvement in weather resistance properties in terms of retention in helium gas barrier and mechanical properties [9, 14].

There is rarely any literature focused on the use of PUNC films for preparation of multi-layered laminates for the envelopes of aerostats and airships. Our group is currently working in this area and utilizing the potential of different PUNC films for developing a high-helium gas barrier and weather-resistant laminates. Different permutations and combinations are being done for designing many multi-layered structures using different PUNC films in different layers and also using different types of fabrics. Few of these laminates are showing excellent properties in terms of retention (more than 90%) in all required properties even after 300 h exposure under accelerated artificial weathering.

6.6 Conclusion

Polyurethane is a very versatile polymer, and different polyurethane nanocomposite-based gas barrier and weather resistance films/coatings have huge potential in developing materials for aerostat and airship envelopes. PU/clay or PU/graphene nanocomposites (film or coating) with improved helium or hydrogen gas barrier properties can be potentially used in the barrier layer of multi-layered coted/laminated structures for aerostat or airship envelopes. On the other hand, different UV shielding nanomaterials (TiO_2, ZnO, CeO_2, etc.) in combination with organic UV additives can be used for developing weather resistance PUNC films and can be potentially used in a weather-resistant layer of multi-layered coated/laminated structures for aerostat or airship envelopes.

References

1. D. Randall, S. Lee, *The Polyurethanes Book*, Wiley, New York, 2002.
2. M. Joshi, B. Adak, B.S. Butola, Polyurethane nanocomposite based gas barrier films, membranes and coatings: A review on synthesis, characterization and potential applications, *Prog. Mater. Sci.* 97 (2018) 230–282. doi:10.1016/j.pmatsci.2018.05.001.
3. J.N. Gavgani, H. Adelnia, M.M. Gudarzi, Intumescent flame retardant polyurethane/reduced graphene oxide composites with improved mechanical, thermal, and barrier properties, *J. Mater. Sci.* 49 (2014) 243–254. doi:10.1007/s10853-013-7698-6.
4. J.O. Akindoyo, M.D.H. Beg, S. Ghazali, M.R. Islam, N. Jeyaratnam, A.R. Yuvaraj, Polyurethane types, synthesis and applications-a review, *RSC Adv.* 6 (2016) 114453–114482. doi:10.1039/c6ra14525f.
5. P. Król, Synthesis methods, chemical structures and phase structures of linear polyurethanes. Properties and applications of linear polyurethanes in polyurethane elastomers, copolymers and ionomers, Prog, *Mater. Sci.* 52 (2007) 915–1015. doi:10.1016/j.pmatsci.2006.11.001.
6. B. Adak, M. Joshi, Coated or laminated textiles for aerostat and stratospheric airship, *Adv. Text. Eng. Mater.* 2018, 257–287. doi:10.1002/9781119488101.ch7.
7. Y. Wang, H. Wang, X. Li, D. Liu, Y. Jiang, Z. Sun, O3/UV synergistic aging of polyester polyurethane film modified by composite UV absorber, *J. Nanomater.* 2013 (2013) 1–7. doi:10.1155/2013/169405.
8. H. Wang, Y. Wang, D. Liu, Z. Sun, H. Wang, Effects of additives on weather-resistance properties of polyurethane films exposed to ultraviolet radiation and ozone atmosphere, *J. Nanomater.* 2014 (2014) 1–7. doi:10.1155/2014/487343.
9. B. Adak, B.S. Butola, M. Joshi, Calcination of UV shielding nanopowder and its effect on weather resistance property of polyurethane nanocomposite films, *J. Mater. Sci.* 54 (2019) 12698–12712. doi:10.1007/s10853-019-03739-7.

10. M. Sabzi, S.M. Mirabedini, J. Zohuriaan-Mehr, M. Atai, Surface modification of TiO_2 nano-particles with silane coupling agent and investigation of its effect on the properties of polyurethane composite coating, *Prog. Org. Coatings.* 65 (2009) 222–228. doi:10.1016/j.porgcoat.2008.11.006.
11. M. Rashvand, Z. Ranjbar, Degradation and stabilization of an aromatic polyurethane coating during an artificial aging test via FTIR spectroscopy, *Mater. Corros.* 65 (2014) 76–81. doi:10.1002/maco.201206544.
12. B. Adak, B.S. Butola, M. Joshi, Effect of organoclay-type and clay-polyurethane interaction chemistry for tuning the morphology, gas barrier and mechanical properties of clay/polyurethane nanocomposites, *Appl. Clay Sci.* 161 (2018) 343–353. doi:10.1016/j.clay.2018.04.030.
13. B. Adak, M. Joshi, B.S. Butola, Polyurethane/clay nanocomposites with improved helium gas barrier and mechanical properties: Direct versus masterbatch melt mixing route, *J. Appl. Polym. Sci.* 135 (2018) 46422. doi:10.1002/app.46422.
14. B. Adak, M. Joshi, B.S. Butola, Polyurethane/functionalized-graphene nanocomposite films with enhanced weather resistance and gas barrier properties, *Compos. Part B Eng.* 176 (2019) 107303. doi:10.1016/j.compositesb.2019.107303.
15. A. Ali, K. Yusoh, S.F. Hasany, Synthesis and physico-chemical behaviour of polyurethane-multiwalled carbon nanotubes nanocomposites based on renewable castor oil polyols, *J. Nanomater.* 2014 (2014) 1–9. doi:10.1155/2014/564384.
16. Y. Cui, S. Kumar, Improving gas barrier performance of polymer nanocomposites with carbon nanotube nanofillers, *Electrochem. Soc.* 8 (2016) 1064–1064.
17. F. Männle, T. Rosquist Tofteberg, M. Skaugen, H. Bu, T. Peters, P.D.C. Dietzel, M. Pilz, Polymer nanocomposite coatings based on polyhedral oligosilsesquioxanes: Route for industrial manufacturing and barrier properties, *J. Nanoparticle Res.* 13 (2011) 4691–4701. doi:10.1007/s11051-011-0435-7.
18. Q. Li, D. Bi, G.X. Chen, Gas barrier and biodegradable properties of poly (l-lactide) nanocomposites compounded with polyhedral oligomeric silsesquioxane grafted organic montmorillonite, *J. Polym. Environ.* 22 (2014) 471–478. doi:10.1007/s10924-014-0653-z.
19. D.K. Chattopadhyay, K.V.S.N. Raju, Structural engineering of polyurethane coatings for high performance applications, *Prog. Polym. Sci.* 32 (2007) 352–418. doi:10.1016/j.progpolymsci.2006.05.003.
20. G. Woods, I.P. Firm, *The ICI Polyurethanes Book*, New York, ICI Polyurethanes and Wiley, 1990.
21. H.F. Mark, J.I. Kroschwitz, *Encyclopedia of Polymer Science and Technology*, New York, Wiley, 2003.
22. G. Oertel, L. Abele, *Polyurethane Handbook: Chemistry*, Raw Materials, Processing, Application, Properties, Hanser Publishers, 1994.
23. A.K. Sen, *Coated Textiles: Principles and Applications*, 2nd ed., Boca Raton, CRC Press, Taylor & Francis Group, 2007.
24. M.F. Sonnenschein, *Polyurethanes: Science*, Hoboken, Technology, Markets, and Trends, Wiley, 2014.
25. U.M. Westhues, *Polyurethanes: Coatings, Adhesives and Sealants*, Hanover, Vincentz Network GmbH & Co KG, 2007.
26. R. Wojcik et al., Super-low-viscosity aliphatic isocyanate crosslinkers for polyurethane coatings, *Mod. Paint Coatings.* 83 (1993) 39.

27. S.L. Bassner, Low viscosity polyisocyanate component and polyol component, US patent 5670599, 1997.
28. A. Goldschmidt, H.J. Streitberger, *BASF Handbook on Basics of Coating Technology*, Hanover, Vincentz Network GmbH & Co KG, 2003.
29. P.G. Lange, *Powder Coatings: Chemistry and Technology*, Hanover, Vincentz Network GmbH & Co KG, 2004.
30. D. Feldman, Polymer barrier films, *J. Polym. Environ.* 9 (2001) 49–55. doi:10.1023/A:1020231821526.
31. N. Mc Crum, C. Buckley, C. Bucknall, *Principles of Polymer Engineering*, New York, Oxford Science, 1996.
32. E. Thostenson, C. Li, T. Chou, Review nanocomposites in context, *J. Compos. Sci. Technol.* 65 (2005) 491–516.
33. J.J. Luo, I.M. Daniel, Characterization and modeling of mechanical behavior of polymer/clay nanocomposites, *Compos. Sci. Technol.* 63 (2003) 1607–1616. doi:10.1016/S0266-3538(03)00060-5.
34. M. Alexandre, P. Dubois, Polymer-layered silicate nanocomposites: Preparation, properties and uses of a new class of materials, *Mater. Sci. Eng. R Reports.* 28 (2000) 1–63. doi:10.1016/S0927-796X(00)00012-7.
35. F.F. Fang, H.J. Choi, J. Joo, Conducting polymer/clay nanocomposites and their applications, *J. Nanosci. Nanotechnol.* 8 (2008) 1559–1581. doi:10.1166/jnn.2008.036.
36. R. Gangopadhyay, A. De, Conducting polymer nanocomposites: A brief overview, *Chem. Mater.* 12 (2000) 608–622. doi:10.1021/cm990537f.
37. M. Joshi, U. Chatterjee, Nanocomposite: An advanced material for aerospace application, in: S. Rana, R. Fangueiro (Eds.), *Advanced Composite Materials for Aerospace Engineering: Processing, Properties and Applications*, Sawston: Woodhead Publishing, 2016.
38. A. S. Khanna, Nanotechnology in high performance paint coatings, *Asian J. Exp. Sci.* 21 (2008) 25–32.
39. M. Joshi, B.S. Butola, U. Chatterjee, G. Singh, Multifunctional multilayered nanocomposite coating for weather resistant aerostat envelope, Indian patent: 201711029449, filed on 23/08/2017, 2017.
40. K. Kalaitzidou, H. Fukushima, L.T. Drzal, Multifunctional polypropylene composites produced by incorporation of exfoliated graphite nanoplatelets, *Carbon 45* (2007) 1446–1452. doi:10.1016/j.carbon.2007.03.029.
41. H. Kim, C.W. Macosko, Processing-property relationships of polycarbonate/graphene composites, *Polymer* 50 (2009) 3797–3809. doi:10.1016/j.polymer.2009.05.038.
42. H. Kim, Y. Miura, C.W. Macosko, Graphene/polyurethane nanocomposites for improved gas barrier and electrical conductivity, *Chem. Mater.* 22 (2010) 3441–3450. doi:10.1021/cm100477v.
43. A. Pattanayak, S.C. Jana, High-strength and low-stiffness composites of nanoclay-filled thermoplastic polyurethanes, *Polym. Eng. Sci.* 45 (2005) 1532–1539. doi:10.1002/pen.20373.
44. G. Choudalakis, A.D. Gotsis, Permeability of polymer/clay nanocomposites: A review, *Eur. Polym. J.* 45 (2009) 967–984. doi:10.1016/j.eurpolymj.2009.01.027.
45. S.S. Ray, *Clay-Containing Polymer Nanocomposites: From Fundamentals to Real Applications*, Amsterdam, Elsevier, 2013.

46. M. Kotal, A.K. Bhowmick, Polymer nanocomposites from modified clays: Recent advances and challenges, *Prog. Polym. Sci.* 51 (2015) 127–187. doi:10.1016/j.progpolymsci.2015.10.001.
47. C.C. Chu, T.E. Fischer, Evaluation of sunlight stability of polyurethane elastomers for maxillofacial use. *I, J. Biomed. Mater. Res.* 12 (1978) 347–359. doi:10.1002/jbm.820120308.
48. C.C. Chu, T.E. Fischer, Evaluation of sunlight stability of polyurethane elastomers for maxillofacial use. *II, J. Biomed. Mater. Res.* 13 (1979) 965–974. doi:10.1002/jbm.820130613.
49. X. Gu, G. Chen, M. Zhao, S.S. Watson, P.E. Stutzman, T. Nguyen, J.W. Chin, J.W. Martin, Role of nanoparticles in life cycle of ZnO/polyurethane nanocomposites thermo-mechanical properties of ZnO/PU films before UV exposure, *NSTINanotech.* 1 (2010) 709–712.
50. G. Shamini, K. Yusoh, Gas permeability properties of thermoplastic polyurethane modified clay nanocomposites, *Int. J. Chem. Eng. Appl.* 5 (2014) 64–68. doi:10.7763/IJCEA.2014.V5.352.
51. Y. Qian, C.I. Lindsay, C. Mac Osko, A. Stein, Synthesis and properties of vermiculite-reinforced polyurethane nanocomposites, *ACS Appl. Mater. Interfaces.* 3 (2011) 3709–3717. doi:10.1021/am2008954.
52. P.K. Maji, N.K. Das, A.K. Bhowmick, Preparation and properties of polyurethane nanocomposites of novel architecture as advanced barrier materials, *Polymer (Guildf)* 51 (2010) 1100–1110. doi:10.1016/j.polymer.2009.12.040.
53. Y.T. Park, Y. Qian, C.I. Lindsay, C. Nijs, R.E. Camargo, A. Stein, C.W. Macosko, Polyol-assisted vermiculite dispersion in polyurethane nanocomposites, *ACS Appl. Mater. Interfaces.* 5 (2013) 3054–3062. doi:10.1021/am303244j.
54. M. Tortora, G. Gorrasi, V. Vittoria, G. Galli, S. Ritrovati, E. Chiellini, Structural characterization and transport properties of organically modified montmorillonite/polyurethane nanocomposites, *Polymer (Guildf).* 43 (2002) 6147–6157. doi:10.1016/S0032-3861(02)00556-6.
55. J.M. Herrera-Alonso, E. Marand, J.C. Little, S.S. Cox, Transport properties in polyurethane/clay nanocomposites as barrier materials: Effect of processing conditions, *J. Memb. Sci.* 337 (2009) 208–214. doi:10.1016/j.memsci.2009.03.045.
56. J.H. Chang, Y.U. An, Nanocomposites of polyurethane with various organoclays: Thermomechanical properties, morphology, and gas permeability, *J. Polym. Sci. Part B Polym. Phys.* 40 (2002) 670–677. doi:10.1002/polb.10124.
57. M.A. Osman, V. Mittal, H.R. Lusti, The aspect ratio and gas permeation in polymer-layered silicate nanocomposites, *Macromol. Rapid Commun.* 25 (2004) 1145–1149. doi:10.1002/marc.200400112.
58. S. Benali, G. Gorrasi, L. Bonnaud, P. Dubois, Structure/transport property relationships within nanoclay-filled polyurethane materials using polycaprolactone-based masterbatches, *Compos. Sci. Technol.* 90 (2014) 74–81. doi:10.1016/j.compscitech.2013.10.015.
59. P. Kaveh, M. Mortezaei, M. Barikani, G. Khanbabaei, Low-temperature flexible polyurethane/graphene oxide nanocomposites: Effect of polyols and graphene oxide on physicomechanical properties and gas permeability, *Polym.-Plast. Technol. Eng.* 53 (2014) 278–289. doi:10.1080/03602559.2013.844241.
60. N. Yousefi, M.M. Gudarzi, Q. Zheng, X. Lin, X. Shen, J. Jia, F. Sharif, J.K. Kim, Highly aligned, ultralarge-size reduced graphene oxide/polyurethane

nanocomposites: Mechanical properties and moisture permeability, *Compos. Part A Appl. Sci. Manuf.* 49 (2013) 42–50. doi:10.1016/j.compositesa.2013.02.005.
61. C. Xiang, P.J. Cox, A. Kukovecz, B. Genorio, D.P. Hashim, Z. Yan, Z. Peng, C.C. Hwang, G. Ruan, E.L.G. Samuel, P.M. Sudeep, Z. Konya, R. Vajtai, P.M. Ajayan, J.M. Tour, Functionalized low defect graphene nanoribbons and polyurethane composite film for improved gas barrier and mechanical performances, *ACS Nano.* 7 (2013) 10380–10386. doi:10.1021/nn404843n.
62. H. Kim, A.A. Abdala, C.W. Mac Osko, Graphene/polymer nanocomposites, *Macromolecules.* 43 (2010) 6515–6530. doi:10.1021/ma100572e.
63. S. Giraud, S. Bourbigot, M. Rochery, I. Vroman, L. Tighzert, R. Delobel, F. Poutch, Flame retarded polyurea with microencapsulated ammonium phosphate for textile coating, *Polym. Degrad. Stab.*, 88 (2005) 106–113. doi:10.1016/j.polymdegradstab.2004.01.028.
64. K. Madhavan, B.S.R. Reddy, Structure-gas transport property relationships of poly (dimethylsiloxane-urethane) nanocomposite membranes, *J. Memb. Sci.* 342 (2009) 291–299. doi:10.1016/j.memsci.2009.07.002.
65. H. Liu, S. Zheng, Polyurethane networks nanoreinforced by polyhedral oligomeric silsesquioxane, *Macromol. Rapid Commun.* 26 (2005) 196–200. doi:10.1002/marc.200400465.
66. S.A. Madbouly, J.U. Otaigbe, Recent advances in synthesis, characterization and rheological properties of polyurethanes and POSS/polyurethane nanocomposites dispersions and films, *Prog. Polym. Sci.* 34 (2009) 1283–1332. doi:10.1016/j.progpolymsci.2009.08.002.
67. K. Madhavan, D. Gnanasekaran, B.S.R. Reddy, Poly(dimethylsiloxane-urethane) membranes: Effect of linear siloxane chain and caged silsesquioxane on gas transport properties, *J. Polym. Res.* 18 (2011) 1851–1861. doi:10.1007/s10965-011-9592-8.
68. U. Chatterjee, B.S. Butola, M. Joshi, High energy ball milling for the processing of organo-montmorillonite in bulk, *Appl. Clay Sci.* 140 (2017) 10–16. doi:10.1016/j.clay.2017.01.019.
69. U. Chatterjee, B.S. Butola, M. Joshi, Optimal designing of polyurethane-based nanocomposite system for aerostat envelope, *J. Appl. Polym. Sci.* 133 (2016) n/a–n/a. doi:10.1002/app.43529.
70. U. Chatterjee, S. Patra, B.S. Butola, M. Joshi, A systematic approach on service life prediction of a model aerostat envelope, *Polym. Test.* 60 (2017) 18–29. doi:10.1016/j.polymertesting.2016.10.004.

7

Weathering and Degradation Behaviour of Materials Used in Envelopes of LTA Systems

Mangala Joshi and Upashana Chatterjee
Indian Institute of Technology, New Delhi, India
Gaurav Singh
Defence Research and Development Organisation, Agra, India

CONTENTS

7.1 Introduction

Successful deployment of LTA systems requires that the envelope meets one of the following two conditions:

- The as-manufactured properties must stay the same throughout the intended system life.
- The design factor of safety must consider predictable decrease in properties.

DOI: 10.1201/9780429432996-7

Too cavalier an approach could jeopardize the safety of the systems, equipment, and personnel, while too cautious an approach would lead to overdesigned systems. Various environmental factors contribute towards the degradation of the envelope materials during their service life such as sunlight, heat, humidity, acid rain, pollutants, and oxygen. The materials are subjected to an outdoors environment wherein temperatures can go 17°C–33°C above ambient temperature depending on the type/colour of material surface, radiation, conduction, wind, and turbulence [1]. These stress factors may degrade the material in isolation, or they may act in tandem or synergistically with each other to change the rate of the degradation. Therefore, weathering is an extremely complex phenomenon.

7.2 Polymers Used in LTA Envelopes

Coated and laminated textiles are mainly used in inflatable systems as a hull material, composed of three main layers: protective layer, strength layer and gas barrier layer. The main components are high performance fibres, i.e. high-tenacity polyester or nylon fibres and other specialty fibres such as polyester polyarylate (Vectran), poly (p-phenylene-2,6benzobisoxazole (Zylon), high-performance polyethylene (Spectra/Dyneema), para aramids (Kevlar), M5, etc., which are used in the strength layer mostly as woven fabrics. The other components are polymers used for coatings, such as polyvinyl chloride (PVC), polyvinylidene chloride (PVDC), nylon, acrylic, and Elastomers. More importantly, polyurethane is a preferred material of choice for protective and gas barrier layers [2]. The films of polyvinyl fluoride (Tedlar®), PTFE used as a protective layer and biaxially oriented polyethylene terephathalate (BOPET) Mylar®, EVOH, Kapton, etc are more popularly used in new-generation aerostat/airship hull materials. The adhesives are used to join the several layers in a laminated fabric and are mostly polyurethane based.

These polymeric materials are specifically formulated with processing additives (stabilizers, plasticizers, fillers, etc.) to improve upon functional performance and compounded into a viscous paste suitable for coating for various applications. Manufacturing of a polymeric coating formulation is one of the most crucial tasks of the coating industry. The chemistry of these polymers, the additives used, and the processing of the overall coating compound itself are vast subjects. Plasticized PVC is commonly used in commercial coated fabrics. It is inexpensive and has good flexibility, but it is highly prone to UV oxidation. Acrylics are also sensitive to UV radiation. It has a tendency to become yellow with time on exposure to direct sunlight. Silicone rubber has the best low-temperature flexibility of all polymeric materials. However, its high gas permeability, low toughness, and low abrasion resistance are problematic. PU has recently attracted greater interest particularly in functional coating applications. It has outstanding overall toughness, high tensile strength, tear resistance, flexibility, abrasion resistance, fair gas permeability, good handling properties, and ease of processibility [3].

7.3 Failure Behaviour of Envelope Materials/Components

The envelope materials are subjected to continued outdoor environment wherein UV from sunlight, temperatures above ambient resulting from solar heating, moisture, rain, oxygen, other pollutants, etc., results in loss in functional and aesthetic properties of the fabric.

The functional parameters that can degrade are:

- Reduction in tensile and tear strength
- Loss of gas impermeability
- Joints failure
- Reduction of Interlamination strength in case of laminated fabric

The aesthetics loss can be in terms of colour fading, yellowing, spots, and specks on envelopes. Apart from outdoor degradation factors, another important factor to be considered are mechanical stresses present in the inflatables which depend upon the diameter of the inflatable. The aerostat material experiences a combination of stresses arising from self-weight, internal pressure, buoyancy load, aerodynamic load, and payload attachments. The location of maximum stress occurs at the maximum diameter and the load patches [4]. The presence of mechanical stresses may accelerate the moisture uptake through micro cavities and increase hydrolytic degradation [5]. This may add to fatigue loading.

There seems to be an absence of complete reporting of the degradation properties of the materials over time. It may be that a large number of permutation and combinations are available as raw materials for designing of envelope material and degradation trends or behaviour of different materials may vary among themselves and also from usage depending on location. It seems that the prediction of envelope lifespan is more experience based. Perhaps, this resulted in similar un-predicted failures in airship envelopes attributed to lack of full understanding of the fabric properties, operating mechanical and environmental stresses, and improper maintenance resulting in degradation that was more rapid than anticipated. Till the mid-1990s, the Skyship series of airships from Airship Industries consisted of polyurethane-coated polyester fabric. The polyurethanes were susceptible to hydrolytic degradation, and of greater concern was the joints of the hull. This necessitated regular painting to avoid coating degradation. In addition, coating degradation resulted in premature UV damage to the base fabric. Painting using incorrect paints and application techniques, and improper or insufficiently frequent painting, led to the premature scrapping of several envelopes. Such experiential-based learning resulted in premature envelope failures; fortunately, none occurred in flight [6].

Nakadate et al. [7] conducted a two-year natural weathering on Zylon-Aluminised Tedlar-PU laminate of ~157 g/m^2 and 993N/cm initial strength. It was found that natural weathering resulted in ~45% loss in tensile strength, and that the maximum strength loss occurs in hot and humid months. Through a supplementary test, it was established that humidity was the primary factor in strength loss.

In a study, PU-coated nylon fabric used in aerostats was exposed to outdoors weathering and accelerated ageing in a xenon arc with lamp on/off and water spray cycles [8]. It was found that after 20 months of outdoor exposure, the tensile strength loss was 36.5%, and the material lost helium retention capability. Further, it was concluded that apart from UV radiations, the moisture also plays a crucial role in the degradation of the nylon base fabric, which was evident as there was nearly twice the tensile strength loss in wet accelerated weathering than outdoors weathering with nearly similar irradiant energies. The effect of moisture is also reflected in text by Gillett [9], which states that even though PU gave good weather resistance than existing materials in temperate climates, its deterioration accelerated in hot and humid climates.

In a study done on PVC-coated fabrics used for tents by Toyoda et al. [10], it was observed that tearing strength loss on environmental exposure is more than tensile strength – 60% vs ~17% in three years and 80% vs ~56% in seven years for tearing and breaking strength loss, respectively.

In a comparative study on architectural fabrics [11] with PVC coating and different top coating/lamination of PVDF, acrylic, and Tedlar, the Tedlar film remained intact, remained thicker than the acrylic and PVDF coatings, retained more than 72% of its initial thickness, and showed excellent gloss retention and no colour change, while PVDF and acrylics failed, after 11,200 kilojoules of exposure, or the equivalent of 9.3 years (refer Figure 7.1).

In a study by Kang et al. [12] done on Tedlar film–Vectran fabric–PU film laminate, it was seen that there was ~25% tensile strength loss when tested at 65°C and nearly 75% gain when tested at –75°C, as seen in Figure 7.2. Also, modulus increased by 47% when tested at –75°C against room temperature.

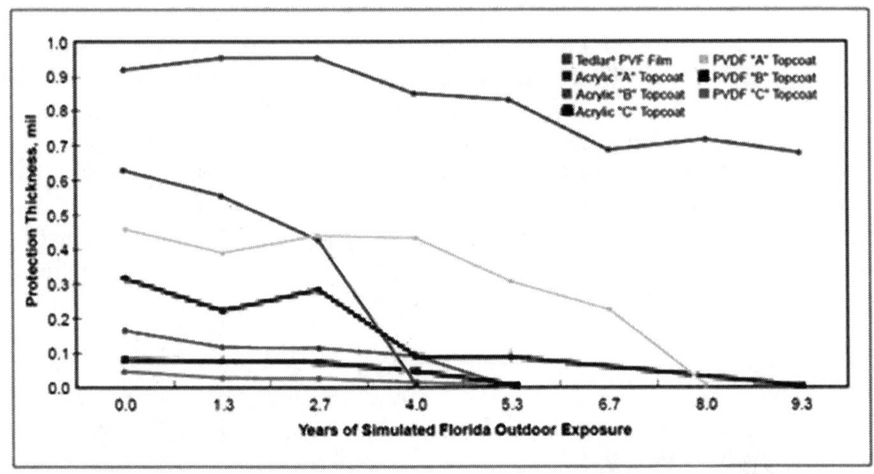

Note: This simulation shows that Tedlar* protective film not only starts out thicker than any of the other protective coatings, but retains superior thickness long after the others eroded completely.

FIGURE 7.1
Thickness loss in top finish materials – Tedlar, PVDF, and acrylic [11].

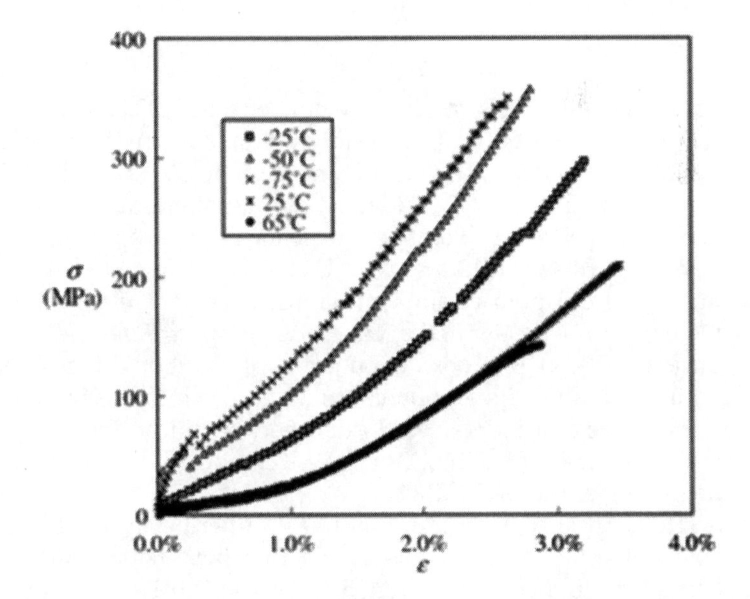

FIGURE 7.2
Effect of temperature on tensile strength of Tedlar film–Vectran fabric–PU film laminate [12].

Hollande et al. [13] tried to use weight loss as a parameter of weathering of thermoplastic polyurethane (PU)-coated fabrics. On comparison of artificial weathering of three PU-coated fabrics in a Weather-o-meter for up to 3500 hours, with and without water spray, with natural weathering in Florida for 42 months, it was found that artificial weathering with water at an acceleration factor between 8 and 10 was found to correlate well with natural exposure. Further, between dry and wet artificial weathering, it was found that the weight loss with wet weathering was nearly five times more, suggesting that presence of water accelerates degradation. The weathering of PU leads to the following [14]:

a. The appearance of a networked layer on the surface, wherein the layer has a near constant thickness by steady formation of degraded-non degraded interface and removal of degraded layer from the exposed surface.
b. Weight loss, wherein the weight loss takes place from the exposed part of the networked layer mostly by volatile compounds which are characteristic of the photo-oxidative degradation of the polyurethane coated fabric, since photo-oxidative degradation results in breakage of urethane bonds and polyether groups of polyurethane. Even though the weight loss was more for ageing with water the network depth and size were much less pronounced than for ageing without water.

Wang et al. [15] conducted a study to see the effect of O3, UV-B, and both O3 and UV-B combined on neat polyester polyurethane (B85A from BASF) film and another added with UV absorbers (0.5% of UV-327 and UV-531 in 1:1 ratio) and nano-ZnO/CeO2 (2% in 3:2 ratio), both surface-treated with an organosilane coupling agent of γ-methacryloxypropyltrimethoxysil ane (KH-570). UV-B lamp (313 nm) with UV radiation of 400 ± 20 μW/cm2 and ozone of 100 ± 2 ppm at ambient temperature and relative humidity of 20% were used for accelerated ageing. UV-Vis spectra, Fourier transformation infrared (FT-IR), photooxidation index, and carbonyl index analysis were used to study the degradation of the films. The results show that that addition of organic and inorganic UV absorbers significantly enhances the weathering resistance of PU film. O3 and UV-B combined shows distinct synergistic degrading effect on PU films.

Eichert [16] conducted long term natural weathering up to 10 years on PVC-coated polyester fabric. It was found that both tensile and tearing strength loss was higher in the weft direction in fabrics having coating thickness of 50 microns or lesser, whereas for fabric with coating thickness of ~230 microns there was negligible tensile strength loss and about 20% tearing strength loss. The location difference also showed that besides UV exposure, air pollution plays a major influence on degradation of coated industrial fabric.

Oprea et al. [17] investigated the effect of different weathering tests on the mechanical behaviour of the polyurethane films. PU films of ~0.5 mm, one precipitated in water and other compacted films as cast were thermally treated at 2 hours at 120°C, synthesised based on polyester diol (adipic acid–ethylene glycol–diethylene glycol), 4,4′-methylenebis phenyl isocyanate (MDI) and dibenzylediisocyanate (DBDI) based aromatic diisocyanates and ethylene glycol as chain extender. These films were subjected to natural weathering of three types:

(a) Soil burial for one month.
(b) Immersion in sea water for five weeks at temperatures of 25°C.
(c) Exposure to sunlight without atmospheric humidity, for 90 days (July–September).

On outdoors exposure the sample colour changed from colourless to yellowish, which increased with increase of exposure time. Even though there was no apparent change in bulk molecular weight after the sunlight exposure vis-à-vis untreated samples, there was ~37% loss in tensile strength and ~50% loss in elongation at break. Therefore, for outdoor applications, the photo-oxidative degradation is the limiting factor.

Since PU can have varying chemistry, the degradation process and the resistance to degradation against weathering can also vary significantly.

PUs prepared with a UV absorber (Tinuvin 400) and hindered amine light stabilizer (Tinuvin 123) after accelerated weathering of 2160 hours in a QUV chamber retained nearly 83% strength compared to 53% for neat PU suggesting that UV additives significantly enhance the UV resistance of PU [18].

Polyvinylidene chloride (PVdC) has low permeability to gases; however, its thermal instability results in degradation at temperatures at or around the melt temperatures [19]. Moreover, PVDC degrades under heat/UV by formation of a conjugated polyene and HCl as a by-product. The conjugated polyenes cause film to change colour from clear to yellow to brown or black. The degradation also reduces the crystallinity, thereby reducing gas barrier properties.

DuPont's datasheet on Tedlar claims that Tedlar® films are made of polyvinyl fluoride and have outstanding toughness, durability, and 20+ year weatherability [20].

In one comparative study [21], nylon, polyester, and Kevlar fabrics were exposed to natural weathering on the roof and accelerated ageing in a xenon weather-o-meter (Table 7.1). It is seen that those polyester goods degrade less than nylon upon natural weathering, but the trend is reversed in the case of exposure to xenon lamp. The strength degradation of Kevlar was inconsistent with large variations in the retained strength.

TABLE 7.1

Effect of Exposure on Roof and Xenon Weatherometer and Time on Strength Retention (%) of Nylon, Polyester and Kevlar [21]

	Nylon		Polyester		Kevlar	
Time, Hrs	Roof	Weather-o-meter	Roof	Weather-o-meter	Roof	Weather-o-meter
48	99.6	91.5	94.3	83.0	85.7	89.4
96	80.8	82.3	93.7	74.5	94.8	58.0
144	69.5	82.3	90.6	65.2	68.4	47.7
192	50.6	82.7	86.2	57.0	50.4	34.3
240	53.1	81.2	86.6	51.8	39.2	30.4
288	52.9	81.2	82.6	47.5	31.3	35.9

7.4 Accidents/Failures of Aerostats/Airships/Blimp Attributed to Material Failures

The accidental failure of envelope hull materials for airships/aerostats suggests that in general, catastrophic tear and joint opening leading to catastrophic tear are the two major failure modes [22–25].

- The fabrics used for the large ZPG3W series (up to 43,000 m^3) airship used Dacron (polyester) base cloths and synthetic rubber coatings, generally neoprene. In 1960, one of the four ZPG3Ws failed during a flight off New Jersey that killed 18 of the crew, resulting in closure of a Navy program. The cause remains in dispute; however, during inspection, a 133-foot tear in the recovered gas bag was observed which indicated that a topside seam 'appeared to have been poorly bonded upon original manufacture, or had deteriorated in service'. The seam, the investigation board said, 'failed in a shearing manner'. Eleven relatives of men killed in the crash sued Goodyear in federal court in Manhattan and argued in 1967 that the disaster had been caused by a defective seam [22].
- In 1993, a Pizza Hut blimp crashed in New York. It is stated that the blimp experienced a seam failure on approach and crashed in the city centre. The rip continued for 38 feet along multiple panels and several gores in both (horizontal and vertical directions) [23, 24].
- At the Gatineau Hot Air Balloon Festival, one airship narrowly avoided landing in the Ottawa River when a seam opened during flight.

- It is reported that in 2012, the Thai Army airship underwent much of the maintenance by replacing helium that had leaked through the seams as they were glued and not heat-sealed [25].

Durney stated that the catastrophic failure of aerostats in high winds tends to occur due to a local failure of the envelope that then propagates into a massive tear [26].

7.5 Causes of Weathering Degradation

In polymers, the decrease in as-manufactured properties or degradation arises due to changes in the molecular structures (including main chain, side chain, substituent groups, or breakage of molecular bonds of any kind) that manifests as diminished physical or chemical properties [27]. The main mechanisms of degradation of polymers in natural weathering are photo-oxidation, thermo-oxidation, and hydrolysis. When the polymer chain length lowers, it manifests as loss in tensile strength, gas barrier, etc.

Many causes have been identified behind the degradation of coatings; however, it should be noted that, usually, the final failure of the coating occurs due to multiple causes [14, 28]. Figure 7.3 provides a schematic diagram where different possible stresses have been shown for the weathering degradation of a typical coating.

The weathering stresses, primarily solar radiation, thermal shock, and moisture, oxygen, and pollutants make complex interactions with the materials. By far, the UV portion of solar radiation is the most destructive environmental factor because it is capable of breaking some of the chemical bonds and thus initiating the degradation reaction.

7.5.1 Weathering and Weathering Tests

The weathering of materials has long been recognized as a natural and universal fact, and the extent of weathering is truly considered to be dependent upon the exposure conditions. Outdoor exposure is a variable function of exposure sites and actual, ever-changing weathering conditions due to the day cycle and to seasonal and year to year variability. The design and simulation of weathering tests must consider all possible variations. The most accurate weathering exposure conditions are therefore those that match exactly the service environment of the product. However, the problem with direct testing is that the period of weathering exposure must be equal to or greater than the required service lifetime of the material which is in reality impossible in maximum cases. An alternative approach is to weather the material

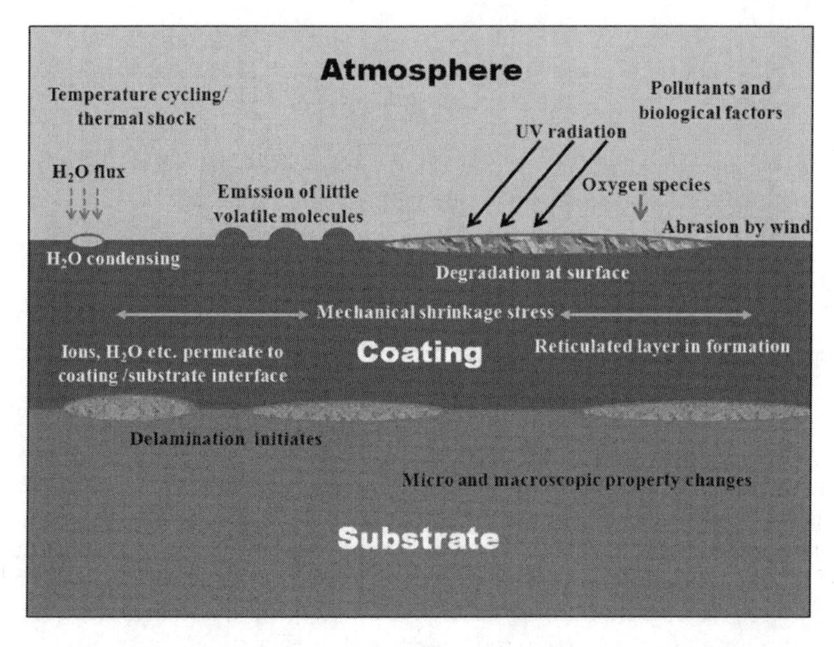

FIGURE 7.3
Schematic diagram of weathering degradation of a polymer-coated substrate.

in extreme weathering conditions and thereafter to relate the rate of degradation with the actual service environment. Table 7.2 lists some of the popular weathering test sites.

Amongst all the natural weathering sites, Florida and Arizona have long been considered as international benchmarks for natural weathering [29, 30]. High-intensity sunlight, high year-round temperatures, and high annual rainfall combine to provide Florida a harsh climate with hot, humid, and high-UV-radiation environment. Arizona exposure testing is widely used for the weathering of materials susceptible to a hot and dry environment. In comparison to Florida, Arizona provides high annual total solar and total UV radiation as well as high air temperatures but lower humidity. Arizona also goes through severe temperature fluctuations that may subject materials to changes in dimension due to expansion and contraction.

Various types of standard sample racks also exist for outdoor exposures. Standard racks are made from anodized aluminium and face to the equator at a fixed angle. The most typical angles of exposures are 0°, 5°, 45°, 90°, or equal to the site latitude (in Florida at 26°, in Arizona at 34°). A specimen can also be exposed at a variable angle which compensates for seasonal differences in the position of the sun. Recently Heikkila et al. [31] demonstrated a new approach to account for seasonal variations in UV exposure by adjusting the timing of the weathering test. They designed a timetable for

TABLE 7.2

List of Some Popular Exposure Sites along with Weathering Specifications

Country	Site Location	Specifications
USA	Miami, Florida	High levels of UV radiation, humidity, and temperature combine to provide a subtropical environment
	Phoenix, Arizona	High levels of UV radiation and temperatures but with low (unlike southern Florida) relative humidity
	Chicago, Illinois	Northern temperate climate characterized by warm summers and cold winters with multiple freezes
	Alberta, Michigan	Northern cold climate with numerous winter freeze days and freeze/thaw cycles in spring and fall, as well as temperate summers
	Medina, Ohio	Northern temperate environment
Canada	Ottawa	Northern cold climate with low average temperature and multiple freeze thaw cycles
France	Sanary	Typical Mediterranean climate
Russia	Moscow	Zone of industrial atmosphere of moderate climate with simultaneous influences of climatic and anthropogenic factors
Australia	Melbourne	Medium levels of temperature and rainfall
	Townsville	Tropical climate with high solar radiation
India	Chennai	Tropical climate with high levels of sunlight, humidity, and temperature
Korea	Seosan	Light industrial and marine environment
China	Guangzhou	Subtropical environment similar to southern Florida
	Hainan	Tropical climate

the outdoor UV exposure of polymeric material using satellite-derived UV climatology. By pre-exposure computation, fixed time increments have been transformed into a schedule with a predicted average accumulation of UV dose. This approach has been applied to create a timetable for an ongoing weathering exposure programme in a network of seven European test sites over the latitude range of 28.5–67.4° N.

Different types of racks can also be used to expose samples, depending on exposure conditions. When the material to be tested is not intended for direct outdoor exposure, under glass exposure methods are used. When a higher test temperature is required to simulate the condition in enclosed spaces, Black Box or Black Box under glass exposure techniques are used. IP/DP (instrumental panel/door panel) test boxes are used mainly for automotive coatings. There are racks designed to perform cyclic sprays (distilled water or salt water) on exposed samples [29]. Sometimes weathering experiments are also done in the actual field where the material is expected to be used. One such interesting investigation was documented by Guseva et al. [32], where the degradation of an aircraft coating was evaluated under an actual service environment. A research programme named Escape Hatch Programme was undertaken by AKZO Nobel together with KLM (formerly Royal Dutch Airlines) for almost eight years beginning in 1994. In this programme, four different pigmented PU-based coatings were applied to each escape hatch of a B-747 aircraft for natural exposure (Figure 7.4). The gloss and colour of the hatches were measured every three months. In that research programme, 15 aircrafts were involved altogether.

Although outdoor exposure seems to provide reliable data, it suffers from several difficulties that have made this method hardly desirable for short-term research and product development. Proper outdoor or field weathering tests are too time consuming and costly. After putting in so much money and effort, there is still a possibility of the samples being unsuitable for instrumental analysis due to heavy contamination during prolonged outdoor exposure. Outdoor conditions have too many uncontrolled and unpredictable variables that make analysis difficult. Thus, researchers have started to focus on accelerating the exposure conditions in a controlled manner. The aim behind artificial weathering is to accelerate the simulated weathering processes for prediction of the service lifetime of a material in an acceptably short time. It was almost a century ago when the first example of so-called artificial photo-ageing was witnessed in the motion picture industry when textile garments worn by actors were found to fade during their exposure to carbon arc light, then used as stage lighting.

For many decades, accelerated outdoor weathering devices are in use where the main light source was nothing but the natural sunlight. Early

FIGURE 7.4
Escape hatch of the B-747 shown with the arrow (Guseva et al. 2003).

versions of this kind of device were designed to track the sun from morning until night in order to maximize the amount of sunlight. Later, light-concentrating devices were developed where mirrors were added to concentrate the sunlight onto the sample for even greater acceleration. A Natural Sunlight Concentrator is a device that automatically follows the sun and at the same time, uses a series of Fresnel mirrors to reflect full-spectrum natural sunlight and concentrate it onto the sample. An Equatorial Mount with Mirrors for Acceleration (EMMA) and an EMMA with distilled water spray for eight minutes out of each hour of operation (EMMAQUA) are such devices developed by ATLAS DSET labs. Thus this kind of system maximizes the amount of sunlight exposure that the sample received by a factor of 8 to 12 times that of the normal exposure.

Accelerated weathering conditions can also be well simulated inside a laboratory environment. This has several advantages over natural outdoor exposure such as precise control of the weathering variables, smaller sample size, elimination of contaminates, and of course, savings in cost, effort, and time [33]. Artificial light sources used in accelerated laboratory weathering instruments are one of the most important exposure elements. The selection criteria should be based on the closest simulation of daylight. Each light source has specific benefits and disadvantages. Xenon arc systems are the most common light sources that can simulate UV and a visible solar spectrum and are widely used by different industries such as textile, polymer, paint, and automotive. Fluorescent UV devices are generally used as a radiation source in tests for relative rank comparisons between samples under specific weathering conditions. This kind of system is useful to vary light or dark cycles and irradiance. A metal halide system is an accepted radiation source where the full spectrum of solar radiation needs to be simulated. Generally, a fluorescent UV source is used for testing components made from several materials and metal halide for research and development purposes. A carbon arc instrument is one specific type of radiation source that emits more intense radiation between 358 nm and 386 nm than natural sunlight. The output of light sources can be corrected by specific filters in order to eliminate UV radiation below 300 nm and excess IR radiation or as per the test requirements. Table 7.3 summarizes different types of common artificial light sources and filters that have been used in artificial weathering instruments.

With artificial accelerated weathering, tests can be conducted easily inside the laboratory, but the validation of data as well as the prediction of service life of the test material, based on the artificial weathering test data, are sometimes questionable. Selection of the proper test program in an artificial weathering instrument is equally important for correlating with actual natural exposure conditions. Jacques [34] described all the procedural variables involved in the design of a weathering test and their relationships in detail, which is beyond the scope of the present discussion.

TABLE 7.3

Common Artificial Light Sources and Filters for Artificial Weathering Instruments [29]

Light sources	Filters
Xenon arc lamps, fluorescent UV lamps, metal halide, carbon arc lamps, indoor actinic source, mercury lamp	Quartz, borosilicate, soda lime, CIRA, Corex

Different natural and artificial exposure sites and instruments are shown in Figure 7.5.

Figure 7.5(a) shows natural weathering of large number samples mounted on fixed frames at an angle of approximately 70° with the ground. Figure 7.5(b) shows large natural exposure field with large number of samples exposed. Figure 7.5(c) shows natural exposure racks with solar concentrator and automated direction control. Figure 7.5(d) shows accelerated weathering laboratory equipment.

Although these weathering devices are sometimes useful in ranking materials through comparison, the validity of correlating the results obtained using such devices with the results obtained for natural outdoor exposure

FIGURE 7.5
(a) and (b) Natural exposure sites. (c) Natural exposure racks with solar concentrator. (d) Artificial weathering laboratory instrument.

are questioned [35]. One reason cited is that the spectral distribution of the lamps differs from that of the solar radiation. A xenon lamp with Pyrex (borosilicate) filters is claimed to produce a spectral distribution closer in intensity to unfiltered terrestrial sunlight than the carbon arc [36–40]. It is also reported that despite the filters, there is significant difference between the radiation distribution in the UV region of the xenon lamp and sunlight [41, 42].

Other factors as possible sources of such disparity in the results are continuous light exposure in the weather-o-meter as against the natural daily cycle of light and dark outdoors, temperature and humidity variations, and difference in the intensity distribution of UV radiation of the xenon lamp in comparison to sunlight. For example, the dissociation energies of CH2-NH and CO-CH2 (nylon bonds) correspond to the energies of wavelengths above 3500 Å [37, 43], where the xenon lamp intensity is lower than natural sunlight radiation; therefore, nylon shows slow degradation rates during weather-o-meter exposure compared to outdoors. The dissociation energies of polyester bonds occur below 3300 Å, where the xenon lamp with borosilicate filters has a slightly higher intensity rate than the sun. Therefore, polyester degrades more in the xenon arc weather-o-meter, in comparison to natural sunlight [44, 45], as xenon lamp radiation intensity in this region is higher than that of the sun. Isakson states that the weather-o-meter is a severe degradation device and should not be used to represent acceleration of natural weathering because it produces large quantities of ozone and has excessive short wavelength energy [46].

Grossman [47] states that moisture penetration has a relatively long-time cycle, 12–16 hours in length; therefore, the rate at which water diffuses inside the polymer is time dependent, and simulators should use water cycles on the order of 8–16 hours with a 40% wetness time. Short cycles of water and UV energy do not allow moisture enough time to carry out its oxidation. He also states that aerated water should be used for spraying samples to simulate actual rainwater, and that temperature should be used as an accelerator.

On the other hand, good correlation between natural weathering and accelerating weathering has been reported for some materials properties [8, 48].

There are a number of standard test methods for measuring the weathering of polymers, which are claimed to be good for comparisons [49–57]. A caveat in ASTM G-151 (Standard Practice for Exposing Non-metallic Materials in Accelerated Test Devices that Use Laboratory Light Sources) states:

> No lab exposure test can be specified as total simulation of actual use conditions in outdoor environment. Results obtained from lab exposures can be considered as representative of actual use exposures only when the degree of rank correlation has been established for specific materials being tested and when type of degradation is same.

Some equivalence of outdoors and accelerated weathering are as follows [58]:

(a) The intensity of the irradiation of an unfiltered metal halide (HMI) lamp compared to an air-mass (AM) 1.5 solar spectrum is about three times as much UVA and seven times as much UVB.

(b) SPART 14 test procedure was originally developed for clear coats in automotive paint systems. An exposure in the SPART 14 test for 1000 h (~6 weeks) is estimated to correspond to about 1.3 years of outdoor testing in Miami, Florida. In the SPART 14 test method, the xenon arc light source is filtered through borosilicate filters and has an irradiance level of 0.5 W/m^2 at 340 nm; this corresponds to an intensity of roughly 1.4 times that of an AM 1.5 spectrum. The test cycle comprises (a) 40 min of light only; (b) 20 min of light with water sprayed on the front surface of the sample; (c) 60 min of light only; and (d) 60 min of no light with water sprayed on the back surface of the sample. Every 14th cycle, the water used to spray the front of the samples is acidic, with a pH of 3.2. The black panel temperature and relative humidity during light periods are 70°C and 75%, respectively. The chamber temperature and relative humidity during the dark periods are 38°C and 95%.

Swallow [59] suggested that due to significant differences between the xenon lamp and sunlight in the UV region, no relationship between the weather-o-meter and outdoor tests can possibly be established. However, within the stipulated conditions in the weather-o-meter, definite configuration of the substrate, and proper recording of the varying outdoors weather conditions, it is possible to arrive at a workable correlation, although it may be an elaborate exercise to perform, and such correlation will not be obviously valid for a different substrate or exposure site.

7.5.2 Evaluation Techniques of Weathering

The characterization of degradation of polymeric materials can be carried out by several subjective and objective means. Various analytical techniques can be used to characterize different degradation events of a coating's lifetime. An outline of series of tests and parameters to monitor the degradation of a coating on exposure to weathering is summarized in Figure 7.6. The accuracy of the analysis of degradation depends on the accuracy and validity of the measured data.

All of these phenomena sufficiently degrades the polymer to drastically increase the film porosity. These can be subjectively assessed through surface cracks, pores, and impermeability and can be objectively measured by SEM, AFM, barrier to gas or liquid, and loss in tensile properties.

Severe cracking occurs resulting in surface degradation and mechanical failure and loss in tensile properties.

Degradation Events	Subjective Measures	Objective Measures
Coating is applied to substrate		
Weathering exposure begins		
Shrinkage stress begin as all plasticizing solvent, etc. evaporates	Weight loss	GSM, film thickness, TGA
Water transport in and out of film leaches out solubles and carries ions to substrate interface	Change in surface chemistry	Chemi-luminescence, surface energy, contact angle
Photo-oxidation of film polymer begins	Chemical changes, change in mol. wt.	FTIR, NMR, XPS, EDX, GPC, GC – MS, viscosity
Measurable gloss changes occur	Change in gloss	Gloss difference
Measurable colour changes occur	Fading of colour, increase in yellowness	Colour difference, yellowness index, UV spectroscopy
Hydrolysis reactions occur	Chemical changes	FTIR, NMR, XPS, GC-MS
Sufficient surface embrittlement occurs	Hardness, surface roughness	Index of hardness, and roughness, AFM, tensile modulus, DMA, contact angle
Reaction products cause delamination at film/substrate interface	Adhesion loss	Strength of adhesion
Cracking/hazing occurs at film surface	Surface micro cracks, loss of strength	SEM, AFM, tensile properties
Sufficient degradation of polymer to drastically increase film porosity	Surface cracks, pores, impermeability	SEM, AFM, barrier to gas or liquid, tensile properties
Severe cracking occurs	Surface degradation, mechanical failure	Tensile properties

Time ↓

FIGURE 7.6
Timeline of a typical polymeric coating: Degradation events and its measures.

7.6 Weathering Degradation: Data Analysis

7.6.1 Basics on Durability and Its Analysis

Durability defines how long an article will last during use. The durability of a polymeric coating depends on its ability to resist the effects of weathering stresses responsible for the deterioration of the performance properties of the ultimate product. This section gives the basic approaches behind the evaluation of durability and estimation of service life of the polymeric coatings against weathering.

The ultimate aim of a weathering test is to provide the most accurate forecast of the performance properties of the product under actual service conditions. Due to the time constraints, it is highly desirable from the industrial point of view to perform tests at higher stress levels, i.e. accelerated levels, and then analyse those short-term accelerated test data to predict the lifetime of the product [60–62]. However, performing this analysis is really a difficult task.

Durability of the polymeric coatings against weathering can be predicted by the generation of the critical photoproduct concentration. For a particular system, when the photo-oxidation mechanism has been recognized fully, the degradation can be quantified by the accumulation of the critical photoproduct. Prediction of durability based on the direct evaluation of a functional property is also possible. However, it should be remembered that the loss of a functional property, in the maximum number of cases, is controlled by more than one dynamic chemical or physical processes. Basically, lifetime is then determined based on an established mathematical relation between the loss of the property of interest and weathering time.

Another obvious disparity that can be noted in the literature related to durability prediction is how to define the failure of a coating. For example, gloss loss is often of prime importance for automobile coatings or exterior house paints, whereas for industrial protective coatings, loss of mass or thickness or reduction of mechanical properties is typically of greater concern. In some special cases, some typical functional property can also play a crucial role in determining coating failure, such as gas barrier property for high-altitude inflatables, hot air balloons, or lighter-than-air (LTA) aircraft systems.

7.6.2 Mathematical Models on Weathering Durability

Several attempts can be found to model degradation of coatings exposed to weathering. The state of the art in modelling techniques of weathering degradation of polymeric coatings can be categorized in two general practices: stochastic (statistical) modelling and mechanistic modelling [62–64].

The mathematical model based on a stochastic approach is generally articulated as reliability theory that fundamentally deals with the probability hypothesis, while mechanistic models are particularly based on knowledge of the underlying degradation mechanisms.

Researchers have made important contributions on modelling of degradation and service life prediction of coatings, although this field still provides enough scope of work. Some typical work, from past to present, on service life prediction of polymeric coatings subjected to weathering, have been categorically discussed here.

In the field of coating, the models based on reliability theory appear to start with the work of Martin [65]. Early in the 1980s, he proposed and partially validated a stochastic model for predicting the service life of polymethyl methacrylate (PMMA) films for coating subjected to photolytic degradation. In his work, the service life was defined as the time after which an unacceptable portion of a nominal population of PMMA films subjected to photodegradation failed. Martin used a Poisson distribution for computing the probability of chain scissions as a function of temperature and intensity of UV radiation. Later, the detailed approaches and protocols for service life prediction models based on reliability theory were assessed critically by Martin [66]. Shah et al. [67] described a more reliable and new approach of predicting the service life in a relatively short time duration. They proposed a method based on the cumulative damage theory, incorporating the statistical treatment of data to provide a realistic output. It was assumed in their methodology that the overall failure of a given material is a cumulative effect of individual deteriorating stresses. The proposed mathematical equations were found to be helpful in analysing the accelerated test data. A reliability model with three stress types, the temperature, UV, and aerosol, was proposed by Guseva et al. [32] for estimating the service life of organic aircraft coatings concerning loss of gloss. They performed accelerated ageing tests on the sample specimens under a newly designed and constructed weathering device, and the results were compared with the data obtained from a unique natural exposure program. Hinderliter and Croll [68] worked on the degradation of the surface of the coating by means of a Monte Carlo model using the solar spectrum photon flux as the initiator for polymer segment scission. Using repeated calculations of random events taking place at the coating surface, a correlation can be established between coating degradation such as gloss loss and any number of input variables such as pigment volume concentration. In this study, the model was designed to predict response behaviour of coatings towards a random process, such as rain or wind action. Later, a different concept was introduced by Hinderliter and Croll [69]. As damage made to a coating can come from the accumulation of a vast number of very small damage events, the Central Limit Theorem (CLT) was found to be a successful modelling technique. Hinderliter and Croll [69] applied CLT to coating property equations which generated additional equations for the prediction of properties of a coating with exposure, such as gloss, colour, fracture toughness, and contact angle. When those equations were fitted to measured data, the outcome was able to provide insight into the mechanisms of degradation processes. The work shows how a wide variety of measurable properties can be modelled through equations based on the CLT. Recently, a cooperative project was completed by the US National Institute of Standards and Technology (NIST) on generating necessary experimental data and the development of a model relating cumulative damage to environmental variables like UV spectrum and intensity, as well as temperature and relative humidity. The parameters of the cumulative damage were estimated from the laboratory data. The adequacy of the model predictions was assessed by comparing with specimens tested in an outdoor environment [63]. In another work, Hinderliter and Croll [70] used simple results to derive algebraic equations to model deterioration trends yet containing a great deal of useful insight. They have considered the degradation of a homogenous coating surface by UV radiation. The rate of reaction at a particular location has been determined only by the probability of adsorbing a photon and the probability that the energy of that photon causes damage (the quantum efficiency or yield). Damage kinetics and damage variation with depth have been modelled for prediction of lifetime.

Mechanistic models can also be used to predict service lifetime of the polymeric coating by identifying the key events in coating degradation [28] and evaluating accelerated coating exposure scenarios. In this regard, as mentioned previously, defining or recognizing coating failure is a very difficult task. Finding clear-cut numerical definitions of a coating failure is the most necessary step towards predicting service life. Towards this objective, Bierwagen [28] proposed a protocol for product failure analysis. The proposed protocol offers a working definition of 'failure' to be used in coating degradation studies which considers the coating performance properties that are crucial for the durability of a coating. However, to date, not many mechanistic models are available in the literature.

In the field of coating, the estimation of in-service weatherability based on quantitative mechanistic models appears to start with the work of Bauer [71]. Based on fundamental studies of the chemistry and physics of failure, he developed a time-to-failure model with some pre-identified specific failure modes. The variation of the key material, process, and weathering exposure parameters were described in terms of distribution functions. By combining the specific failure model with in-service variations in the key parameters, finally, he estimated the in-service failure rates as a function of material and weathering variables. Although in Bauer's work, the hypothetical mechanisms were illustrated in detail, it was not validated on a practical coating case. Recently, Wood and Robien [72] attempted to establish a model on weathering degradation of thermoplastic coating where the constant mass loss period was simulated as a function of pigment volume concentration of the coating. They found that the predicted value of constant mass loss rate was three to four times higher than the measured value and proposed several explanations for this observation. Apart from these studies, no other important mechanistic model has been reported so far.

Very little of the literature is available on modelling of weathering of aerostat/airship envelope materials. However, modelling on similar use materials such as solar panels, glazings, etc., can be used for better appreciation of applicable conditions and methods. Several models were evaluated for use in the description of physical property degradation of materials used in solar panels. These include simple exponential, normal or half normal, log normal, gamma, extreme value, and Weibull [73].

Hampton listed out merits and demerits of various statistical models for modelling property change due to weathering as given in the following [35]:

(a) The exponential model was found to be inadequate when predicting the probability of a material failure within a time Δt as failure rate is not a function of time.
(b) Both normal and log normal models are inaccurate because they indicate an increasing and then a decreasing failure rate.
(c) The Weibull model with certain assumptions could predict the loss in properties of polymers as a function of time in different climatic areas. It was found to fit with 95% confidence intervals.

Jorgensen et al. [58] tried to model the accelerated ageing data at elevated stress conditions to assess the durability of polymers. They conducted a preliminary accelerated ageing test on polycarbonate and PVC films, identifying the key functional parameter that degrades and then associating it with the predominant degradation mechanism such as (a) photo-oxidation, (b) thermal oxidation, and (c) combined photo-oxidation and hydrolysis. They then compared it with reference or in-service material with a known time frame and used Arrhenius time-temp transformation to substitute for long-time exposures at lower stress levels.

A number of empirical relations have been proposed to determine the rate of degradation of polymers due to sunlight, without and with temperature [29, 74–78].

(a) Linear (without temperature effects)

$$P = P_0 + bD$$

where D is the ultraviolet radiation dose, b is a constant, and P_0 is the initial property value.

(b) Linear with temperature effects (i.e. Arrhenius relationship)

$$P = P_0 + De^{(-E_a/RT)}$$

where E_a is the activation energy, R is the gas constant, and T is absolute temperature.

(c) Power law (without temperature effects)

$$P = P_0 + bD^n$$

where n is a constant.

(d) Exponential (without temperature effects)

$$P = P_0 + Ae^D$$

where A is a constant.

However, Maxwell et al. [5] conclude that weathering is impossible to simulate, being very complex due to natural fluctuations in ultraviolet radiation, humidity, temperature, other environmental factors, and interaction of these factors.

Very recently Chatterjee et al. [79] proposed a very systematic approach to predict the in service weatherability of a model aerostat envelope. A reliability model with two stress types (UV radiation and temperature) has been achieved to predict the service life of a model aerostat envelope material concerning the gas permeability through the material. Accelerated ageing tests at higher stress levels have been performed to obtain a life stress relationship from which service life of the material has been determined at the use level conditions. Validation of the proposed model has also been performed using the actual field test data. It will be worthy to apply this approach in various fields of application to predict the service life of newly developed materials before their commercialization.

7.7 Conclusion: Towards a Weathering Resistant LTA Envelope

Weathering properties of the PU coatings can be varied significantly by playing with the structural chemistry of the PU polymer. The degree of phase separation of hard and soft segments plays an important role in determining the physical properties of PU coatings. Chattopadhyay and Raju [3] clearly pointed out several factors which affect the properties of thermoplastic PU coatings. The composition and chemical nature of soft and hard segments, lengths, and molecular weights of the two segments as well as the sequence of length distribution, anomalous linkages (branching, crosslinking) all determine the morphology in the solid state and of course the durability of the coatings. Stabilization against UV radiation practically means elimination or retardation of primary photochemical reactions. In principle, stabilization of polymers can be done in three basic ways: (a) by preventing the radiation from reaching the polymer (UV screeners); (b) by preferential absorption of radiation by some molecule which can then dissipate that into harmless radiation energy (UV absorbers); (c) by addition of a compound which can stop the auto-oxidation process by trapping the excited state energy of free radical by H donation (antioxidants and radical scavengers).

In today's research, nanotechnology has come out as an efficient and powerful strategy to upgrade the structural as well as functional properties of polymers. Applying nanomaterials to high-performance coatings is a recent concept for newly developed coating systems that has shown better performance than the conventional one [80]. Incorporation of nanomaterials for weather-resistant coatings is a very new concept and has huge potential in terms of research significance.

References

1. Test Method for Environmental Engineering Considerations and Laboratory Tests. 2000. US Department of Defense. Appendix-C of MIL-STD-810F, 1 January; Part one, C-2.
2. Sen, A. K. 2007. *Coated Textiles: Principles and Applications*, CRC Press.
3. Chattopadhyay, D., & Raju, K. 2007. Structural engineering of polyurethane coatings for high performance applications. *Progress in Polymer Science*, 32(3), 352–418.
4. Hunt, J. D., 1982. Structural analysis of aerostat flexible structure by the finite-element method. *Journal of Aircraft*, 19(9), 674–678.
5. Maxwell, A. S., Broughton, W. R., Dean, G., et al. Review of accelerated ageing methods and lifetime prediction techniques for polymeric materials, Report of Engineering and Process Control Division, National Physical Laboratory Teddington, Middlesex, Report no. DEPC MPR 016, March 2005: 1–84.
6. Brooke, L., Wakefield, D. S., & Bown, A. 2008. *The development history of inflated lifting body form LTA vehicle hulls*. In *7th International Airship Convention, Friedrichshafen*, 9–11 October 1–6.
7. Nakadate, M., Maekawa, S., Kurose, T., & Kitada, T. (2011). *Investigation of long term weathering characteristics on high strength and light weight envelope material zylon*. In *11th AIAA Aviation Technology, Integration, and Operations (ATIO) Conference, including the AIAA Balloon Systems Conference and 19th AIAA Lighter-Than Air Technology Conference*, Virginia, USA (p. 6938).

8. Pal, S. K., Thakare, V. B., Singh, G., & Verma, M. K. (2011). Effect of outdoor exposure and accelerated ageing on textile materials used in aerostat and aircraft arrester barrier nets. *Indian J. Fibre Text. Res.*, 36, 145–151.
9. Islam, S., & Bradley, P. 2012. Materials, In: K. G. Alexander, ed., *Airship Technology*, 2nd Ed., Cambridge Aerospace Series, 139–148.
10. Toyoda, H., Wu, Y., & Torii, T. *Deterioration with time of the PVC coated fabrics for tent warehouses: some experimental results.* In *International Symposium Textile Composites In Building Construction, Part I*, 23–25 June 1992, Lyon (France), Ed. Pluralis, Paris, 243–252.
11. A Comparative Study: Architectural Fabric Top Finish Performance. 1999; Cited Aug. 18, 2017; Available from: https://cdn2.hubspot.net/hubfs/481608/Top_Finish_Whitepaper.pdf
12. Kang, W., et al. (2006). Mechanical property characterization of film-fabric laminate for stratospheric airship envelope. *Composite Structures* 75(1–4), 151–155.
13. Hollande, S., & Laurent, J.-L. 1998. Weight loss during different weathering tests of industrial thermoplastic elastomer polyurethane-coated fabrics. *Polymer Degradation and Stability*, 62, 501–505.
14. Hollande, S., & Laurent, J.-L. 1999. Degradation process of an industrial thermoplastic elastomer polyurethane-coated fabric in artificial weathering conditions, *Journal of Applied Polymer Science*, 73, 2525–2534.
15. Yanzhir, W. et al. 2013. O3/UV synergistic aging of polyester polyurethane film modified by composite UV absorber, *Journal of Nanomaterials, Hindawi Publishing Corporation*, [Online]. Cited Aug 18, 2017; Available from: https://www.hindawi.com/journals/jnm/2013/169405/
16. Eichert, U. 1994. Residual tensile and tear strength of coated industrial fabrics determined in long-time tests in natural weather conditions, *Journal of Industrial Textiles*, 23, 311–327.
17. Oprea, S., & Oprea, V. 2002. Mechanical behavior during different weathering tests of the polyurethane elastomers films, *European Polymer Journal*, 38, 1205–1210.
18. Jana, H., & Bhunia, R. N. 2010. Accelerated Hygrothermal and UV ageing of thermoplastic polyurethanes, *High Performance Polymer*, 2(1), 3–15.
19. Kaas, R., & Mueller, C. (2003). U.S. Patent No. 6,514,626. Washington, DC: U.S. Patent and Trademark Office.
20. DuPont™ Tedlar® Polyvinyl Fluoride (PVF) Films General Properties 2014; [Online]. Cited Aug 18, 2017; Available from: https://www.dupont.com/content/dam/dupont/amer/us/en/tedlar-pvf-films/public/documents/EI00241-Dupont_TedlarGeneralProperty-Digital.pdf
21. Harris, G. 1982. *A study into the degradation of nylon, Kevlar and polyester fabrics when exposed to varying amounts of ultra violet in the laboratory and in natural environment and the effects of varying degrees of heat on the degradation of these fabrics. SAFE Association, 18th Annual Symposium*, San Diego, CA Oct. 1980.
22. http://articles.philly.com/2003-05-21/news/25460291_1_crew-blimp-navy-program (2003, accessed September 12, 2015).
23. National Transportation Safety Board. https://web.archive.org/web/20141129014643/https://www.ntsb.gov/aviationquery/brief.aspx?ev_id=20001211X12779&key=1 (accessed Oct. 17, 2021)

24. Mc Fadden, R. D. 1993. *Blimp Crash-Lands on Roof of a Building in Manhattan* [Online]. Cited Aug. 18, 2017; Available from: http://www.nytimes.com/1993/07/05/nyregion/blimp-crash-lands-on-roof-of-a-building-in-manhattan.html
25. https://asiatimes.com/2017/09/limp-blimp-ignites-thai-military-corruption-debate/
26. Durney, G. P. 2000. *Concepts for Prevention of Catastrophic Failure in Large Aerostats*, In *Proceedings of the AIAA International Meeting and Technical Display on Global Technology*, Baltimore, Maryland, May 6–8, 1980.
27. Slater, K. 1991. Textile degradation. *Textile Progress* 21.1(2).
28. Bierwagen, G. 1987. The science of durability of organic coatings: A foreword. *Progress in Organic Coatings*, 15(3), pp. 179–195.
29. Wypych, G. 2015. *Handbook of UV Degradation And Stabilization*. Elsevier.
30. Tracton, A. 2005. Coating calculations. In Tracton, A.A., ed., *Coatings Technology Handbook*. CRC Press, pp. 71–76.
31. Heikkila, A., Tanskanen, A., Karha, P., & Hanhi, K. 2007. Adjusting timing of weathering test to account for seasonal variations in UV exposure. *Polymer Degradation and Stability*, 92(4), pp. 675–683.
32. Guseva, O., Brunner, S., & Richner, P., 2003. Service life prediction for aircraft coatings. *Polymer Degradation and Stability*, 82(1), pp. 1–13.
33. Schulz, U. 2009. *Accelerated Testing: Nature and Artificial Weathering in the Coatings Industry*. Vincentz Network.
34. Jacques, L. 2000. Accelerated and outdoor/natural exposure testing of coatings. *Progress in Polymer Science*, 25(9), pp. 1337–1362.
35. Hampton, H. L., & Lind, M. A. 1978. Weathering characteristics of potential solar reflector materials: A survey of the literature. No. PNL-2824. Battelle Pacific Northwest Labs., Richland, WA (USA).
36. Rosato, D. V., & Schhwartz, R. T. 1968. *Environmental Effects on Polymeric Materials*. Vol 1, Environments Interscience, New York.
37. Hirt, R. C., & Searle, N. D. 1967. Energy characteristics of outdoor and indoor exposure sources and their relation to the weatherability of plastics. *Applied Polymer Symposia* 4.
38. Norton, J. E., Kiuntke, H. O., & Connor, J. D. 1969. New developments in water cooled xenon lamps for light-fastness and weathering tests (Part I). *Canadian Textile Journal* 86(19), 50–54.
39. Singleton, R. W., Kunkel, R. K., & Sprague, B. S. 1965. Factors influencing the evaluation of actinic degradation of fibers. *Textile Research Journal* 35(3), 228–237.
40. Singleton, R. W., & Cook, P. A. C. 1969. Factors influencing the evaluation of actinic degradation of fibers: Part II: Refinement of techniques for measuring degradation in weathering. *Textile Research Journal* 39(1), 43–49.
41. Wall, M. J., & Frank, G. C. 1971. A study of the spectral distributions of sun-sky and xenon-arc radiation in relation to the degradation of some textile yarns: Part I: Yarn degradation. *Textile Research Journal* 41(1), 32–38.
42. Wall, M. J., Frank, G. C., & Stevens, J. R. 1971. A study of the spectral distributions of sun-sky and xenon-arc radiation in relation to the degradation of some textile yarns: Part II: spectral distribution studies. *Textile Research Journal* 41(1), 38–43.
43. Benson, S. W. 1965. III-Bond energies. *Journal of Chemical Education* 42(9), 502.

44. Marcotte, F. B., et al. 1967. Photolysis of poly (ethylene terephthalate). *Journal of Polymer Science Part A-1: Polymer Chemistry* 5(3), 481–501.
45. Stephenson, C. V., Lacey, J. C., Jr, & Wilcox, W. S. 1961. Ultraviolet irradiation of plastics III. Decomposition products and mechanisms, *Journal of Polymer Science* 55(162), 477–488.
46. Isakson, K. E. (1972). Use of infrared specular reflectance in study of ultraviolet degradation of polymer films, *Journal of Paint Technology*, 44(573), 41–62.
47. Grossman, G. W. 1975. *Weathering paints and plastics in the laboratory*. In *Proceedings of SPE Miami Valley Section Regular Technical Conference, Coloring and Decorating of Plastics*. Cincinnati, Ohio: s. n., Sept. 23–24.
48. Anagnostou, E., & A. F. Forestieri 1977. Real time outdoor exposure testing of solar cell modules and component materials. No. NASA-TM-X-7365. National Aeronautics and Space Administration Cleveland, Ohio: Lewis Research Cen Ter.
49. ISO 4582: 2017. Plastics – Determination of Changes in Colour and Variations in Properties and Exposure to Daylight under Glass, Natural Weathering of.
50. ISO 877:1994. Plastics – Methods of Exposure to Direct Weathering, to Weathering Using Glass-Filtered Daylight, and to Intensified Weathering by Daylight Using Fresnel Mirrors.
51. ISO 4892-2:2013. Plastics – Methods of Exposure to Laboratory Light Sources – Part 2: Xenon-Arc Lamps.
52. ISO 9060:2018. Solar Energy — Specification and Classification of Instruments for Measuring Hemispherical Solar and Direct Solar Radiation.
53. ASTM D1435-99. Standard Practice for Outdoor Weathering of Plastics.
54. ASTM D1499. Standard Practice Filtered Open-Flame Carbon-Arc Type Exposures of Plastics.
55. ASTM D2565. Standard Practice for Xenon Arc Exposure of Plastics Intended for Outdoor Applications.
56. ASTM D4329. Standard Practice for Fluorescent UV Exposure of Plastics.
57. ASTM D4364. Standard Practice for Performing Outdoor Accelerated Weathering Tests of Plastics Using Concentrated Sunlight.
58. Jorgensen, G., et al. 2003. *Durability of Polymeric Glazing Materials for Solar Applications*. No. NREL/CP-520-34702. National Renewable Energy Lab. (NREL), Golden, CO.
59. Swallow, J. E. 1975. *Effects of Dyes and Finishes on the Weathering of Nylon Textiles*. No. RAE-TR-74179. Royal Aircraft Establishment Farnborough (England).
60. Herling, R. J. 1996. *Durability Testing of Nonmetallic Materials, ASTM STP 1294*, American Society for Testing and Materials, Scranton, PA, USA.
61. Martin, J. W., Saunders, S. C., Floyd, F. S., & Wineburg, J. P. 1996. *Methodologies for Predicting the Service Lives of Coating Systems*, Federation of Societies for Coatings Technology Blue Bell, PA.
62. Martin, J. W., Ryntz, R. A., Chin, J., & Dickie, R. 2008. *Service Life Prediction of Polymeric Materials: Global Perspectives*, Springer.
63. Trigo, I. V., & Meeker, W. Q. 2009. A statistical model for linking field and laboratory exposure results for a model coating. In Martin, J.W., Ryntz, R. A., Chin, J., & Dickieed, R., eds. *Service Life Prediction of Polymeric Materials*, Springer, 29–43.

64. Kiil, S. 2012. Model-based analysis of photoinitiated coating degradation under artificial exposure conditions. *Journal of Coatings Technology and Research*, 9(4), 375–398.
65. Martin, J. W. 1984. A stochastic model for predicting the service life of photolytically degraded poly (methyl methacrylate) films. *Journal of Applied Polymer Science*, 29(3), 777–794.
66. Martin, J. W. 1993. Quantitative characterization of spectral ultraviolet radiation-induced photodegradation in coating systems exposed in the laboratory and the field. *Progress in Organic Coatings*, 23(1), 49–70.
67. Shah, C. S., Patni, M. J., & Pandya, M. V. 1994. Accelerated aging and life time prediction analysis of polymer composites: A new approach for a realistic prediction using cumulative damage theory. *Polymer Testing*, 13(4), 295–322.
68. Hinderliter, B. R., & Croll, S. G. 2005. Monte Carlo approach to estimating the photodegradation of polymer coatings. *Journal of Coatings Technology and Research*, 2(6), 483–491.
69. Hinderliter, B. R., & Croll, S. G. 2006. Simulations of nanoscale and macroscopic property changes on coatings with weathering. *Journal of Coatings Technology and Research*, 3(3), 203–212.
70. Hinderliter, B. R., & Croll, S. G. 2010. Predicting coating failure using the central limit theorem and physical modeling. *ECS Transactions*, 24(1), 1–26.
71. Bauer, D. R. 1997. Predicting in-service weatherability of automotive coatings: A new approach. *Journal of Coatings Technology*, 69 (864), 85–96.
72. Wood, K. A., & Robien, S. 2009. A quantitative model for weathering-induced mass loss in thermoplastic paints. In J.W. Martin et al., ed.*Service Life Prediction of Polymeric Materials*, Springer: Boston, MA, 457–474.
73. Clark, J. E., & Slater, J. A. 1969. Outdoor Performance of Plastics III: Statistical Model for Predicting Weatherabi1ity, NBS Report 10 116, October 30.
74. Brown, R. P., et al. 1995. A review of accelerated durability tests. Versailles Project on Advanced Materials and Standards (VAMAS Report no. 18).
75. Brown, R. P., & Greenwood, J. H. 2002. *Practical Guide to the Assessment of the Useful Life of Plastics*. iSmithersRapra Publishing.
76. White, J. R., & Turnbull, A. 1994. Weathering of polymers: Mechanisms of degradation and stabilization, testing strategies and modelling. *Journal of Materials Science* 29(3), 584–613.
77. Chatterjee, U., Butola, B. S., & Joshi, M. 2016. Optimal designing of polyurethane-based nanocomposite system for aerostat envelope. *Journal of Applied Polymer Science* 133(24).
78. Clark, J. E., & Clark, J. E. 1969. *Outdoor Performance of Plastics: III. Statistical Model for Predicting Weatherability*. US Department of Commerce, National Institute of Standards and Technology.
79. Chatterjee, U., Patra, S., Butola, B. S., & Joshi, M. 2017. A systematic approach onservice life prediction of a model aerostat coating. *Polymer Testing*, 60, 18–29.
80. Khanna, A. 2008. Nanotechnology in high performance paint coatings. *Asian Journal of Experimental Science*, 21(2), 25–32.

8

Developments in Weather-Resistant Polyurethane Coatings for Inflatables

Mangala Joshi and Upashana Chatterjee
Indian Institute of Technology, New Delhi, India

CONTENTS

8.1 General

Tethered helium inflatable systems are receiving renewed attention in the scientific communities. Unlike fixed-wing aircraft or helicopters, these systems are 'lighter than air' (LTA), typically use helium gas to stay aloft and are tethered using a mooring system operated from a fixed location. Any damage on the inflatable envelope that affects the barrier property and gives an escape to the helium gas causes an immediate failure of the whole structure, leading to loss of a huge amount of money. From the manufacturing point of view, design of new and improved materials for this kind of structure that consistently survive in the harsh atmospheric condition is a critical issue. A typical inflatable aerostat structure consists of a strength layer generally made up of woven textiles, with high strength-to-weight-ratios, to provide a flexible high-strength base to the structure and a protective layer which acts as a gas barrier layer for maintaining the inflated condition of the structure for a prolonged time. In such cases, the lifetime of the ultimate inflatable structure is primarily determined by the protective properties of the polymeric layer. Use of conventional polymers (polyvinyl chloride, acrylic, etc.) in these applications can be very limited because

DOI: 10.1201/9780429432996-8

of the poor weathering resistance, low flex fatigue, poor adhesion to substrate or high permeability to air or gases. As an alternative material for the protective layer of the aerostat envelope, polyurethane (PU), has recently attracted a lot of interest [1–3]. However, similar to other polymer materials, exposure of PU to aggressive environments (mainly to UV radiation) for a prolonged time causes changes in its physical, chemical and mechanical characteristics and even loss of use value [4].

8.2 Mechanism of Photo-Stabilization of PU

Initial studies to establish the segmented structure of PU were made in the 1960s. The work of Cooper and Tobolsky [5] as well as Schollenberger [6] demonstrated that segmented PUs consist of hard segments microphase separated from soft segments having relatively low glass transition temperature (Tg). The degree of phase separation of hard and soft segments plays an important role in determining the physical properties of PU coatings. Chattopadhyay and Raju [1] have clearly pointed out several factors which affect the properties of thermoplastic PU coatings. The composition and chemical nature of soft and hard segments, lengths and molecular weights of the two segments as well as the sequence of length distribution and anomalous linkages (branching, crosslinking) all determine the morphology in the solid state and, of course, the durability of the coatings.

Stabilization against UV radiation practically means elimination or retardation of primary photochemical reactions. In principle, there are three basic ways in which stabilization of polymers can be done:

(a) By preventing the radiation from reaching the polymer (UV screeners)
(b) By preferential absorption of radiation by some molecule which can then dissipate that into harmless radiation energy (UV absorbers)
(c) By addition of a compound which can stop the autooxidation process by trapping the excited state energy of free radicals by H donation (antioxidants and radical scavengers).

There are some other specific groups of UV stabilizers such as quenchers, peroxide decomposers and nucleating agents. Table 8.1 summarizes some conventional UV stabilizers [7, 8].

One of the most important conditions for effective stabilization is that the incorporated additives must persist in the polymer matrix throughout the expected lifetime of the product. There can be physical loss through migration, blooming, evaporation or leaching. The key variables for physical loss

TABLE 8.1

Conventional UV Stabilizers for Polymers

Type	Function	Example
Light screeners	Act as a screen or shield between radiation and polymers	Carbon black
UV absorbers	Absorbs UV light	Benzophenone, benzotriazole, triazines, oxanilides, cyanoacrylates
Radical scavengers	Scavenge or trap excited radicals by hydrogen donation	Hindered amine light stabilizers, mercaptans, quinines, polynuclear hydrocarbons
Antioxidants	Inhibit the oxidation processes	Hindered phenols, hindered amine light stabilizers
Quenchers	React with excited state singlet/triplet to result in non-reactive species	Metal chelates
Peroxide decomposers	Donate electron to form peroxide anion	Alkyl xanthates, N,N′-disubstituted dithiocarbamates and dithiophosphates (secondary antioxidants)
Nucleating agents	Reduce the chain mobility and diffusivity of attacking sites	Metal salts, cyclic bis-phenol phosphates, dibenzylidene sorbitol dicarboxylates

of UV stabilizers on weathering are solubility of the additives in the polymer matrix as well as in water, their diffusion co-efficient, vapour pressure, etc.

(a) *UV screeners*: These compounds prevent the penetration of UV radiation by making an opaque covering or coating on the polymer and thereby restricting the total amount of degradation to a thin surface layer. The efficiency of this kind of screener depends on the type and size of the particles as well as the degree of dispersion of the particles within the polymer. One of the most common and efficient example of UV screeners for polymers is carbon black.

(b) *UV absorbers*: UV absorbers protect against photodegradation by competing with the polymer for absorption of UV radiation.

When the UV absorber molecule absorbs UV energy, an electronic rearrangement occurs to form the tautomeric form of that molecule which reverts back to the original form through a non-radiative decay process (Figure 8.1). This process is repeated indefinitely. An ideal UVA should be very light stable and should have high absorption over the UV range from 290 to 400 nm. Generally, UV absorbers are based on one of five compounds: benzophenone, benzotriazole, triazines, oxanilides and cyanoacrylates.

A strong disadvantage of UV absorbers is the fact that to provide good protection, they need a certain absorption depth (item thickness).

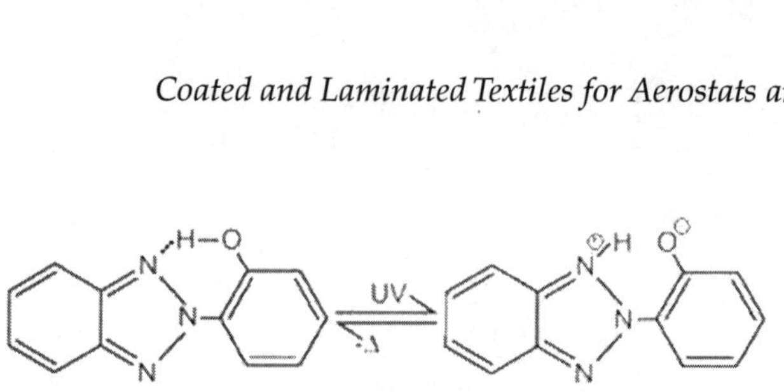

FIGURE 8.1
Mechanism of tautomerism of UVAs.

Therefore, the protection is only moderate in case of polymer surfaces and of thin items such as films or fibers.

(c) *Antioxidants and radical scavengers*: Antioxidants can stop the autooxidation process by trapping the excited state energy of free radicals by hydrogen donation. Radical scavengers are antioxidants capable of trapping free radicals. Hindered amine light stabilizers (HALS) are examples of this type of UV stabilizers.

Several theories have been advanced to explain the mechanism of stabilization by HALS. The most widely accepted explanation involves efficient trapping of free radicals with subsequent regeneration of active stabilizer moieties, represented in Figure 8.2.

In today's research, nano technology has come out as an efficient and powerful strategy to upgrade the structural as well as functional properties of polymers. Applying nanomaterials to high-performance coatings is a recent

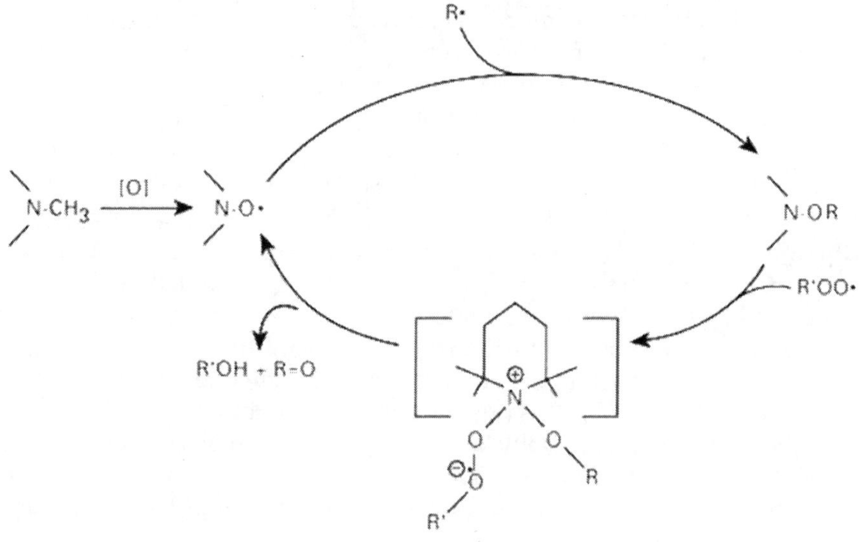

FIGURE 8.2
Radical trapping mechanism of HALS.

concept for newly developed coating systems that has shown better performance than the conventional one [9]. Having exceptionally high surface area-to-volume ratio, nanomaterials give rise to exceptional properties in the new products. Nano additive or the filler has at least one dimension in the nanometer scale. Thus, owing to a very high surface area, the dispersion of only a little amount of nanomaterials within the polymer matrix can lead to a tremendous increase in interfacial contacts between the polymer and the filler which ultimately leads to the superior properties of the new product with only a negligible increase in weight. Figure 8.3 shows the schematic diagram of various types of nano additives or fillers with nanoscale dimensions.

Nano additives also have several other advantages in comparison to conventional organic UV protective additives in coatings. First of all, conventional additives are typically used for one particular purpose, such as reducing or retarding UV degradation processes. On the other hand, the additives based on nanoparticles can impart improved UV resistance along with other functional properties such as higher mechanical properties, thermal stability, electrical properties, gas barrier properties, flame retardancy, etc. Moreover, there are many drawbacks of conventional organic additives such as high loss rate due to continuous conversion to radicals, leaching migration with low molecular weight stabilizers (especially HALS), poor mobility with high-molecular-weight stabilizers, yellowing with phenolic antioxidants, etc. On the other hand, the efficiency of these conventional additives is also affected by their 'active' presence in the system. Physical loss of organic UV stabilizers is very common and occurs through volatilization, solubility, diffusion and leaching during its service. On the contrary, migration and physical loss of nanoparticles from polymeric materials is truly a difficult task owing to their superior interfacial interaction [8].

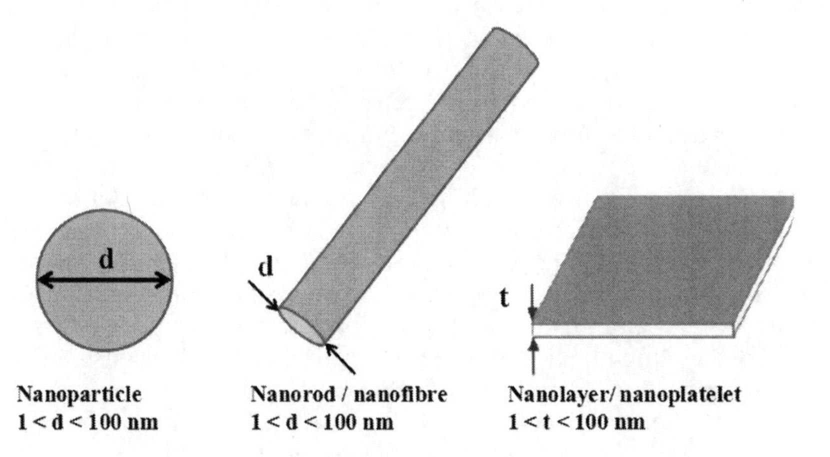

FIGURE 8.3
Scheme of various types of nanofillers or fillers with nanoscale dimensions.

8.3 Studies on Photo-Stabilization of PU

8.3.1 Structural Engineering of Neat PU

Weathering properties of the PU coatings can be varied significantly by playing with the structural chemistry of the PU polymer. Rek et al. [10] studied the effect of different kinds of soft segments (polyols) on the mechanical, chemical and structural properties of photodegraded PU. They synthesised three series of PU elastomers, using poly(ethylene butylene adipate) glycol, poly(caprolactone) glycol and poly(tetramethylene ether) glycol with 4,4′-diphenylmethane di-isocyanate as the hard segment and 1,4-butandiol as the chain extender. All three types of PUs indicated photo-oxidative degradation after 500 h of irradiation under quartz-mercury vapour lamp (150 W) at room temperature. The intensity of degradation was found to be dependent on the chemical nature of the polyols and was found to be less in polyester PU than in polyether PU. The tensile strength and tear strength were observed to decrease in all the samples after irradiation. The extent of the changes in mechanical property was found to correlate well with the extent of photo-oxidative degradation of the same samples. Due to the photochemical degradative reactions, urethane sequences and aromatic structures decreased by simultaneous appearance of new groups like amine, quinoid and carbonyl. In PU samples having poly(buthylene ethylene adipate) glycol, an azo structure also was formed, while the polyether type of PU produced aldehyde, carboxyl and hydroxyl structures. It was concluded that the degradation of PUs occurred through urethane group scission in polyester PU, and through the scission of polyether group in polyether PU.

Bajsic et al. [11] studied the role of the concentration of hard segments (urethane) and molecular weight of soft segments (polyol) in PU resin. PU elastomers were prepared from 4.4′-dipheylmethane di-isocyanate (MDI) and poly(oxytetra-methylene) glycol (PTMO) of 1000 and 2000 molecular weight at NCO:OH ratios of 2:1 and 4:1 respectively. Laboratory irradiation at room temperature of the solution casted PU films was carried out for 100 h using a quartz-mercury vapor lamp (400 W). It was found that the molecular weight of soft segments as well as the NCO:OH ratios have a significant influence on the weathering behaviour of the prepared elastomers. The photo-oxidative degradation was found to be more prevalent in the PU with lower hard segment concentration and with higher soft segment molecular weight.

In early 1980s, Brauman et al. [12] first found and reported that the substitution of the aliphatic di-isocyanate methylenebis(4-cyclohexyl isocyanate) for MDI can give a polymer more resistant to light-induced discoloration and degradation. Recently, Wang et al. [13] carried out a detailed photochemical degradation study on a series of PU elastomers having MDI-, TDI- and HDI-based isocyanates. An UV-B lamp (310 nm radiations) was used as a radiation source. By analysing the ATR FT-IR spectrum of the samples irradiated up to 412 h, they concluded that the photodegradation mechanism of the PU might be the scission of carbamate and the formation of amino, alkyoxyl and alkyl radicals that successively react. For TDI and MDI based PUs, their colour easily became yellow under UV exposure as expected since there are aromatic isocyanate structures that can easily transform into chromophores containing quinoid structure. However, less or no discoloration was observed in case of the HDI-based PU, on exposure to UV radiation.

Wilhelm and Gardette [14] investigated the photochemical behaviour of PU having aliphatic-based hard segments and polyester-based soft segments at two different wavelengths of irradiation. At longer (> 300 nm) wavelengths, degradation involved formation of coloured species derived from primary hydroperoxides. It was shown that long wavelength irradiations provoke an induced oxidation of urethane. This reaction was initiated by hydrogen atom abstraction from the methylene groups at the alpha position to nitrogen atoms. On irradiation at short wavelength (254 nm), degradation occurred through a photo-fries type reaction, and the results provide evidence for a dual mechanism for photo-oxidation: induced oxidation and direct photolysis of the urethane groups following three possible photo-scissions that involve the homolysis of carbon–nitrogen and carbon–oxygen bonds. In another study, Wilhelm and Gardette [15] analysed photo-oxidative degradation of an aliphatic diisocyanate-based PU having polyether as a soft segment. It was confirmed that the oxidation of this type of PU results from the sensitivity of the polyether soft segments to oxidative degradation. Oxidation of the ether groups produces formates, which further decompose to give ethyl and methyl formates.

8.3.2 PU with Conventional UV Resistant Additives

Although a large number of different types of compounds are commercially known to be useful UV stabilizers, only a few classes are effective in stabilization of PU. Some of the recent experimental works on stabilization of PU against degradation using conventional stabilizers are discussed next.

Formerly, Thapliyal and Chandra [16] reviewed in detail several investigations of UV stabilizers that provide a comparatively stable PU matrix against degradation. They concluded that triazole-type stabilizers can be good; however, hindered amine light stabilizers (HALS) along with UV absorbers can be the most effective stabilizers for PU elastomers.

Singh et al. [17] investigated three types of stabilizers (0.5 wt %); a benzotriazole UV absorber (Tinuvin P or Tinuvin 900), a hindered amine light stabilizer (Tinuvin 292) and a phenolic antioxidant (Irganox 1010) to suppress the extensive yellowing that occurs in aromatic PU clear coatings based on MDI on exposure to weathering. Irganox 1010 was also incorporated into the

TABLE 8.2

Changes in Yellowness Index Values of Aromatic PU Clear Coating with Irradiation Time [17]

Sample	0 h	100 h	200 h	300 h	400 h
Neat film	21.50	64.50	83.70	93.64	96.12
Tinuvin P	43.10	66.50	74.29	77.63	79.33
Tinuvin 292	32.10	66.61	63.50	65.70	70.12
Tinuvin P + Tinuvin 292	24.77	50.52	69.33	65.37	69.84
Irganox 1010	24.32	44.10	63.04	76.96	84.95
Tinuvin 900	21.29	66.91	78.06	79.13	88.98
Tinuvin 292 + Tinuvin 900	28.40	48.56	56.12	71.03	73.75
Irganox 1010 + TiO_2	29.54	80.97	90.45	94.62	97.67

formulation in combination with TiO_2 (rutile) pigment. A synergistic mixture of benzotriazole and hindered amine light stabilizers was also used. The results are presented in Table 8.2.

As discussed previously, the yellowness developed in the aromatic PU clear coats on exposure is due to the transformed products of MDI with conjugated p-electrons, which absorb in the visible spectrum. From their study, it was found that the combination of benzotriazole UV absorber (Tinuvin P) and piperidine hindered amine (Tinuvin 292) produces the best results as the UV absorber (Tinuvin P) acts as an effective light filter towards the short wavelength radiation and HALS scavenges any free-radicals formed in the coating. The synergistic mixture (Tinuvin P + Tinuvin 292) also has good compatibility with PU due to more alkyl groups in the chain.

In most cases, standard low molecular weight organic additives are used to protect commercial PU from the solar radiation-induced degradation. However, as discussed previously, one of the basic problems associated with such additives is that they may partly evaporate or get washed away from the polymer matrix. But, if the molecule of the additive is attached to the PU chain by a covalent bond, no such physical losses are expected. Such self-stabilizing polymers have been a subject of intensive study for the last few decades [3, 18]. Recently, such an approach has been used by Podešva et al. [19]. Three new low-molecular-weight stabilizers, bearing in their molecules a tri(dimethylsiloxane) chain end-capped with primary OH groups, have been synthesized and characterized. One of the stabilizers was based on a sterically hindered phenolic antioxidant, and the others were on sterically hindered piperidines. Each of these stabilizers, when added to the reaction mixture containing a polybutadiene diol, a diisocyanate and a catalyst, for the preparation of elastomeric PUs, due to the presence of the OH group, was joined by a covalent bond at one or both chain ends of the PU formed and acted simultaneously as a molecular weight regulator. They concluded a long-term persistence of the stabilizers into the modified PUs due to a negligible loss of the antioxidants by evaporation or washing away.

Recently, Jana and Bhunia [20] studied accelerated hygrothermal and UV ageing of thermoplastic PU synthesized from 4,4′-diphenyl methane diisocyanate, polyether polyol and 1,4 butane diol, with and without HALS (Tinuvin 123) and UV absorber (Tinuvin 400). The commercially available HALS and UV absorber were used in a combination of 1:2 w/w ratio to protect the PU from UV degradation. The accelerated hygrothermal and UV ageing were performed at different humidity and aging temperatures for 720, 1440 and 2160 h of exposure, in an environmental chamber and a QUV chamber respectively.

8.3.3 PU with Nanomaterials

Incorporation of nanomaterials for weather-resistant coatings is a very new concept, and the research done in this area is also rather limited to date. There is a good review by Pandey et al. [21] on degradability of polymers incorporated with nanomaterials which specifically mention that, unfortunately, no work had been done till now on durability of nanocomposites based on PU. Another review by Kumar et al. [9] discussed the effect of nanoscale particles on polymer degradation and stabilization. They studied photo-oxidation of different polyolefins systems such as polypropylene [22], polyethylene [23] and EPDM [24], polyamides and polycarbonates nanocomposites filled with layered silicate nanoclay. Although nanoclays are thought to have a UV screening effect, it has also been reported that the nanocomposites degrade faster than the pristine polymers [25] because of the adsorption of additives onto the clay or the degradation of the alkyl-ammonium cation exchanged in montmorrilonite and the catalytic effect of iron impurities of the organo-montmorrilonite. Iron (Fe^{3+}) is able to catalyse the decomposition of the primary hydroperoxide formed by photo-oxidation reaction of polymers. But in the case of the polyester/clay nanocomposite, the degradation was found to be higher for the neat polyester than the clay-filled polyester for longer exposures [26]. Material service life was found to be significantly increased for polyester nanocomposites with nanoclay. They have also discussed degradation of polymer nanocomposites incorporating other nanomaterials such as CNT, nano TiO_2 and nano ZnO.

In this section, the current status of worldwide research on nanomaterials for weather resistant coatings, particularly on PU is reviewed. Chen et al. [27] studied the photo-oxidation of PU with TiO_2 nanoparticles, and they suggested that with increase in TiO_2 nanoparticle content, the photosensitivity of PU/anatase TiO_2 nanocomposite increases, while the photostability of PU/rutile TiO_2 nanocomposites increases.

A typical FT-IR spectra of PU and its nanocomposites having 2 wt % of nanoparticles, measured after 10 min of UV irradiation under a Xenon arc

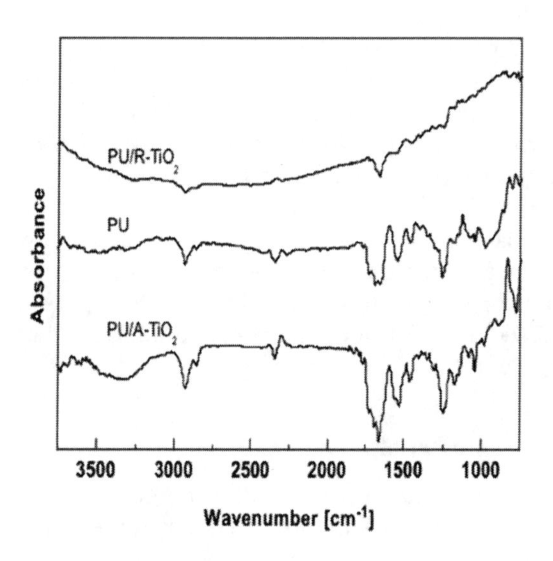

FIGURE 8.4
FT-IR differential spectra of PU and its nano composites with anatase (A) and rutile (R) TiO_2 nanoparticles exposed to UV radiation [27].

lamp, is shown in Figure 8.4. It can be seen that all the absorption bands of PU/anatase TiO_2 composite were greater than those of PU and PU/rutile TiO_2, and moreover, all the absorption bands of PU were greater than those of PU/rutile TiO_2. It was concluded that the rutile TiO_2 nanoparticles offer a stabilization function for PU, while anatase TiO_2 enhances photosensitivity of the matrix polymer.

Mirabedini et al. [28] worked further on the effect of rutile TiO_2 nanoparticles on PU. Surface modification of rutile TiO_2 nanoparticles was done with aminopropyltrimethoxysilane (APS) before incorporation in PU and this method was found to control photocatalytic activity of nanoparticles, and thus to improve further the weathering behaviour of the PU nanocomposite coatings.

It was also found that direct addition of unmodified nanoparticles into the coating system results in poor dispersion into the polymer matrix, and thus weathering performance suffers. It has been discussed that the surface modification of TiO_2 nanoparticles increases the steric hindrance between the nanoparticles and also improves their wet-ability and compatibility into the PU matrix. Significant improvement was found in tensile properties of the coatings filled with modified TiO_2 nanoparticle, as shown in Table 8.3.

Nano ZnO particles are also considered to provide UV protection to the polymeric coatings. Rashvand et al. [29] investigated the effect of nano ZnO particles on the photo-degradation of the aromatic PU-based coatings. It was concluded from their study that the presence of nano ZnO particles could

TABLE 8.3

Tensile Properties of Nanocomposite Films after 1000 h of QUV Chamber Exposure [28]

	Ultimate Tensile Strength (MPa)		Modulus (MPa)		Elongation (%)	
Sample	**Before QUV**	**After QUV**	**Before QUV**	**After QUV**	**Before QUV**	**After QUV**
Neat PU	9.76	2.34	350	59.5	4.31	1.64
TT-0.5[a]	10.70	3.05	390	97.5	4.98	1.89
TT-1[a]	12.54	3.50	444	150.9	5.80	1.97
TT-2[a]	14.00	4.62	490	343.0	6.80	2.23
UT-2[b]	12.90	1.55	468	257.7	6.47	1.49

[a] TT-0.5, TT-1 and TT-2 coating samples containing, 0.5, 1 and 2 wt % treated TiO_2 nanoparticles respectively.

[b] UT-2 coating sample containing 2 wt % untreated TiO_2 nanoparticles.

protect the PU matrix against the deteriorating effects of UV radiation to a certain extent. In their study, surface roughness of the coatings, as a measure of degradation, was measured using a Surface Roughness Tester, and it was evident that the surface roughness of the unfilled or blank coating was much more than the nano-ZnO-stabilized coating (Figure 8.5).

Among different metal oxide nanoparticles, nano cerium oxide (CeO_2) has gained much scientific importance recently as an effective UV absorber. CeO_2 nanoparticles have band gap energies ranging from 2.9 eV to 3.5 eV depending on their particle sizes (quantum confinement effect) and show a UV cut-off threshold at around 370 nm, similar to that of nano TiO_2. These two oxides are both semi-conductive in nature with a similar band gap energy and exhibit the same absorption mechanism under UV radiation. An electron-hole pair is created due to the absorption of a photon with a higher energy than the band

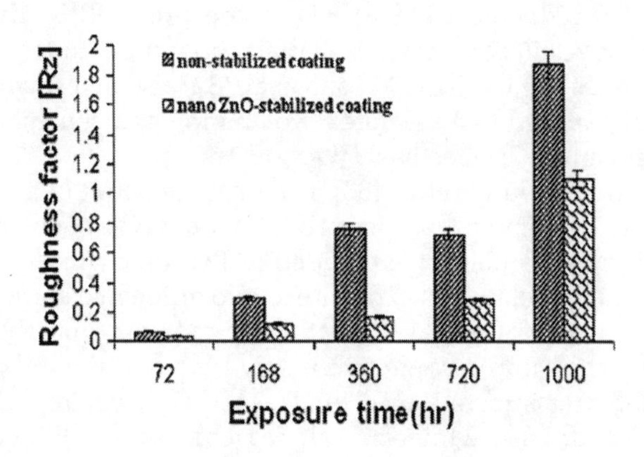

FIGURE 8.5

Variation of the surface roughness factor as a function of time of exposure in QUV chamber [29].

gap. In the case of nano TiO_2, these holes and electrons migrate to the surface of the particles and eventually react with oxygen, water or hydroxyls groups to form free radicals which can actually intensify or accelerate the degradation process of the protective coatings. This process is known as photocatalysis. In contrast, nano CeO_2 absorbs UV without being photoactive as the electron-hole pair recombines very fast before it can migrate to the particle surface because of the crystal defects and oxido-reduction reaction. Besides, nano TiO_2 has fewer localized electrons (3d orbital) than that of nano CeO_2 (4f orbital), resulting in Ce–O bonding being more ionic than Ti–O bonding, and thus, the formation of the electron-hole pair is also much lower in case of nano CeO_2 than that of nano TiO_2. Due to the combination of these two phenomena, nano CeO_2 does not show any photo-catalytic effect; rather, it presents superior transparency in the visible spectrum due to having a lower refractive index.

Monfared et al. [30] have reported an interesting study on the influence of nano CeO_2 particles on UV protective properties of PU. In this work, surface-modified nano CeO_2, prepared separately, was incorporated into PU to obtain better dispersion of the particles. PU with nano CeO_2 showed less deterioration in comparison to the blank PU when exposed to 700 h of accelerated artificial weathering. PU clear coats with increasing nano CeO_2 content showed excellent resistance toward discoloration in a QUV exposure test as compared to the pure PU (Figure 8.6) clear coat. These differences are more visible during first 150 h of the exposure cycle.

In addition to effective UV absorption by nano CeO_2, the variations in structural and mechanical performances of PU/nano CeO_2 coatings up on weathering have also been discussed by Monfared and Mohseni [31]. DMTA studies revealed that the Tg of films systematically increased with increase in nano CeO_2 content. Due to decreased rate of photo-oxidation and chain scission reactions, the storage moduli of PU films containing nano CeO_2 were higher than that of the neat PU after UV exposure. ATR FT-IR spectra also supported these results by showing depreciation in photodegradation reactions in PU containing nano CeO_2. Saha et al. [32] also investigated PU/nano CeO_2 coatings exposed to accelerated weathering and concluded that CeO_2 nano particles can act as effective UV absorbers.

A recent study by Nuraje et al. [33] discussed the effect of graphene nanoflakes on the weathering properties of PU coatings. Graphene nanoflakes in different weight percentages were added to PU top coatings, and the coatings were evaluated relative to exposure to two different experimental conditions: one was a QUV-accelerated weathering cabinet, while the other was a corrosion test. The samples were alternately placed in the UV chamber and salt fog chamber at intervals of 12 h. The UV chamber replicated the dry atmospheric conditions, while the salt fog chamber replicated wet atmospheric conditions. This process was followed for 20 d. The tests confirmed the hypothesis that the addition of graphene does in fact improve resistance

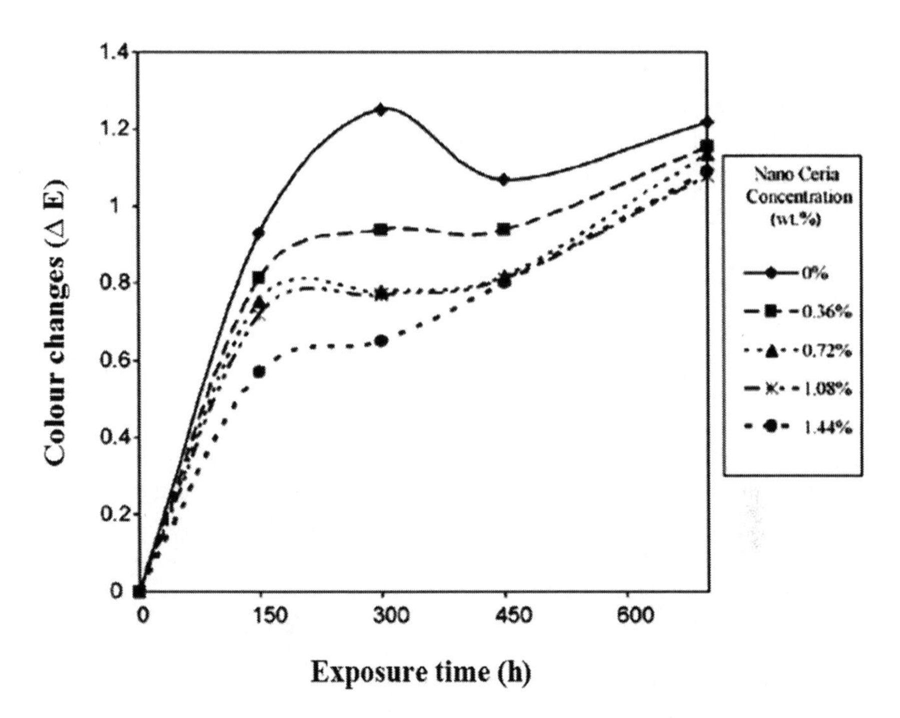

FIGURE 8.6
Colour change of PU clear coats with increasing nano CeO_2 content as a function of exposure time [30].

against UV degradation and corrosion. The PU coating containing 2 wt % graphene showed greatly improved performance as compared to the standard PU coating, since graphene provides hydrophobicity, absorbs incident light and improves mechanical robustness of the coatings.

Wang et al. [34] have reported modified thermoplastic PU films for the weathering layer of a high-altitude balloon or airship by using a composite additive containing nano ZnO, nano CeO_2, UV-531, UV-327 and the hindered amine-622. The comprehensive UV/ozone aging properties of PU films have been studied in a self-designed UV/ ozone aging test device. The results have shown better efficiency in respect to weathering resistance.

Zhang et al. [35] have presented a very interesting and green approach towards the improvement in the yellowing tendency of PU using surface-modified nanocrystalline cellulose (NCC) along with several other advantages, such as environmental compatibility and biodegradability. In that study, the yellowing of the NCC/PU composite was found to be decreased by 57.7% with 1.5 wt % surface-modified NCC.

More recently, the studies done by Chatterjee, Butola and Joshi [36, 37] have enlightened on the durability of TPU-based nanocomposite coatings, particularly targeting for inflatable aerostat applications. In their study, a series of nanocomposite coating systems have been evaluated for efficiency to UV resistance, helium gas impermeability and also the losses in both properties with weathering. An optimized coating formulation containing a UV stabilizer as well as nanomaterials (e.g. nanoclay and graphene) was also identified [37]. Both sided coating with the optimized mixture formulation of aliphatic TPU seems to have much improved quality, and its multi-layered oriented prototype performed even better in some cases. When compared with natural weathering, the degradation behaviour of the samples was found to be quite similar with artificial exposure. A reliability model with two stress types, the temperature and UV irradiance was applied successfully for calculating the service life for three types of TPU-based coatings targeting inflatable systems [37]. Currently, this dynamic research group supervised by Professor Joshi is vigorously engaged in doing research on nanomaterial-based weather-resistant coatings and films for inflatables; that has been reflected in their progressive publications [38–40].

A summary of some of the important research done on weathering of PU and its stabilization is given in Table 8.4.

8.4 Futuristic Approach

The major disparity that should be noted in the literature related to weathering degradation and stabilization of PU is the chemistry of the PU system studied. Obvious differences are also present in various sample forms (films, coated metal panel, coated fabric, etc.; their thickness and processing) that were exposed for the weathering study. Even the weathering methods and parameters (UV intensity, temperature etc.) used for the degradation study were different too. So, an overall conclusion on the degradation behaviour of PU and its performance properties could not be drawn from the previous studies. In this respect it will be promising to study the degradation behaviour of a particular PU system of interest, systematically, in detail. In view of stabilization of PU based coatings, little research has been done that incorporates nanomaterials, and it will be worthwhile to focus on further studies with nanomaterial based coatings. Research should also proceed to study the multifunctional properties of coatings (UV resistance, tensile properties, barrier properties, etc.) and their degradation trend that have particular significance towards high-performance applications. Unfortunately, no significant work particularly dedicated to service life prediction of PU-based coating has been found in the literature. One of the main reasons could be the versatility in the PU structure which makes degradation behaviour of each type significantly different from another. However, some of these issues have been

TABLE 8.4

Weathering of PU and Its Stabilization

Work done by	Studied on	PU Type	Weathering	Main Findings
• **Weathering studies on neat PU**				
Rek et al. [10]	Effect of different kinds of polyols	Polyester, poly caprolactone and polyether with MDI	Artificial (quartz-mercury)	Degradation of polyester < polycaprolactone < polyether-based PU
Bajsic et al. [11]	Role of mol. wt. of soft segments and conc. of hard segments	PTMO (mol wt. 1000/ 2000) + MDI, NCO:OH = 2:1 and 4:1	Artificial (quartz-mercury)	• Higher soft segment molecular weight and lower hard-segment concentration • Higher degradation
Wilhelm and Gardette [14]	Photochemical behaviour	Aliphatic polyester	Artificial (SEPAP 12–24 unit and mercury)	Degradation started from oxidation of urethane functions
Wilhelm and Gardette [15]	Photochemical behaviour	Aliphatic polyether	Artificial (SEPAP 12–24 unit and mercury)	Degradation started from oxidation of the ether groups which produced formates
Wang et al. [13]	Effect of different di-isocyanates	MDI/TDI/HDI + polyester polyol	Artificial (ultraviolet-B lamp 310 nm)	• For TDI and MDI, colour easily became yellow • For HDI, also degraded but with less or no discoloration
Work done by	**Additives used**			**Main findings**
• **Weathering studies on PU with UV protective additives**				
Singh et al. [17]	Benzotriazole UVA (Tinuvin P or Tinuvin 900), HALS (Tinuvin 292), a phenolic antioxidant (Irganox 1010) and TiO_2			Benzotriazoles and HALS separately and in combination are effective

(*Continued*)

TABLE 8.4 (Continued)

Weathering of PU and Its Stabilization

Work done by	Additives used	Main findings
Podešva et al. [19]	Synthesized new low-molecular-weight stabilisers, based on a sterically hindered phenolic antioxidant and sterically hindered piperidines, bearing in their molecules a tri(dimethylsiloxane) chain and end-capped with primary OH groups	Due to the presence of the OH group, stabilizers were joined by a covalent bond at one or both chain ends of the PU that acted simultaneously as a self-stabilized polymer
Jana and Bhunia [20]	HALS (Tinuvin 123) and UVA (Tinuvin 400) in 1:2 w/w ratio	Stabilized PU maintained 83% strength retention even after 2160 hrs
Work done by	**Nanomaterials**	**Main findings**
• **Weathering studies on PU with nanomaterials**		
Chen et al. [27]	TiO_2 nanoparticles, anatase and rutile form	Rutile TiO_2 nanoparticles offer a stabilization function for PU, while anatase TiO_2 enhances photosensitivity
Mirabedini et al. [28]	Surface-modified rutile nano TiO_2 using aminopropyltrimethoxysilane (APS)	Better dispersion and weathering performance than the coating with unmodified nano TiO_2
Rashvand et al. [29]	Nano ZnO particles	Rate of photo-induced degradation of the nano-ZnO incorporated coating was much less than the neat
Nuraje et al. [33]	Graphene nanoflakes	Coating containing 2 wt % graphene showed greatly improved performance since graphene provides hydrophobicity, absorbs incident light, and improves mechanical robustness of the coatings
Wang et al. [34]	Nano-ZnO, nano-CeO_2, UV-531, UV-327, and the hindered amine-622	• Can be useful for the weathering layer of high-altitude balloon or airship • Better efficiency to resist UV/ozone ageing
Chatterjee, Butola and Joshi [36, 37]	UV stabilizer as well as nanomaterials (e.g. nanoclay and graphene)	• High performance durable coating particularly for inflatable aerostat application • Coating system performs efficiently to UV resistance, helium gas impermeability and also the losses in both the properties with weathering

addressed in the recent studies done by Chatterjee, Butola and Joshi [36, 37] to improve the overall performance of the PU-based high-performance protective coatings for inflatable applications.

References

1. Chattopadhyay, D. & Raju, K., 2007. Structural engineering of polyurethane coatings for high performance applications. *Progress in Polymer Science*, 32(3), pp. 352–418.
2. Mater, C. E. & Kinnel, M. J., 1992. Laminate material particularly adapted for hull of aerostats. US patent 5118558.
3. Datta, R. N., Huntink, N. M., Datta, S. & Talma, A. G., 2007. Rubber vulcanizates degradation and stabilization. *Rubber Chemistry and Technology*, 80(3), pp. 436–480.
4. Boubakri, A., Guermazi, N., Elleuch, K. & Ayedi, H., 2010. Study of UV-aging of thermoplastic polyurethane material. *Materials Science and Engineering: A*, 527(7), pp. 1649–1654.
5. Cooper, S.L. & Tobolsky, A. V., 1966. Properties of linear elastomeric polyurethanes. *Journal of Applied Polymer Science*, 10 (12), pp. 1837–1844.
6. Schollenberger, S. C., 1959. Simulated vulcanizates of polyurethane elastomers. US patent 2871218.
7. White, J. & Turnbull, A., 1994. Weathering of polymers: mechanisms of degradation and stabilization, testing strategies and modelling. *Journal of Materials Science*, 29(3), pp. 584–613.
8. Kumar, A. P., Depan, D., Singh Tomer, N. & Singh, R. P., 2009. Nanoscale particles for polymer degradation and stabilization—trends and future perspectives. *Progress in Polymer Science*, 34(6), pp. 479–515.
9. Khanna, A., 2008. Nanotechnology in high performance paint coatings. *Asian Journal of Experimental Science*, 21(2), pp. 25–32.
10. Rek, V., Braver, M., Jocic, T. & Govorcin, E., 1988. A contribution to the UV degradation of polyurethanes. *Die Angewandte Makromolekulare Chemie*, 158(1), pp. 247–263.
11. Bajsic, E. G., Rek, V., Sendijarevic, A., Sendijarevic, V. & Frish, K, 1996. The effect of different molecular weight of soft segments in polyurethanes on photooxidative stability. *Polymer Degradation and Stability*, 52(3), pp. 223–233.
12. Brauman S. K., Mayorga G. D. & Heller J., 1981. Light stability and discoloration of segmented polyether urethanes. *Annal Biomedical Engineering*, 9(45), pp. 45–58.
13. Wang, L.Q., Liang, G. Z., Dang, G. C., Wang, F., Fan, X. P. & Fu, W. B., 2005. Photochemical degradation study of polyurethanes as relic protection materials by FTIR-ATR. *Chinese Journal of Chemistry*, 23(9), pp. 1257–1263.
14. Wilhelm, C. & Gardette, J. L., 1997. Infrared analysis of the photochemical behaviour of segmented polyurethanes: 1. Aliphatic poly (ester-urethane). *Polymer*, 38(16), pp. 4019–4031.

15. Wilhelm, C. and Gardette, J. L., 1998. Infrared analysis of the photochemical behaviour of segmented polyurethanes: Aliphatic poly (ether-urethane). *Polymer*, 39(24), pp. 5973–5980.
16. Thapliyal, B. & Chandra, R., 1990. Advances in photodegradation and stabilization of polyurethanes. *Progress in Polymer Science*, 15(5), pp. 735–750.
17. Singh, R., Tomer, N. S. & Bhadraiah, S. V., 2001. Photo-oxidation studies on polyurethane coating: effect of additives on yellowing of polyurethane. *Polymer Degradation & Stability*, 73(3), pp. 443–446.
18. Crawford, J. C., 1999. 2(2-hydroxyphenyl)2H-benzotriazole ultraviolet stabilizers. *Progress in Polymer Science*, 24(1), pp. 7–43.
19. Podešva, J., Kovarova, J., Hrdlickova, M. & Netopilik, M., 2009. Stabilization of polyurethanes based on liquid OH-telechelic polybutadienes: Comparison of commercial and polymer-bound antioxidants. *Polymer Degradation and Stability*, 94(4), pp. 647–650.
20. Jana, R. N. & Bhunia, H., 2010. Accelerated hygrothermal and UV aging of thermoplastic polyurethanes. *High Performance Polymers*, 22(1), pp. 3–15.
21. Pandey, J. K., Reddy, K. R., Pratheep, A. K. & Singh, R., 2005. An overview on the degradability of polymer nanocomposites. *Polymer Degradation and Stability*, 88(2), pp. 234–250.
22. Mailhot, B., Morlat, S., Gardette, J. L., Boucard, S., Duchet, J. & Gerard, J. F., 2003. Photodegradation of polypropylene nanocomposites. *Polymer Degradation and Stability*, 82(2), pp. 163–167.
23. Morlat, S., Mailhot, B., Gonzalez, D., and Gardette, J. L., 2005. Photooxidation of polypropylene/montmorillonite nanocomposites. 2. Interactions with Antioxidants. *Chemistry of Materials*, 17(5), pp. 1072–1078.
24. Morlat, S., Mailhot, B., Gardette, J. L., Silva, C. D., Haidar, B. & Vidal, A., 2005. Photooxidation of ethylene-propylene-diene/montmorillonite nanocomposites. *Polymer Degradation and Stability*, 90(1), pp. 78–85.
25. Zaidi, L., Kaci, M., Bruzaud, S., Bourmaud, A. & Grohens, Y., 2010. Effect of natural weather on the structure and properties of polylactide/Cloisite 30B nanocomposites. *Polymer Degradation & Stability*, 95(9), pp. 1751–1758.
26. Goldman, A. Y., Montes, J. A., Barajas A., Beall G. & Eisenhour D. D., 1998. *Effect of aging on mineral-filled nanocomposites. Annual Technical Conference Proceedings, Society of Plastics Engineers*, pp. 2415–2425.
27. Chen, X. D., Wang, Z., Liao, Z. F., Mai, Y. L. & Zhang, M. Q., 2007. Roles of anatase and rutile TiO_2 nanoparticles in photooxidation of polyurethane. *Polymer Testing*, 26(2), pp. 202–208.
28. Mirabedini, S., Sabzi, M., Zohuriaan Mehr, J., Atai, M. & Behzadnasab, M., 2011. Weathering performance of the polyurethane nanocomposite coatings containing silane treated TiO_2 nanoparticles. *Applied Surface Science*, 257(9), pp. 4196–4203.
29. Rashvand, M., Ranjbar, Z. & Rastegar, S., 2011. Nano zinc oxide as a UV-stabilizer for aromatic polyurethane coatings. *Progress in Organic Coatings*, 71(4), pp. 362–368.
30. Monfared, A. S., Mohseni, M. & Tabatabaei, M. H., 2012. Polyurethane nanocomposite films containing nano-cerium oxide as UV absorber. Part 1. Static

and dynamic light scattering, small angle neutron scattering and optical studies. *Colloids and Surfaces A: Physicochemical and Engineering Aspects*, 408, pp. 64–70.
31. Monfared, A. S. & Mohseni, M., 2014. Polyurethane nanocomposite films containing nano-cerium oxide as UV absorber; Part 2: Structural and mechanical studies upon UV exposure. *Colloids and Surfaces A: Physicochemical and Engineering Aspects*, 441, pp. 752–757.
32. Saha, S., Kocaefe, D., Boluk, Y. & Pichette, A., 2013. Surface degradation of CeO_2 stabilized acrylic polyurethane coated thermally treated jack pine during accelerated weathering. *Applied Surface Science*, 276, pp. 86–94.
33. Nuraje, N., Khan, S. I., Misak, H. & Asmatulu, R., 2013. The addition of graphene to polymer coatings for improved weathering. *ISRN Polymer Science*, 2013(1), pp. 1–8.
34. Wang, H., Wang, Y., Liu, D., Sun, Z. & Wang, H., 2014. Effects of additives on weather-resistance properties of polyurethane films exposed to ultraviolet radiation and ozone atmosphere. *Journal of Nanomaterials*, 2014, pp. 1–8.
35. Zhang, H., Chen, H., She, Y., Zheng, X., & Pu, J., 2014. Anti-yellowing property of polyurethane improved by the use of surface-modified nanocrystalline cellulose. *BioResources*, 9(1), pp. 673–684.
36. Chatterjee, U., Butola, B. S. & Joshi, M., 2016. Optimal designing of polyurethane-based nanocomposite system for aerostat envelope. *Journal of Applied Polymer Science*, 133(24), pp. 1–9.
37. Chatterjee, U., Butola, B. S. & Joshi, M., 2017. High energy ball milling for the processing of organo-montmorillonite in bulk, *Applied Clay Science*, 140, pp. 10–16.
38. Adak, B., Butola, B. S., & Joshi, M., 2018. Effect of organoclay-type and clay-polyurethane interaction chemistry for tuning the morphology, gas barrier and mechanical properties of clay/polyurethane nanocomposites, *Applied Clay Science*. doi:10.1016/j.clay.2018.04.030.
39. Adak, B., Joshi, M., Butola, B. S., Polyurethane/clay nanocomposites with improved helium gas barrier and mechanical properties: Direct versus master-batch melt mixing route. *The Journal of Applied Polymer Science* (2018). doi:10.1002/app.46422.
40. Adak, B., Joshi, M., Butola, B. S., 2019. Polyurethane/functionalized-graphene nanocomposite films with enhanced weather resistance and gas barrier properties. *Composites Part B: Engineering* 176, 107303. doi:10.1016/j.compositesb.2019.107303.

9

Testing and Evaluation of LTA Systems

B. S. Butola, S. Parasuram, and Neeraj Mandlekar
Indian Institute of Technology, New Delhi, India
Gaurav Singh
Defence Research and Development Organisation, Agra, India

CONTENTS

DOI: 10.1201/9780429432996-9

9.1 Introduction

The envelope of LTA systems plays a significant role in providing the essential lift while containing the buoyant gas, and it offers an aerodynamically suitable shape for static and dynamic stability, structural strength, and rigidity to the inevitably large structure. Similar to the envelope design and material development, establishing test requirements and estimating its service life are also equally challenging tasks. For example, materials of the different parts of the aerostat structure – i.e., hull, ballonet, and fins – must be tested in several directions (warp, weft, bias) at different environmental conditions (hot, cold, humid, high UV), which involves the envelope material being subjected to rigorous tests for assessing immediate functional parameters and long-term properties before its conversion into a large aerodynamic body [1, 2]. This makes the testing of LTA materials more challenging. Subsequently, the transformation of envelope material into an envelope requires optimum sealing conditions, minimum handling, abrasion, folding, and excellent workmanship. Some of the fundamental properties that are to be tested for LTA envelop qualification are as follows [3].

- Weight
- Base fabric breaking strength
- Tensile strength
- Seam tensile strength at elevated temperature
- Tear strength
- Creep
- Peel strength
- Flex resistance
- Helium permeability
- Solar absorptivity and emissivity

To satisfy all these properties, extensive testing of envelope material starts right from the testing of raw materials, i.e. the selection of yarn for conversion into base fabric and picking the polymer/film used for coating/lamination.

The acceptance test matrix for raw materials and the envelope material is given in Tables 9.1–9.3. Besides the acceptance criteria, these materials are subjected to several qualification tests that are customized as per guidelines under MIL-810-G, such as solar radiation (sunshine), high/low temperature, temperature shock, rain, flex, relative humidity, salt fog, sand, dust, and creep.

The aforementioned specifications on envelope material and certification requirements help to define the material development and qualification process. Moreover, the strength of the material is considered one of the critical testing parameters. The uni-axial tensile strength is probably the most often-used property for the evaluation of a hull material mechanical performance. Another critical testing parameter for hull material is its tear resistance after sustaining damage. In recent times, considerable efforts have been made to investigate the tear-resistance performance of various types of laminated and coated fabric materials, as failure to contain tear under a critical length could lead to catastrophic failure of the entire structure. The sample size and the geometry of the notch applied to the sample have a significant effect on the tear resistance. Several methods have been developed: the central cut slit tear test, the tongue tear test, the wounded (slit) burst test, and the trapezoidal tear test.

The LTA systems, such as aerostat and airship materials, are primarily exposed to the outdoor environment during their intended service life. A wide range of fabrics, polymers, and organic as well as inorganic materials

TABLE 9.1

Test Methods for Testing of Yarns

Test Parameter	Test Method
Denier	ASTM D 1907-01
Breaking Strength	ASTM D 2256
Elongation at Break	ASTM D 2256
Tenacity gpd	ASTM D 1907-01
Twist	ASTM D 1422, ASTM D 1423
Ply	ASTM D 1423

TABLE 9.2

Test Methods for Testing of Base Fabrics

Test Parameter	Test Method
Breaking Strength	ASTM D 5035
Elongation at break, %	ASTM D 5035
Ends/Picks per cm	ASTM D 3775
Mass	ASTM D 3776
Thickness	ASTM D 1777
Bow and Skewness	ASTM D 3882

TABLE 9.3

Test Methods for Testing of Envelope Materials

Test Parameter	Test Method
Breaking Strength	ASTM D 751
Elongation at break	ASTM D 751
Tear Strength – Cut Slit	ASTM D 751, ASTM D1424
Helium Permeability	ASTM D 1434
Coating Adhesion / Peel Strength	ASTM D 1876
Inter Lamination Strength of Different Layers	ASTM D 1876
Seam Strength	ASTM D 1876
Mass	ASTM D 751
Thickness	ASTM D 751
Low Temperature Flex/Bent Test at (–) 30°C	ASTM D 2136
Surface Finish – Interior	Visual Inspection
Surface Finish – Exterior	Visual Inspection
Seam Helium Permeability	ASTM D 1434
Seam Creep Test	ASTM D 2990
Resistance to Ozone	ASTM D 1149
Solar Absorptivity	ASTM E 903
IR Emissivity	ASTM E 408

are used in such systems. Various environmental factors contribute towards the degradation of these materials during their service life. Testing of these environmental parameters and their direct correlation with the actual environmental degradation is another challenging area. Generally, the three critical factors for environmental degradation are ultraviolet (UV) radiation, humidity, and heat. The polymeric material normally degrades when it is exposed to these environmental influences. Primarily, UV irradiation has a more severe impact on surface degradation of an envelope material, causing irreversible changes in the chemical structure of polymeric films, which affect both the physical properties – loss of gloss, yellowing, cracking – and the mechanical properties – loss of tensile strength, brittleness, changes in glass transition temperature (Tg), etc. [4]. Several studies have been carried out on the envelope material to know the extent of degradation during exposure to natural sunlight and accelerated ageing environments. It was evident that natural weathering gives a more realistic representation of the variation in product performance during use compared to artificial weathering.

This chapter focuses on testing procedures and characterization techniques adopted for LTA hull assessment. Standard procedures exploited for evaluation of the strength layer, the gas barrier layer, and the protective layer have been discussed in detail. Besides, the chapter also emphasizes the estimation of the service life of the LTA material during practical application.

It is globally accepted that environmental factors such as sunlight, humidity, heat, oxygen, and ozone significantly contribute towards the degradation of LTA envelope material. For this purpose, testing exposure under natural weathering and accelerated artificial weathering are discussed in detail.

9.2 Testing and Characterization

9.2.1 Tensile Test

High tensile strength is a crucial parameter required for envelope material in order to bear the pressure during operation of LTA systems at high altitudes. The tensile strength of the laminate samples is determined using the guidelines provided in the ASTM standard. Uni-axial tensile testing is probably the most often reported property for envelope material. The methodology behind uniaxial testing for fabrics is standardized by ASTM standard D-5035-06, 'Standard Test Method for Breaking Force and Elongation of Textile Fabrics (Strip Method)' [5]. This test is done with type 1R specimens, which are 1.0-inch-wide ravelled strips. Each sample is mounted securely in the clamp, as shown in Figure 9.1, taking care that the long dimension is as parallel as possible to the direction of the applied force. The test is performed using a constant rate of extension, typically at a rate of 12 inches/300 mm per minute.

Komatsu et al. from the Japan Aerospace Exploration Agency (JAXA) evaluated the uni-axial tensile strength of more than 30 high-specific-strength laminated fabrics for LTA system applications. Based on the comparison of strength results, the laminated fabric with Zylon fabric as the core layers exhibited the most satisfactory mechanical performance. Further study on the temperature effects on mechanical properties also verified that the high specific strength of Zylon-based envelope material could be well maintained at both high and low temperature. More comprehensive and in-depth studies on the tensile property of the laminated and coated envelope material were also conducted to help better understand their behaviour towards the mechanical loading in varied conditions [7]. Kang et al. conducted a uniaxial tensile test for Vectron-fabric-reinforced envelope material to attain the experimental result of tensile properties and then compared it with the results predicted by the geometrically nonlinear finite element analysis [8].

The failure mode of uni-axial tensile testing of Kevlar-fabric-based hull material was studied by observing SEM images of fractured specimens. Two distinct failure modes of the envelope material were identified; namely, interface failure and fibre bundle fracture. The failure mode of de-bonding between fibres and polymer films correlated well with the force-displacement curve obtained [9]. As can be seen from Figure 9.2(a), a significant orthotropy

FIGURE 9.1
Uniaxial tensile test with strip specimen [6].

is observed for the load-displacement curve of the envelope material. Potentially owing to the fact that slightly higher initial crimps were introduced to weft yarns, the maximum tensile force in the weft direction is about 13.4% lower than that in the warp direction. However, the authors' statement that weft yarn sustained more damage than warp yarns from the weaving process may not be entirely correct. Two significant nonlinear regions on the force-displacement curves in both warp and weft directions are shown in Figure 9.2(b). This may be attributed to the occurrence of debonding between the fabric substrate and the polymeric adhesives/coating, which was also found by Chen et al. [10]. The stress-strain curve is essential for characterizing the mechanical behaviour of airship hull materials. Meng et al. [9] developed the Monte Carlo simulation method based on the Ising model to analyse the stress-strain curve of their hull material samples, achieving a good fit of the simulated curves to the experimental data.

For the specific laminated fabric developed in their study, Chen et al. also proposed a stress-strain model based on the uniaxial tensile testing, as shown in Figure 9.3. The stress-strain curve consists of three linear regions and two

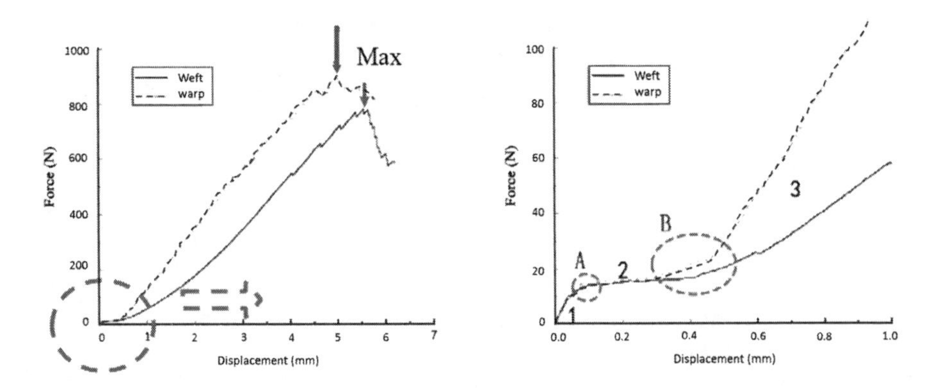

FIGURE 9.2
Measured force-displacement curves of the uniaxial tensile tests; (a) global curve and (b) magnification at the low strain range [9].

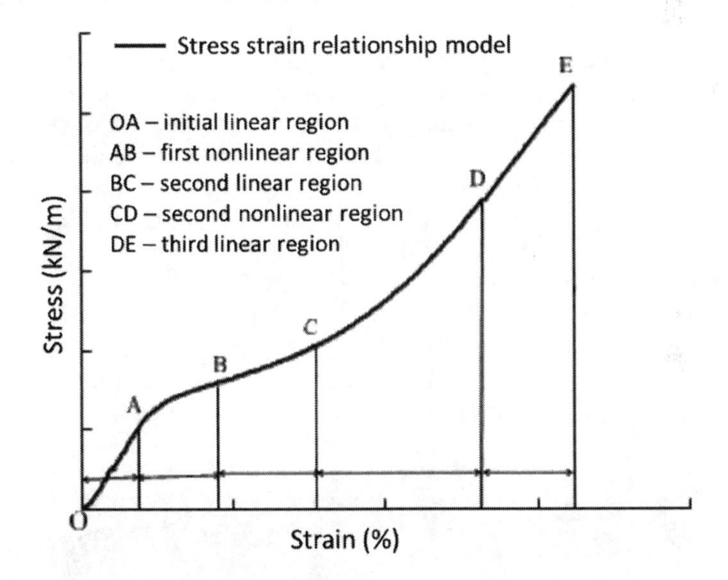

FIGURE 9.3
The stress-strain model of the coated fabric. Adapted from [10].

nonlinear regions [10]. Initially in region OA, the laminate sample behaves as an undamaged, linear elastic material, indicating that the fabric and coating are extended together. In the first and second nonlinear regions (AB and CD), the coating may be debonded from the fabric, and the fabric behaved as a nonlinear material. There is also a second linear stage BC, where the behaviour of the laminate material is primarily controlled by yarns. The yarns elongate linearly again with a smaller slope (elastic modulus) than that of the

OA region. At the end region DE, fabric elongates linearly again after point D, and at point E, most of the yarns get fractured [10].

In these five regions, the stress-strain relationships were expressed as:

$$OA, BC, DE : \sigma = \alpha \cdot \varepsilon + b$$

$$AB : \sigma = \alpha : \text{In}\varepsilon + b$$

$$CD : \sigma = \alpha \cdot e^{(b \cdot \varepsilon)}$$

where σ is the stress, ε is the strain, and α and b are the best-fit constant parameters that were determined by the experimental data [10].

9.2.2 Gas Permeability

One of the most critical parameters to be assessed while studying the performance of the LTA systems is the gas retention ability of the envelope hull material. The service life of LTA systems is primarily decided by the extent of increase in the gas transmission rate (GTR) of the envelope material over time. LTA systems, as the name suggests, attain their lift by use of a gas having a lower density than the surrounding air. Generally, LTA systems have adopted the use of hydrogen (H), helium (He) and hot air to attain flotation. Figure 9.4 gives the comparison of the lift capacity of different gases for a fixed volume. It is evident from the figure that hydrogen provides the maximum lift, given that its density is the least.

Early models of airships employed hydrogen as the buoyant gas; they were later replaced with helium taking into consideration the high flammability of hydrogen gas when mixed with ordinary air even at a low volume ratio of 4% (hydrogen to air). Helium makes an ideal buoyant gas for LTA systems, having just 7% lower lift capacity than hydrogen and being inert

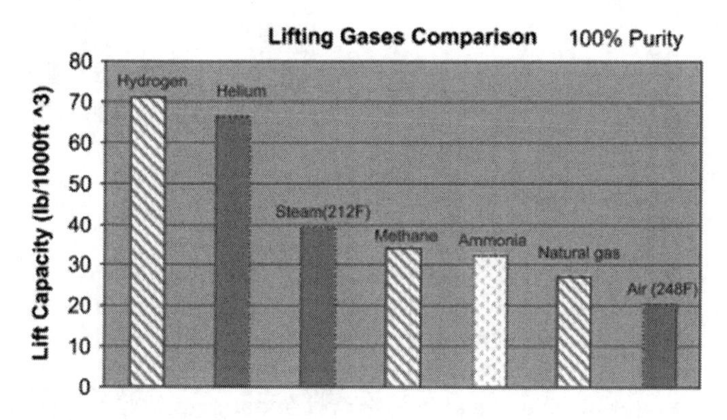

FIGURE 9.4
Schematic of lifting gases comparison [11].

TABLE 9.4

The Kinetic Diameter of Common Gaseous Molecules [13]

Gas Molecule/Atom	Diameter (nm)
He	0.26
H_2	0.29
Nitrous oxide (N_2O)	0.32
Carbon dioxide (CO_2)	0.33
Oxygen (O_2)	0.35
Nitrogen (N_2)	0.36
Methane (CH_4)	0.38
Propane (C_3H_8)	0.43
Butane (C_4H_{10})	0.50

and incombustible [11, 12]. However, the primary drawback in using helium is its high cost and limited availability. In addition to the lifting capacity provided by the gas, the size of the permeate molecule is also an important parameter to be considered. Assuming a molecule takes the shape of a sphere, the kinetic diameter is the smallest effective diameter of the molecule. Table 9.4 gives the kinetic diameter of few gaseous molecules/atoms; it can be seen that the kinetic diameter does not necessarily correlate with molecular weight. The factors such as steric hindrance and polarity of the molecule decide the kinetic diameter [13]. From this discussion, it is evident that helium is an ideal candidate to be used as a lifting gas in LTA applications. Still, its high cost and low kinetic diameter make it particularly necessary to have an effective gas barrier layer in the envelope system to provide adequate resistance to transfer/leakage of the lifting gas.

9.2.2.1 Gas Transport Mechanism

The generally accepted theory of gas transmission through polymeric materials is the activated diffusion mechanism. Gas molecules tend to initially get adsorbed or become soluble on the surface and then diffuse through it. This diffusion process takes place from a region of higher concentration to lower and continues until there exists no concentration difference between either side of the polymeric membrane. The factors that influence the transport phenomenon include the nature of the polymer, specifically the free volume within the polymer matrix, and the segmental mobility of the chains which in turn is dependent on the degree of crystallinity, degree of crosslinking, and extent of unsaturation [14]. The size of the permeate molecule, temperature, and humidity also have a considerable effect on the diffusion process [15].

Permeation of gas through a polymeric system is a complex phenomenon involving multiple stages:

(a) Diffusion of permeant from the upstream atmosphere to the polymeric film/coating.
(b) Adsorption of permeant on the surface of polymeric film/coating.
(c) Diffusion of permeant through the film/coating. This step is generally the slowest and hence the rate-determining step in gas permeation.
(d) Desorption of the permeant from the surface of the film/coating.
(e) Diffusion of permeant from the film and into the downstream atmosphere.

Figure 9.5 illustrates the permeation process through a homogenous polymeric film of thickness l, permeant concentration c with $c_1 > c_2$ and the permeant pressure p with $p_1 > p_2$. A combination of Henry's law of solubility and Fick's first law of diffusion can be used to deduce a relationship that governs the permeation process [17]. By Fick's first law of diffusion, the permeant gas flux is equal to

$$J = -D.\Delta C \tag{9.1}$$

where ΔC is the difference in concentration (mol/cm^3) across the membrane of thickness l. D is the diffusion coefficient/diffusivity (cm^2/s), and J is the diffusion flux (mol cm^{-2} s^{-1}).

Considering one-dimensional diffusion through the polymeric film

$$J = -D(\Delta C / 1) \tag{9.2}$$

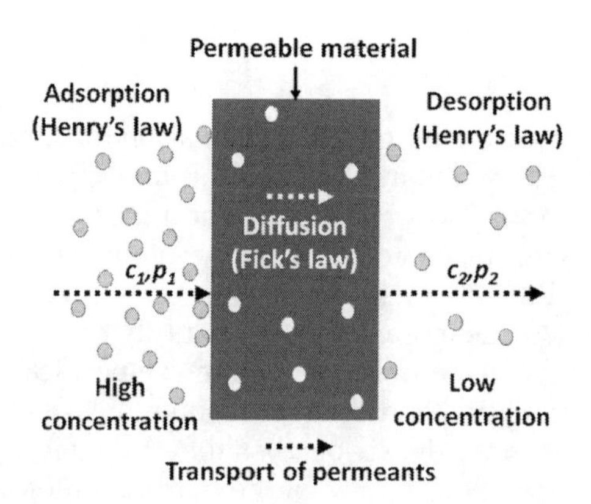

FIGURE 9.5
Schematic showing the general gas permeation mechanism through polymeric film, where p and c represent the pressure and concentration of the permeant [16].

Under the steady-state condition, gas partial pressure and gas concentration obey Henry's law. Since it is relatively easier to measure the pressure instead of concentration, ΔC can be replaced by S. ΔP, where S is the coefficient of solubility ($mol \cdot cm^{-3} \cdot Pa^{-1}$) which denotes the amount of permeant in the polymeric film and ΔP the difference in pressure.

Equation (9.2) becomes

$$J = -D(S.\Delta P / 1) \tag{9.3}$$

The product D. S is indicated as the permeability coefficient (or constant) or permeation coefficient or only as permeability (P). The ratio of P/l is generally indicated by q and is named as permeance [15].

If S is independent of the concentration, the gas permeability coefficient P (in $mol \cdot Pa^{-1} \cdot cm^{-1} \cdot s^{-1}$), can be expressed as follows:

$$P = -J1 / \Delta P = D.S. \tag{9.4}$$

9.2.2.2 Method to Measure Permeation

In general, most methods to study gas permeation follow the same procedure, in which one side of the specimen – polymeric film/coated or laminated fabric – is exposed to the testing gas, the test gas can either be static at a particular pressure, or a continuous stream of gas can be maintained throughout the test. On the other side of the specimen, the chamber is maintained at low pressure (vacuum), and the increase in pressure due to permeation of the test gas through the specimen can be recorded. In some cases, the gas diffused through the specimen is taken away by a stream of carrier gas and fed into a sensor. The relative difference between the permeant and carrier gas mixture and the pure carrier gas is calculated. Given the advancement in pressure sensors and the ability to accurately measure minute rise in pressure, commercially available permeability measurement systems which are predominantly used in industries at present employ the direct method and measure the increase in pressure in the lower chamber [18]. As the permeant diffuses through the film sample and accumulates in the lower chamber, the pressure gradually increases. Once the flow is allowed to stabilize, the pressure-time curve will tend to appear more linear. The gas permeation rate is calculated from the rate of pressure increase.

The transmission rate of the buoyant gas (hydrogen or helium) through the polymeric film/coated or laminated fabric is generally obtained by carrying out tests in accordance with the ASTM D1434 procedure; the general schematic of the test set-up is as follows [19]. The test is mostly carried out at an ambient temperature of 25°C and relative humidity of 50%; the test

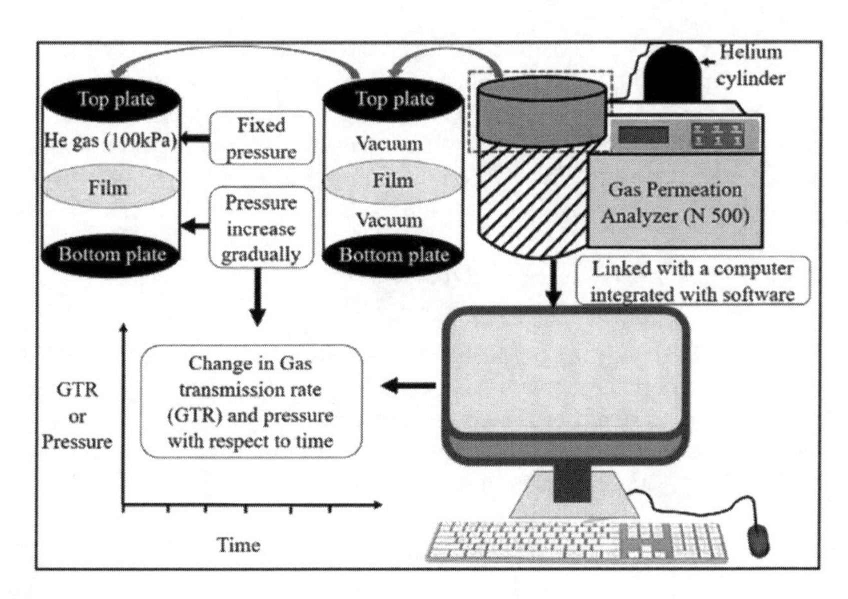

FIGURE 9.6
Gas permeability testing setup [20].

conditions can vary depending on the final intended application of the material. It needs to be noted that factors such as temperature and humidity considerably influence the gas transmission through the polymeric film or coating, and care must be taken to condition the specimen appropriately before performing the test in line with the ASTM standard (Figure 9.6).

The sample is clamped securely to develop a semi barrier between the two chambers following which both the chambers are evacuated. Subsequently, the top chamber is filled with the buoyant gas (helium or hydrogen) at a specific high pressure (100 kPa). The gas transfer takes place through the sample following the 'adsorption-diffusion-desorption phenomenon' described earlier. The flow is allowed to stabilize, and the instrument measures the GTR by a differential pressure method.

Given that the scope of the chapter is primarily centred around testing, a brief summary of materials used as the gas barrier layer and the GTR at ambient temperature and pressure is described here. Loss in helium due to leakage can lead to a reduction of lift and thus decrease the operational capability of the LTA system. It would also significantly increase the operating expenses to replenish the lifting gas to maintain the necessary altitude. Given the high cost of helium and limited availability, it is paramount to have a gas barrier layer in the enveloped system that provides adequate resistance to permeation of helium. Earlier airships employed the use of polymeric and elastomeric coating for enhancing the gas retention property; modern airships of late have sided towards incorporating speciality polymeric films for

TABLE 9.5

Desirable Environmental Protective and Gas Holding Polymeric Materials [22]

Material	Permeability	Weatherability	Flex Fatigue	Adhesion to Fabric/Film	Heat Sealability
PVF (Tedlar)	Good	Excellent	Good	Poor	No
PTFE (Teflon)	Good	Excellent	Good	Poor	(> 500 Yes °F)
Polyurethane	Fair	Good	Excellent	Excellent	Yes
PVC	Fair	Good	Good	Excellent	Yes
PVDC (Saran)	Excellent	Poor	Fair	Fair	Yes
Nylon	Excellent	Poor	Excellent	Fair	Yes
Polyester (Mylar)	Good	Fair	Fair	Fair	No

the same [21]. In a review article published by researchers at TCOM, a pioneer in developing aerostat and airship hull material, the potential materials which can be used as gas retention layer were summarized [22] (Table 9.5).

In addition to excellent gas retention property, it is desirable for the polymeric film to possess excellent flex fatigue resistance, good weather resistance, and shear stiffness. Of all the commercially available speciality films, polyester is the preferred candidate for the gas barrier layer; the High Modulus polyester film Mylar (DuPont Teijin) is extensively used due to its low gas permeability and relatively high strength and stiffness. Few other commercially available films that have also been explored for application as the gas barrier layer are polyvinylidene chloride (PVDC, Saran), polyimide (PI) film, Kapton [21], and ethylene vinyl alcohol copolymer (EVOH) [7, 24]; but these commercially available films do have drawbacks such as poor low-temperature flexibility, adhesion, and weatherability, which are essential attributes for applications in LTA systems (Table 9.6). EVOH, despite having an exceptional gas barrier property due to a high degree of crystallinity, is very sensitive to moisture, caused by the interaction between the water molecules and the polar hydroxyl groups in EVOH, which limits its application [23].

9.2.3 Tear Strength

Tearing strength is an essential attribute, as it the most common mode of failure in inflatables. It has been experimentally demonstrated that any tear beyond a critical tear length, i.e. small crack or cut that can lead to degradation of mechanical performance, would cause catastrophic failure. Therefore, tear strength or tear resistance is a critical property of the envelope material used in LTA systems. Tearing is a complex loading environment in which there is a combination of tensile and shearing forces that act upon an individual or small group of yarns within a fabric. Tearing strength is, therefore,

TABLE 9.6

Gas Permeability of Different Gases Through Polymeric/Elastomeric Films at Standard Testing Conditions [16]

Polymer/elastomer	Permeability ($cm^3 \cdot mm/m^2 \cdot day \cdot atm$)			
	Nitrogen (N_2)	Oxygen (O_2)	Carbon dioxide (CO_2)	Helium (He)
Polyvinyl fluoride (PVF)	0.1	1.3	4.4	59.1
Polyvinylidene fluoride (PVDF)	3–3.5	0.55–5.2	2.2–30	59–86
Polytetrafluoroethylene (PTFE)	69–129	178–255	487–720	–
Fluorinated ethylene-propylene copolymer (FEP)	33–125	101–290	648–838	–
Vinylidene fluoride-hexafluoropropylene copolymer	4.67–22	71–95	209–508	771
Polyvinylidene chloride (PVDC)	0.02–0.12	0.03–0.04	0.47–3.2	–
Biaxially oriented polyester (BoPET)	0.18–0.39	1–2.4	5.9–9.8	71
Nylon 6	0.28–0.35	0.57–1.3	1.8–5.9	45.7
Nylon 6,6	0.28	0.3–1.36	3–4.6	59.1
Ethylene vinyl alcohol copolymer (EVOH)	0.003–0.01	0.01–0.05	0.03–0.08	4–14
Liquid crystal polymer (Vectran)	0.03–0.5	0.05–0.1	–	–
Low-density polyethylene (LDPE)	39–79	69–274	394–959	–
Polyurethane (PU)	9–297	3–1067	175–2014	36–2340
Silicon rubber	17,280	19,685	118,110	–

not the intrinsic property of an envelope but an outcome of fabric-coating-weathering interaction. Several efforts have been taken to investigate the tear strength and the tear resistance performance of various types of coated and laminated structures [25–27]. A number of studies have been conducted to know the effect of tear length, orientation, pressure, and diameter of inflatable, etc. [28, 29]. The tear strength itself has no definite relationship with the actual tear propagation [30]; therefore, several studies have been carried out to establish a relationship between tearing strength and tear propagation in order to ascertain safe tearing limits of envelope material.

Numerous tear test methods for fabrics have been proposed, such as Tongue [31] (ISO 13937-2), Wing rip (ISO 13937-3), Double rip tear (ISO 13937-4), trapezoid [32], central cut slit [33], and pressurized cylinder tear test [34]. Among the various tear methods for fabric, the cut slit tear method is considered as an accepted test method as it better simulates the tearing phenomena of a damaged inflatable envelope than other standard tear methods (tongue/trapezoid) for the airship hull material. Similarly, uniaxial, bi-axial, dome-shaped bursting, and pressurized cylinder-based experimental techniques have been used to study the factors affecting tear propagation and to estimate critical tear length in LTA systems.

9.2.3.1 Central Cut Slit Tear Testing

This method is most effective for measuring the tear strength of a fabric and the material capability towards the tearing resistance after the initial damage. It was initially developed by the U.S. Navy in the 1950s as an acceptance test for the airship hull materials. Later it was modified and specified in MIL-C-21189, 'Cloth Laminated, ZPG2 and ZPG2W type Airship Envelope'. The Federal Aviation Administration adopted this test method in FAA P-8110-2, 'Airship Design Criteria', and documented as Appendix A in 1995.

9.2.3.1.1 Description of the Cut Slit Tear Test

This method is used to determine the tearing strength of the coated or laminated fabric. The fabric sample is 102 mm (4 in.) wide × 152 mm (6 in.) long having a 32 mm (1¼ in.) wide razor cut slit across the centre of the sample at right angles to the longest dimension (Figure 9.7).

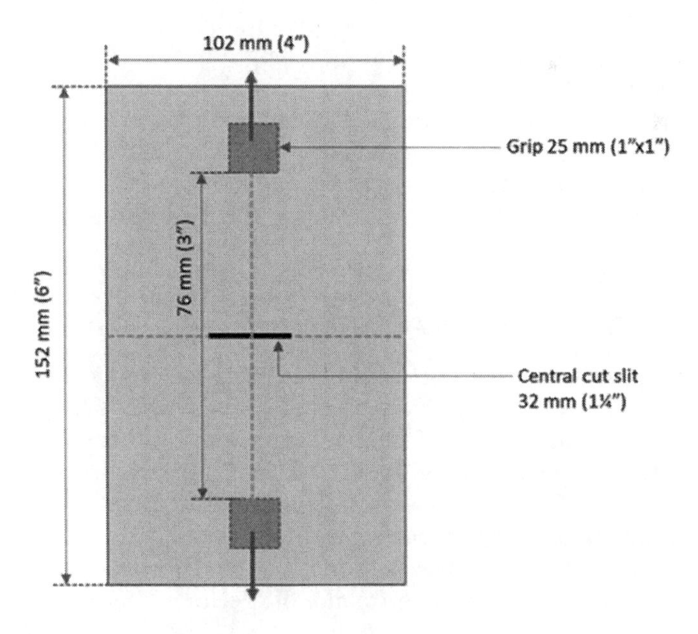

FIGURE 9.7
Central cut slit tear test sample dimension.

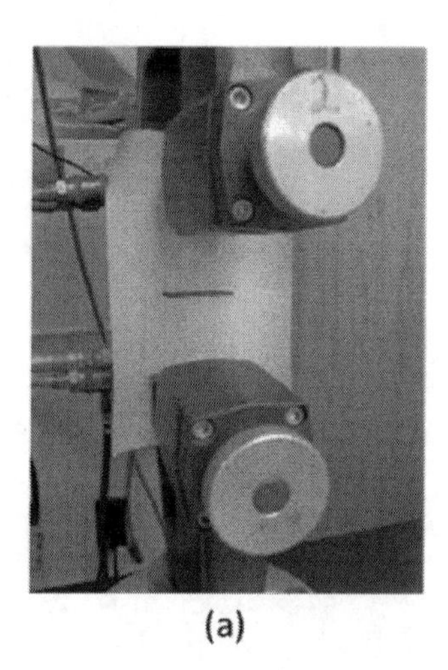

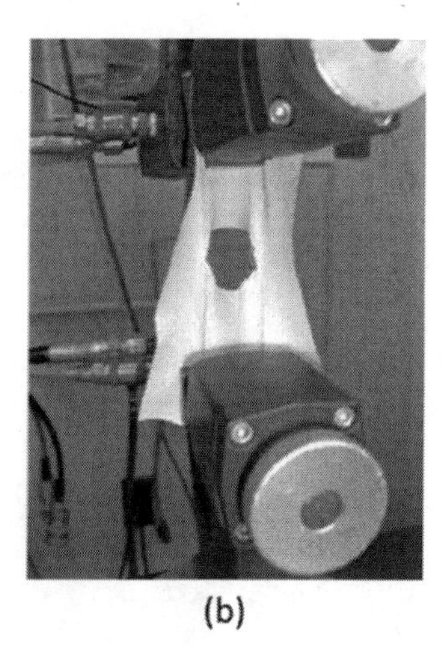

(a) (b)

FIGURE 9.8
Central cut slit tear test: (a) test setup; (b) test in progress. [30].

The specimen is placed symmetrically into clamps of a universal testing machine (Figure 9.8(a)) with the longest direction parallel to the direction of load application. The clamps must be 25 mm (1 in.) wide and must grip the yarns that are cut. At the start of the test, the distance between the clamps (gage length) must be 76 mm (3 in.), with the slit an equal distance from each clamp. Breaking force is applied to the sample at a rate of 305 mm/min (12 in./min) (Figure 9.8(b)). The tearing strength is determined as the average load of the highest recorded peaks of five specimens recorded in N/cm.

Cut slit tear testing may be accepted globally to measure the tear strength of certain coated and laminated fabrics, yet the tear strength simply has no direct relationship with the actual tear propagation characteristics of the airship hull material [30]. Therefore, much effort has been made to establish the relationship.

9.2.3.2 Tear Propagation of Uni-Axial Tear Test

Regardless of the inability to better simulate the real stress field on the hull material, the uni-axial tear test is still an effective way to evaluate the tear strength. Wang et al. conducted uniaxial tearing experiments on ultra-high molecular weight polyethylene fibre (UHMWPE)-based coated fabric, 170 gsm, 1004 N/cm BS (warp) and 804.4 N/cm BS (weft). They experimented with different initial cut slit length at different angles, and the different stretching rate was applied. It was found that the stretching rate is not significant to determine the tear performance of specimens in a uni-axial tensile test. However, the initial crack length does make a difference concerning the maximum tearing force. The difference in initial crack orientations affected

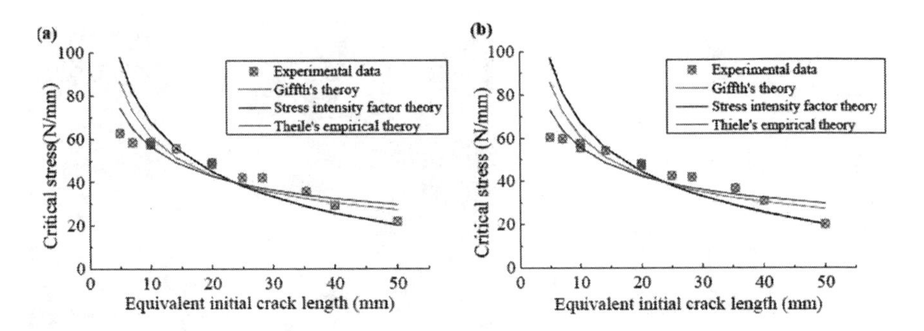

FIGURE 9.9
Comparison between experimental data and the theoretical values for uni-axial specimens: (a) the warp specimens; and (b) the weft specimens [35].

the tear propagation by changing the number of cutting yarns in the loading direction.

To predict the tear propagation upon initial crack, Wang et al. developed analytical models based on the uni-axial testing data. By analysing three of the most critical methods for tear propagation modelling, i.e. Griffith energy theory, the stress intensity factor theory, and Thiele's empirical theory, it was found that the stress intensity factor theory gives the best correlation with the test data from uni-axial tear test, shown in Figure 9.9.

9.2.3.3 Tear Propagation of Bi-Axial Tear Test

In order to better simulate the real stress field on the hull material, bi-axial tear tests were performed on the cruciform specimen with their arms aligned to warp and weft direction of the fabric, as shown in Figure 9.10. The crack was made at the centre with an angle of φ to the warp direction. The length of the slit was relatively short due to the restriction of specimen size. The specimen had a cross-area of 160×160 mm^2, and an effective cantilever of 160 mm. Chen et al. [29] found from the bi-axial tearing analysis that the crack length, crack, orientation, and stress ratio can be considered as the parameters to affect the tearing strength and tearing stress-displacement curves. It is also stated that stress parallel to the crack can build the barrier for tear propagation, and therefore, for the same crack length, the bi-axial test can show a greater tearing strength than the uni-axial test.

9.2.3.4 Tear Propagation of Pressurized Cylinder Test

Using an air-pressurized cylinder to generate the bi-axial tensile field is the most practical method for the development of airships [3, 30]. This method closely simulates the actual stress field, and longer tear length can be tested compared to the bi-axial tensile test [29, 36]. In this test, the bi-axial load is

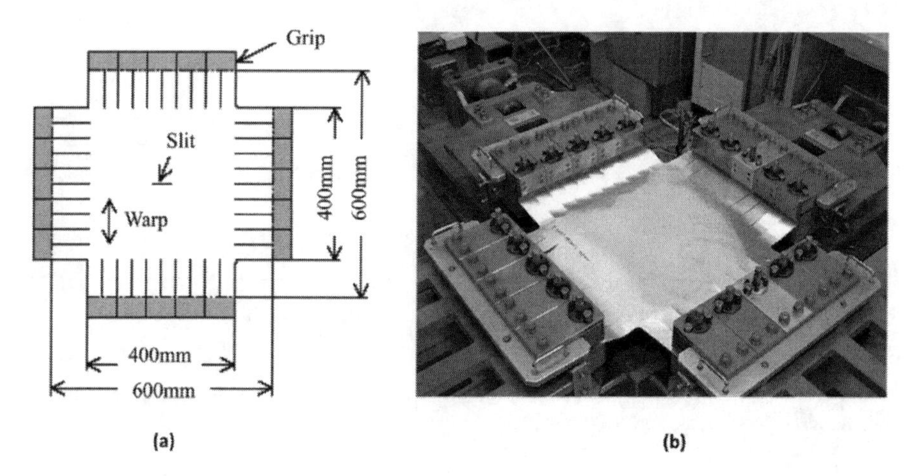

FIGURE 9.10
Tear strength measurement in a bi-axial tensile test: (a) test specimen; (b) test set-up [29].

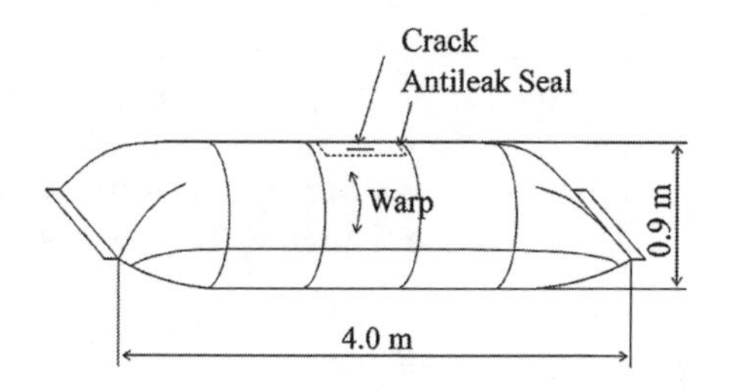

FIGURE 9.11
Pressurized cylinder tear test for airship envelope [30].

applied in stress-control mode. Figure 9.11 shows a generic sketch and a picture of the real specimen produced. The specimen is a cylinder closed at both ends, and the slit is covered from inside by an expandable film to prevent air leakage. The specimen was set in a stand, and the pressure is increased at a constant speed. The slit was recorded by a video camera. The gas filling was stopped immediately when the tear started to propagate, and then the cylinder was depressurized.

9.2.4 Seam Sealing Strength

LTA systems such as aerostats and airships are constructed by joining multiple sections of the base material. It is essential to design a structurally efficient joint to ensure effective transfer of the load from one piece of the base material to another. To maintain structural stability of the whole hull structure, the seam must provide tensile strength that is greater or at least equal to the strength of the base material [22]. Standard joint configurations include

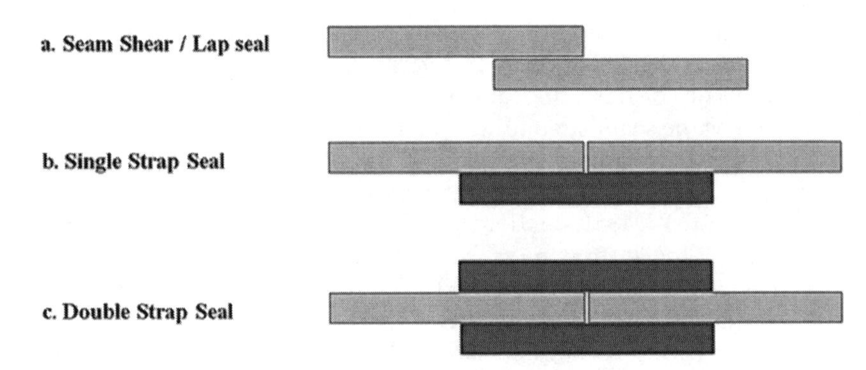

FIGURE 9.12
Common joint configurations of envelope materials.

lap joint, single strap joint, and double strap joint (Figure 9.12), and the various sealing technologies that are used in the sealing process include radio frequency (RF) or high frequency (HF), adhesive sealing, and impulse heat sealing. These topics were extensively discussed in Chapter 2 on manufacturing of airships and aerostats; thus, the emphasis here is on testing and evaluation.

Factors that influence joint strength are joint design, seaming technology, and adhesive selection. In addition to the strength requirement, severe operating conditions such as sub-zero temperature and thermal cycling are also detrimental to the seam integrity in the long term. The seam must absorb stresses and have high failure strength to the extreme service temperatures varied during a day. The strength or dimensional stability of joints of the inflatable structure is often a limiting factor, irrespective of the strength of an inflatable envelope material.

There are several mechanisms by which the base material adheres to one another, such as:

1. *Mechanical interlocking* – where the adhesive interlocks around the irregularities or pores of the substrate, forming a mechanical anchor.
2. *Adsorption* – the attractive forces may be physical, that is, physical adsorption by van der Waal's forces and H-bonding, or by chemical bonding (chemisorption).
3. *Electrostatic attraction* – when the adhesive and substrate have substantial differences in their electronegativity, leading to the formation of an electrostatic double layer at the interface.
4. *Diffusion* – when the adhering materials have macromolecular mobility and mutual solubility leading to an interlayer that consists of inter-diffused molecules or segments from each layer [37].

Given the number of molecular mechanisms by which the coated fabrics adhere to one another, it is difficult to determine the strength of the joint purely on the properties of the fabric, adhesive, and polymers. Hence, mechanical testing such as the seam strength test by a conventional uni-axial tensile test and a T peel test is carried out to quantify the joint strength of the seam.

The seam strength test is generally carried out in accordance with ASTM D5035 and ASTM D4851 standards by loading the strip-shaped specimen uniaxially with the seam joint at the centre of the gauge length [39, 40]. A typical specimen schematic as per ASTM D4581 is shown in Figure 9.13. The crosshead displacement is maintained at 300 mm/min, and generally, five specimens are tested in both the warp and weft directions. Proper specimen preparation technique and clamping need to be adopted to ensure repeatability. The accepted mode of failure in the seam strength test is fibre fracture failure mode; that is, failure taking place on the base material away from the

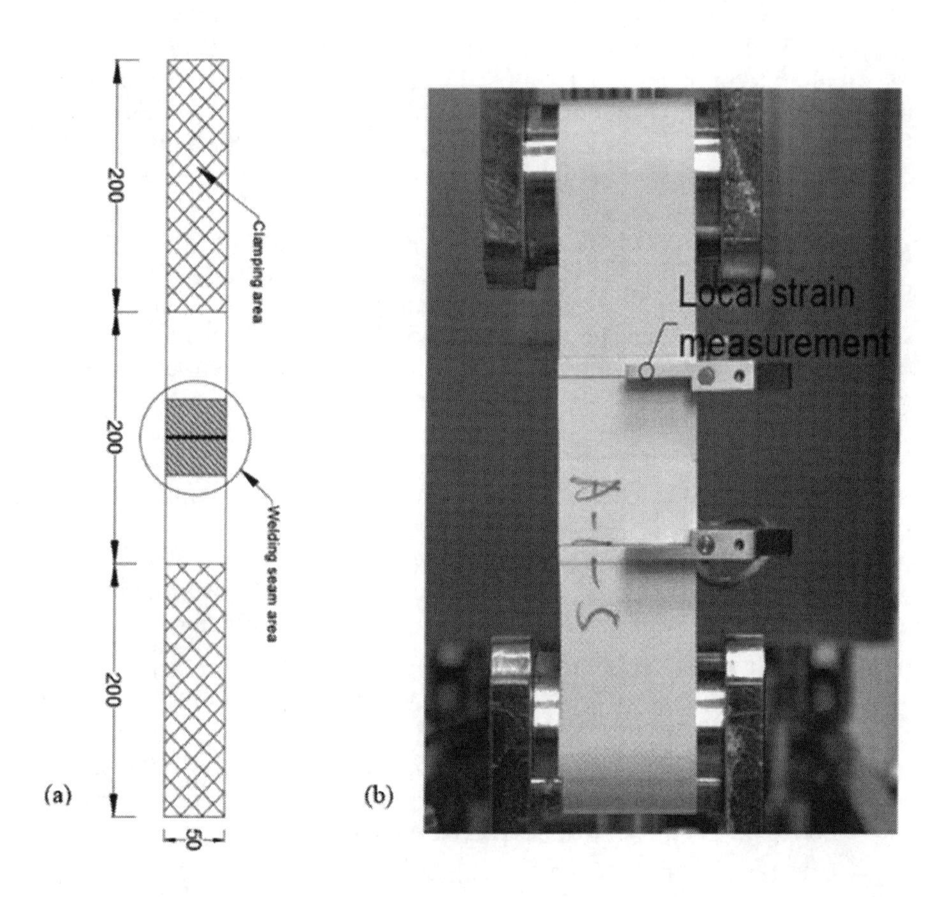

FIGURE 9.13
(a) Schematic of seam strength specimen. (b) Specimen loaded under uniaxial tension [38].

seamed area. A weak seam joint would show delamination, seam failure, or breakage of base material at the edge of the seam.

The second method adopted to study the seam efficiency is the T-peel test carried out in accordance with ASTM D1876 standard [41]. This method essentially gives the relative peel resistance of adhesive bonds between the joining base fabric adherents employing a T-shaped specimen. To perform the T-peel test, a Universal testing machine is employed with a suitable loading range while ensuring adequate clamping is maintained to make sure no slippage occurs throughout the test. The different modes of failure that could arise in a T-peel test are: (a) intralaminar fracture, where the adhesive layer fails, indicating weak strength of adhesion, the cause for which include insufficient heat at the welding interface resulting in poor curing of the adhesive; (b) interlaminar fracture, where the adhesive layer is well cured and durable, but the fracture occurs at the interface between the fabric and the adhesive; (c) fibre fracture, which occurs when the strength of the weld zone exceeds the parent material, and the failure occurs at the base material surrounding the weld zone. The third mode would show the highest fracture force [37] (Figure 9.14).

In one study, Tang and Keefe [43] analysed the seam strength of a double tape polyester seam comprising of two-polyester base fabric, a polyvinyl fluoride cover strap, and a structural tape made of the same polyester base fabric. They tested 42 specimens of dimension 50.8 mm width and 152.4 mm length at a crosshead displacement speed of 50 mm/min in accordance with ASTM D5035 standard to determine the breaking force and elongation. The average breaking strength of the seam was around 20% lower than that of the base laminate, with considerable samples failing at the edge between

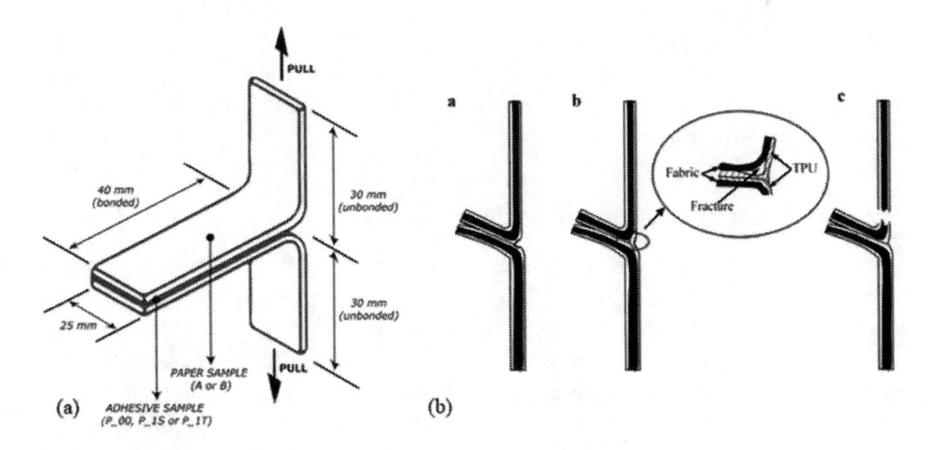

FIGURE 9.14
(a) Schematic representation of a T-peel test specimen. (b) Different failure modes: intralaminar fracture (fracture a), interlaminar fracture (fracture b), and fibre fracture (fracture c) [42].

the structural tape and the base fabric. They recommended two structural modifications to improve the seam strength – one was to reduce the stiffness and geometric discontinuity at the edge of the structural tape by increasing the width of the cover tape, the other was to use a nylon-based structural tape to reduce the stiffness discontinuity. Komatsu et al. [7] carried out extensive mechanical testing on Vectran-, Kevlar-, and Zylon-based airship envelope material and concluded that in the envelope structures of high-specific strength fabrics, the joint strength and creep resistance are far more critical factors to ensure structural stability than tensile strength. A few articles in which the researchers have tried to optimise the sealing process to obtain high seam strength include the work done by Theller [44], where the heat sealability of plastic film was studied in bar sealing applications and reported that the interfacial temperature and dwell time are the primary factors which control the heat-seal strength. Pressure normal to the sealing surface had little effect above the level required to flatten the web for good contact. In a study by Hollande et al. [45], thermoplastic polyurethane (TPU)-coated nylon fabric was sealed with a high-frequency (27.12 MHz) welding machine and impulse heat-sealing machine. The fabric was sealed using various HF machine parameters such as weld time, hold time, and m/c power settings and compared with impulse heat-sealing temperature at the same pressure of 5 kg/cm^2. It was concluded that the melting point of hard chain segments (180°C) must be exceeded to create true cohesion between the TPU-coated fabrics. At temperatures below the melting point of hard segments, there were entanglements only between the flexible segments of the two fabric surfaces in contact which produces a superficial weld at room temperature that readily peeled off at 70°C at relatively modest loads. Johnson et al. [46] explored the use of UV-visible reflectometry and thermography as potential non-destructive examination methods for assessing the quality of such seams in structures comprised of composite laminate fabric panels. From this discussion, it is evident that the seam strength is an essential parameter to be tested to ensure the structural integrity of airship and aerostat envelopes.

9.2.5 Resistance to Ozone and Atomic Oxygen

Airships operate in the stratospheric region of the earth's atmosphere, which extends from the tropopause at 10 km to an altitude of about 50 km from the earth's surface, and this region comprises up to 90% of the earth's ozone. Ozone is a strong oxidant which breaks down molecular chain structures in polymers, and this coupled with UV radiation can drastically deteriorate the hull material and can lead to a loss in lifting gas retention property of the barrier layer and loss in strength of the load-bearing fabric layer. The test to study the deterioration of envelope material in an ozone environment is generally carried out as per the ASTM D1149 standard under dynamic or static strain conditions [47]. The generally adopted test methods are Procedure A in which specimens of dimension 10 ± 0.03 mm width and 100 ± 25 mm length are subjected to a maximum dynamic tensile strain of 25% ± 3% and a fixed amplitude of 0.5 Hz, whereas in Procedure B rectangular specimens of dimension 25 mm width by 150 mm length and 2 mm thickness is subjected to a static tensile strain of 20%. The standard test parameters are ozone partial pressure of 50 ± 5 mPa (equivalent to 50 pphm at 100 kPa atmospheric pressure) and standard temperature of 40°C ± 1°C.

In one of the earlier studies, Maekawa et al. subjected a Zylon-based high-performance airship envelope material to artificial weathering under UV radiation at 180 W/m^2 for 100 hours and Ozone at 50 ± 5 ppm for 24 hours which is roughly five times the concentration that is experienced at an altitude of 20 km. Excellent retention in mechanical strength was noticed despite exposure to a high concentration of ozone content; this was attributable to the presence of Tedlar as the top protection layer, which acts as a useful ozone resistance element. It was concluded that the use of Tedlar as the top protection layer acts as a useful ozone resistance element. Maekawa et al. [48] concluded that the use of Tedlar film as the protection layer shielded the load-bearing Zylon strength layer, but longer exposure duration under high ozone concentration was required to simulate the actual working conditions and a combination of UV and ozone would have given a better understanding of weathering resistance. Wang et al. [49], in a study published in 2013, used a combination of inorganic (nano-ZnO/CeO_2) and organic (UV-531/UV-327) UV absorbers to develop modified polyester polyurethane films by a spin-coating technique and subjected them to accelerated ageing. Self-designed equipment was used to expose the specimens to UV radiation in an ozone atmosphere at a relative humidity of 20% under ambient temperature conditions. FTIR technique was extensively used to study and compare the synergetic ageing effect of ozone and UV radiation on the specimens, and the results showed a significant reduction in the degradation of TPU films with the addition of UV absorbers.

To quantitatively access the extent of accelerated UV radiation/ozone ageing on the modified polyester PU films, parameters such as Photo-oxidative Index (P.I) and Carbonyl index (C.I) were adopted in this study. Considering C-H stretching peak as the internal standard, P.I can be calculated according to the following formula:

$$\mathrm{P.I} = \left(\frac{A_{-\mathrm{OH},-\mathrm{NH}}}{A_{-\mathrm{CH_3}}}\right)_{\mathrm{exposed}} - \left(\frac{A_{-\mathrm{OH},-\mathrm{NH}}}{A_{-\mathrm{CH_3}}}\right)_{\mathrm{initial}}$$

where $A_{-OH,-NH}$ is the area of the absorption band associated with –OH, –NH stretching vibration (3650–3100 cm^{-1}) and A_{-CH_3} is the area of the absorption band associated with the –CH stretching band (3000–2700 cm^{-1}).

C.I can be calculated to quantitatively analyse the extent of broadening in the Carboxyl stretching peak using the following formula:

$$C.I = \left(\frac{A_{-C=O}}{A_{-CH_3}}\right)$$

where $A_{-C=O}$ is the area under the peak (1780–1600 cm^{-1}) assigned to C=O stretching vibration and A_{-CH_3} is the area of the absorption band associated with –CH stretching band (3000–2700 cm^{-1}) considered as the internal standard.

It was noticed that P.I of the TPU films increased slightly with an increase in exposure time for both UV radiation and ozone environment exposed separately. However, a considerable increase in the value of P.I was shown for a sample subjected to a combination of UV radiation plus ozone simultaneously, suggesting a synergetic ageing effect. UV light contributed by providing enough energy for TPU ageing, and ozone acts as the oxidant for inducing degradation resulting in photo-oxidation of the aromatic functional groups and direct photolytic cleavage of the urethane group producing a diquinone imide, which is attributed to the accelerated ageing in specimens subjected to a combination of UV radiation plus ozone. A similar trend was observed in C.I with the increase in exposure time, as the area of the C=O absorbance peak increases with time. This suggests the generation of new carbonyl species, which could include the formation of quinine-imides structure (yellow). The values for C.I and P.I were considerably lower for modified TPU when compared to neat TPU, indicating enhanced photo-oxidation resistance with the incorporation of the composite UV absorbers as they absorb the UV light and reduce TPU degradation (Figures 9.15–9.17).

In another study, Wang et al. [50] studied the effects of antioxidants (Antioxidant-1010) and composite stabilizers (UV-327, UV-327 and hindered amine-622) on the weather resistance behaviour of polyurethane films, by subjecting films to both UV radiation and ozone atmosphere using a self-designed UV-ozone ageing test setup.

The specimens were subjected to an ozone concentration of 100 ppm and a UV radiation intensity of 400 ± 20 $\mu W/cm^2$ at a wavelength of 313 nm at a relative humidity of 20%. The results indicate that the presence of composite stabilisers could reduce the UV transmittance considerably, and the P.I and C.I of the PU films were lower in case of the composite stabilizer and Antioxidant-10101 modified films. In comparison, the Antiodiant-1010-modified PU film showed the best performance, followed by composite stabilizer-modified PU films under UV/O_3 ageing conditions.

Atomic oxygen, abundantly available in the stratosphere, is highly reactive and has shown to exhibit intense erosion towards a wide range of polymers. Zhai and Euler [22] reported the Trition Oxygen Resistant (TOR)

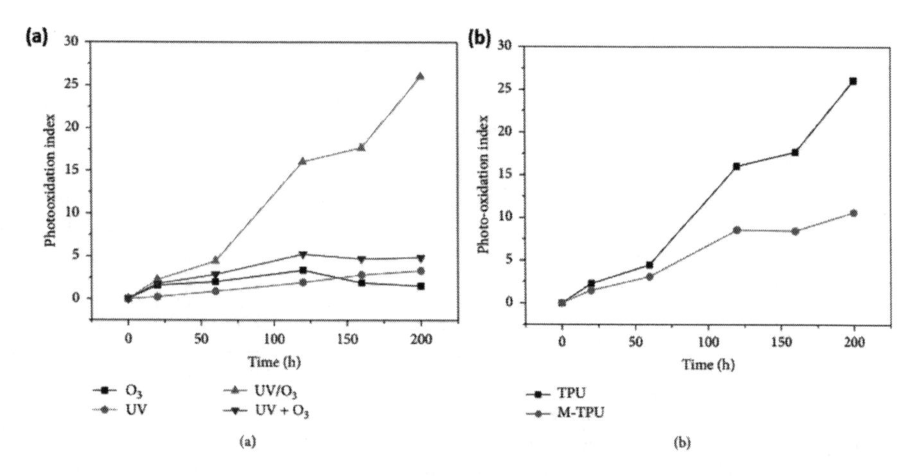

FIGURE 9.15
Photooxidation Index (a) of thermoplastic polyurethane (TPU) film subjected to different conditions and (b) of TPU and M-TPU films exposure to O_3/UV environment [49].

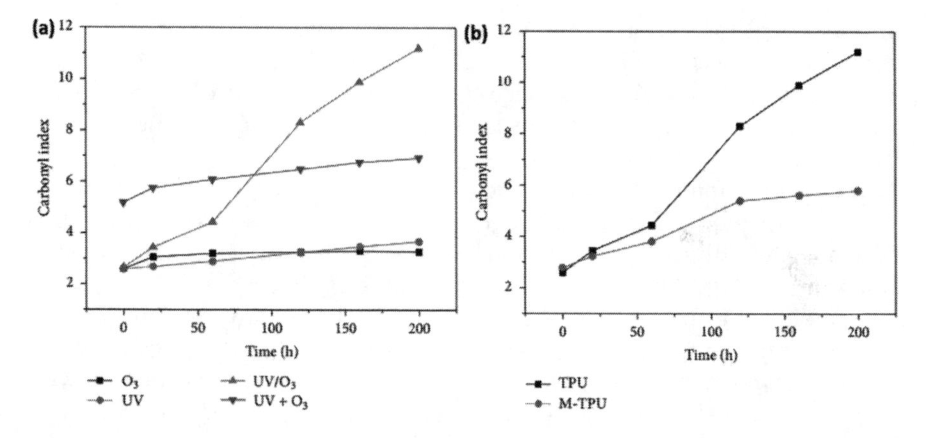

FIGURE 9.16
Carbonyl Index (a) of thermoplastic polyurethane (TPU) film subjected to different conditions and (b) of TPU and M-TPU films exposure to O_3/UV environment [49].

polymer as a potential material for use in the protection layer of high-altitude airship envelopes owing to its ability to form an outer oxidized layer that not only protects against abrasion but also shows signs of reforming if damage to the surface does not completely penetrate through the material's core. The TOR polymer has shown to provide resistance to atomic oxygen, ozone, and UV radiation, making it an ideal candidate to be used in high-altitude airships. NASA tested TOR in low earth orbit, and it showed potential to work as weather resistance films and coating in harsh environmental conditions.

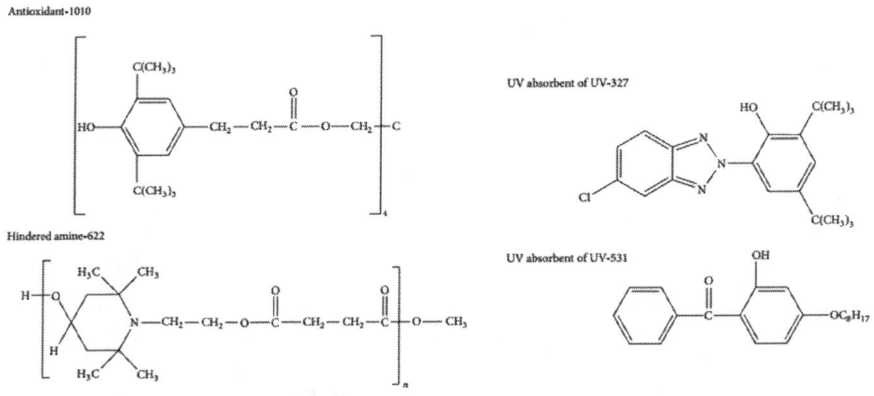

Antioxidant-1010

Hindered amine-622

UV absorbent of UV-327

UV absorbent of UV-531

FIGURE 9.17
Chemical Structures of Antioxidant-1010, UV absorbent of UV-327 and UV-531, and hindered amine-622. [50].

9.2.6 Solar Absorptivity and IR Emissivity

It is essential to study the thermal behaviour of stratospheric airship structures as a whole and the thermal response of envelope materials in particular, as the variations in the service temperatures that the LTA systems are subjected to on a daily basis induce significant challenges to the vehicle's altitude control and power consumption [51].

The thermal interactions that take place between the stratospheric airship and the surrounding are illustrated in Figure. 9.18. The external thermal fluxes include direct solar radiation, diffused solar radiation, reflected solar radiation (ground and clouds), the convective flux between wind and airship, and the infrared radiation flux emitted by the airship envelope into the surroundings. In particular, two parameters that are widely used to study the thermal characteristics of stratospheric airships are solar absorptivity (α) and IR emissivity (ε). Solar absorptivity essentially is a measure of how much radiation is absorbed by the body, and IR emissivity is the effectiveness of the material to emit thermal radiation compared to a black body. The need to control the solar absorbance of envelope structure is paramount, as it is the primary source of heat energy for the airship, and failure to do so could lead to degradation of the energy storage compartment and solar photovoltaic cells. LTA systems are generally large aircraft structures containing a gas volume of several thousand cubic metres, so even a small change in temperature between the lifting gas and the outer atmosphere can cause a substantial change in volume resulting in buoyant variations and would also compromise the structural stability of the envelope materials [52].

To measure the solar absorptivity (α), first spectral absorptance is determined as per the ASTM E903 standard [53] with the help of a spectrophotometer between 2450 and 250 nm wavelengths. The solar absorptivity is

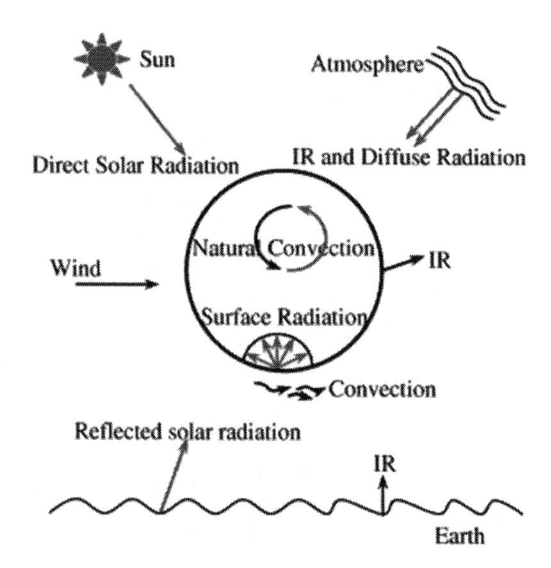

FIGURE 9.18
Thermal interaction of a stratospheric balloon with its surroundings [51].

calculated by integrating the α-values over the wavelength range and weighted against the standard spectral irradiance E(λ) taken from ASTM E490 [54]. Infrared reflectance is measured in accordance with the ASTM E408 standard with the help of a Gier-Dunkle infrared reflectometer against standard black-body surface [55]. The IR emissivity (ε) is calculated from the measured reflectances using $\varepsilon = 1 - R^*$; R^* is the average reflectance value. Ideally, envelope materials used in LTA systems need to have low absorptivity and high IR emissivity to minimize temperature fluctuation induced due to the thermal interactions between stratospheric airships and the surrounding. Factors that influence the absorptivity and emissivity of envelope material include thickness, surface preparation, coating formulation and manufacturing technique. Few studies that concentrated on studying the thermal characteristics of LTA systems include Fulton et al. [56], in their article on developing thermal control coatings for high-altitude airships suggested the use of silver for its low absorptivity and high reflectance for minimal coating thicknesses and SiO_2 as an emittance coating material for obtaining optimal thermal performance of the envelope system. Zhai and Euler [22], in their article on material challenges for LTA systems compared different material combination such as Quartz over silver, White Tedlar, and Silvered Tedlar with their lab developed metallic finished sample. It was found that the thermal performance of the lab specimen was better than the other three in terms of the ratio of absorptivity to emissivity, and the White Tedlar gave the highest value indicating that the presence of a metallic backing is more efficient than just a white non-metallic surface.

9.3 Weatherability and Polymer Degradation

'Weathering is the adverse response of a material or product to the environment, often causing unwanted and premature product failure' [57]. It causes irreversible structural and mechanical changes in the polymeric material and directly affects the durability. In practice, the three main environmental factors are solar radiation (UV light), heat (temperature), and humidity (water) contributing towards the degradation of laminated and coated envelope material of the LTA system during their service life. These factors, in combination with secondary effects such as airborne, pollutant, microbial, and acid rain, act together to cause 'weathering' [58].

9.3.1 Factors Responsible for Weathering

9.3.1.1 UV Radiation

Solar radiation that reaches the earth's surface consists of wavelengths ranging from 295 to 3000 nanometres (nm). The terrestrial sunlight is commonly separated into three main wavelength ranges: ultraviolet (UV), visible (VIS), and infrared (IR). Wavenumbers between 295 and 400 nm are the ultraviolet (UV) portion of the solar spectrum, making up around 4–7% of the total radiation. Ozone in the stratosphere absorbs and essentially eradicates all UV radiations below 295 nm, but this amount is considered negligible by most of the experts [59]. UV rays are the most harmful radiation in terrestrial sunlight, which can initiate degradation processes in polymeric materials. For a polymer subjected to UV radiation, photodegradation is the primary destructive mechanism. UV radiation causes photo-oxidative ageing, which results in the breakage of polymer chains, produces free radicals, and reduces the molecular weight of polymers, resulting in a loss of surface gloss and the significant deterioration of many material properties with exposure time [60, 61].

9.3.1.2 Temperature

The temperature of materials exposed to solar radiation impacts the effect of the radiation. Photochemical reactions are usually accelerated at higher temperatures. Besides, temperature determines the rate of subsequent reaction steps. A number of factors control the temperature of a material exposed to natural sunlight. For example, specimen surface temperature is a function of ambient temperature, solar irradiation, specimen solar absorptivity, and surface conductance. Therefore, in the presence of sunlight, the surface temperature of a material is generally higher than the temperature of the air. Solar absorptivity in both the visible and infrared regions is closely related to colour, varying from about 20% for a white surface to over 90% for the black surfaces; thus, materials of different colours will reach different temperatures on exposure. This surface temperature dependency on colour can have a secondary effect on material as well.

9.3.1.3 *Water (Moisture)*

All materials used for outdoor applications are exposed to moisture. The moisture absorption by synthetic materials and coatings from humidity and direct wetness is an example of physical effects. As the surface layer absorbs moisture, volume expansion is produced that places stress on the dry layers of the subsurface. Rain, which periodically washes dirt and pollutants from the surface, has an effect on the long-term rate of deterioration that is determined more by its occurrence than its volume. For instance, it makes a great difference if a certain amount of rain falls on material for a few minutes than a sudden shower or the type of sprinkle lasting several hours. The depth of penetration into the material, and thus the influence on the weathering behaviour, is more prominent in the second case than in the first.

Water can also be directly involved in the degradation reaction from a chemical perspective. The 'chalking' of titanium dioxide (TiO_2) in pigmented coatings and polymers is a good example. Contact with water in any phase can accelerate the rate of oxidation. Moisture may also act as a pH adjuster, particularly when considering the effect of acid rain, which may cause etching of many paints and coatings.

9.3.1.4 *Secondary Effects*

The secondary effects of the atmosphere or weather that may cause degradation cannot be underestimated. Gases and pollutants in the atmosphere, especially in the form of acid rain, may cause entirely new reactions. In highly industrialized areas, acid rain is the primary element driving the weathering process that affects a wide range of materials. Blowing dirt and dust may have effects on the weathering process without reacting with the actual molecular structure of the material.

9.3.2 Effect of Environmental Factors on Critical Material Properties

All these environmental factors act in combination to degrade materials. The influence of these factors is determined by three main testing techniques; natural weathering, accelerated natural weathering, and accelerated artificial weathering. Each technique has its specific application along with the pros and cons. There is no simple answer that which is the better technique to test. Depending on the application, either method can be effective. All the weathering tests are based on regular monitoring and characteristic of the aforementioned weathering factors.

Natural weathering: This process is also called static weathering or an outdoor direct exposure method. It involves placing samples on inclined racks oriented toward the sun. Sites used for this type of testing are usually in tropical areas, as high temperatures, UV intensity, and humidity are needed for maximum degradation. Often the benchmark sites are Miami, Florida, and Arizona, which possesses all three characteristics. This process is the most authentic way of testing because it is a real-world and real-time process. Despite harsh conditions, testing takes several years before significant results are achieved.

Accelerated natural weathering: To speed up the weathering process, accelerated natural testing can be applied. It involves direct exposure to the sun; various devices are designed specifically to simulate and accelerate the sunlight exposure. One method uses mirrors and lenses to amplify available UV radiation. A black box under glass exposures is used to test materials intended for interior automotive conditions. Maximum acceleration of the ageing process is obtained by exposure on a Fresnel reflecting concentrator panel rack, which uses photo-receptor cells to maintain alignment with the sun and ten mirrors to reflect sunlight onto the test specimens. A variety of environmental chambers are also used in association with industry standards.

Laboratory accelerated weathering: This process has been used for a long time with increasing importance due to the development of more weatherable materials and the need to determine in a short time the effects of natural exposures over prolonged periods. The process is accelerated through the use of specially designed weathering chambers. Gas-discharge lamps (e.g. xenon arc lamps), electric arc (carbon), or fluorescent lamps are used to simulate/accelerate the effects of sunlight. These devices have the unique requirement of accelerating as well as simulating the effect of the natural environment.

9.3.3 Natural Weathering

Natural weathering is also known as direct or outdoor weathering, which is typically accomplished by mounting the material on some type of exposure racks (Figure 9.19), made from anodized aluminium and having it face the sun directly at a fixed angle. The typical sample size is 15 x 30 cm, and the common angles of exposure are near horizontal (usually 5°), 45° and vertical (90°). Direct weathering tests follow ISO 877 (accelerated natural weathering for plastics), ISO 2810 (natural weathering of coatings), and ISO 10-B03 (effect of natural weathering on textiles).

FIGURE 9.19
Natural weathering test racks [57].

Thermoplastic polyurethane (TPU) is versatile polymeric material, often used as a weather-resistant coating polymer in the design of inflatable structural materials. However, TPU films/coating, like other polymeric materials, are also susceptible to degradation when exposed to aggressive environments (e.g. UV radiation). Recently, the photo-oxidative degradation of different thermoplastic polyurethane, which includes such processes as photo-oxidation, chain scission, crosslinking, and secondary reactions, occurring by free radical mechanisms, were well discussed in a recently published review by Fengwei Xie et al. [62].

The service life of any ultimate inflatable structure is determined primarily by the protective layer. Thermoplastic polyurethane (TPU) and its derivative have been exploited as a protective layer due to its stability against UV radiation [63]. In this context, Pal et al., a research group in ADRDE, India, studied the degradation behaviour of PU-coated nylon fabric upon exposure to the direct outdoor environment [64]. Coated fabric samples developed for an aerostat envelope were mounted on custom-designed frames directly exposed to sunlight. Degradation data in terms of tensile and gas barrier properties were evaluated at every three-month interval for three years. It was observed that PU-coated nylon fabric was sufficiently resistant to outdoor weathering of one year (10.5% loss). However, its strength fell rapidly thereafter, and the strength loss after 20 months was 36.5% (Figure 9.20). The maximum strength loss in winter suggests that moisture plays a major role

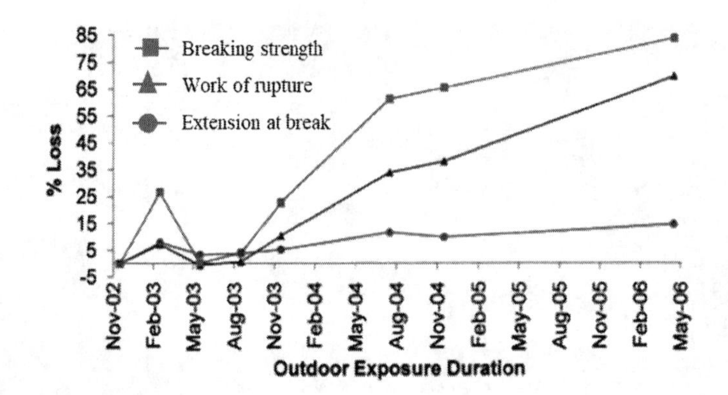

FIGURE 9.20
Effect of outdoor exposure on mechanical properties of TPU coated nylon fabric [64].

in the degradation of the material compared to sunlight. After some critical time exposure, a decrease in strength was proportional to the time of exposure. Even though the strength loss was linear with time, the hydrogen gas permeability increased drastically on prolonged outdoor exposure beyond a critical duration. This suggests that gas permeability should be taken as a critical factor in determining the life of the aerostat system.

The severity of natural weathering depends on the climate type, Shi et al. [65] studied six different natural exposure sites with distinct climate types, including Qionghai (typical hot-humid climate), Sansha (island hot-humid climate), Chennai (savanna hot-humid climate), Jeddah (xerothermic climate), Sanary-sur-Mer (Mediterranean climate), and Hoek van Holland (warm-temperate climate). Results showed that Sansha had the highest severity value while Hoek van Holland had the lowest. As shown in Figure 9.21, the extent of degradation degree at Qionghai, Sansha, and Chennai was higher than that at other sites, resulting in a deterioration of the optical properties

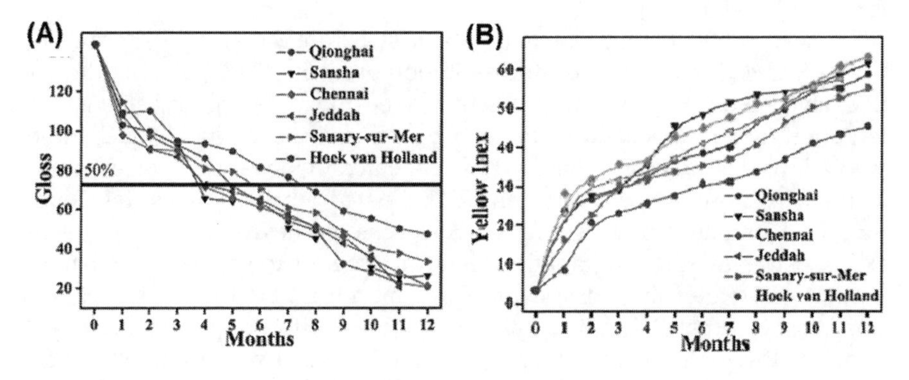

FIGURE 9.21
Optical properties evaluation of polystyrene samples: (A) loss in gloss; (B) Yellow index [65].

and severe damage to the sample surface. In the xerothermic climate, the optical properties decreased drastically during the process of natural weathering, while only slight change in hydroxyl index and carbonyl index was observed, and the micrographic surface did not differ much from the surface of un-weathered materials.

9.3.4 Accelerated Natural Weathering

In spite of the optimization of weathering factors by site selection, type of backing, exposure angle, and follow-up of a long period of outdoor testing to establish the durability of materials, it is possible to further accelerate natural weathering tests by using a specially designed device or chamber to simulate the sunlight exposure; for example, black box chambers for higher panel temperatures and longer wetness time than open or backed exposures and Fresnel reflectors and concentrators for increased irradiance. *Black box glass exposures* are explicitly designed to stimulate and accelerate the effect of the interior automotive condition. Air temperature in such a device may exceed 80°C under condition of outside ambient air temperature and solar irradiance. A layer of glass filters out the most damaging portion of sunlight, i.e., the short wavelength UV. It also protects the specimen from direct rainfall, pollutants, and dust [66]. The maximum acceleration of natural weathering is obtained by exposure to Fresnel reflectors and concentrator devices such as EMMM (developed by Atlas) and Q-trac (developed by Q-lab). These devices use ten flat mirrors to concentrate the natural sunlight onto the sample at an intensity of approximately eight times that of global daylight and about five times the global radiation in the UV portion of the spectrum (Figure 9.22). Because the test method exposes samples to the full spectrum of natural concentrated sunlight, it is one of the most realistic accelerated weathering tests available. The parameters of the test device are governed by ISO 877 performing for plastic materials [67] and ASTM G90 for Nonmetallic Materials [68]. A good correlation between Fresnel-reflector exposures and standard outdoor tests varies based on the stability ranking of polymeric materials [69].

Fresnel reflecting systems have been used to study the weathering resistance of different polymer mortars to harsh environments. In an experiment, specimens were exposed to the natural and accelerated environment during one year at a maritime site in the middle of Portugal. Acceleration factors, based on a one-year time scale, were determined and employed to predict the flexural strength of the materials under a long-term outdoor exposure at a natural maritime site [70]. A Q-trac sunlight concentrator testing device was used to study the impact of different weathering environments [71]. Samples were exposed for six months in Arizona, parameters for this exposure method were set according to the ASTM G90 standard test method 'Standard Practice for Performing Accelerated Outdoor Weathering of Nonmetallic Materials

FIGURE 9.22
Fresnel reflecting solar concentrators. (Atlas Material Testing Technology).

Using Concentrated Natural Sunlight' was used. This device is equipped with ten mirrors enabling the concentration and reflection of the total solar spectrum. Samples were sprayed with water three times a day for eight minutes. When expressing colour changes in terms of length of exposure, and/or quantity of MJ UV/m^2 received by sample, Arizona exposure caused the most severe degradation of both paints (Figure 9.23). These conditions are more severe than those encountered during normal end uses: one day of exposure in Arizona is more degrading than one day in a QUV or Florida or Quebec.

9.3.5 Laboratory Accelerated Weathering

Natural weathering exposure is a variable function of the exposure site and the actual ever-changing weathering conditions due to the day cycle, seasonal, and year-to-year variability. Natural conditions have too many uncontrolled and unpredictable variables that make analysis difficult. Thus, the focus has shifted to artificial weathering where the exposure conditions are maintained in a controlled manner. In artificial weathering, acceleration can be achieved via two different approaches: by enhancing exposure parameters' overuse level conditions and the time-lapse approach that repeats episodes with higher effect at shorter frequency. Due to the time constraints, it is highly desirable to investigate short-term data of about 50 responses at accelerated levels and analyse the data effectively, leading to the most accurate forecast of the lifetime of a material.

Xenon arc laboratory weathering experiments have been performed with different filter combinations to study the effect of the light source and its intensity on the weathering of various engineering thermoplastics [72]. The effects of irradiance level and dark cycle were determined in order to establish the

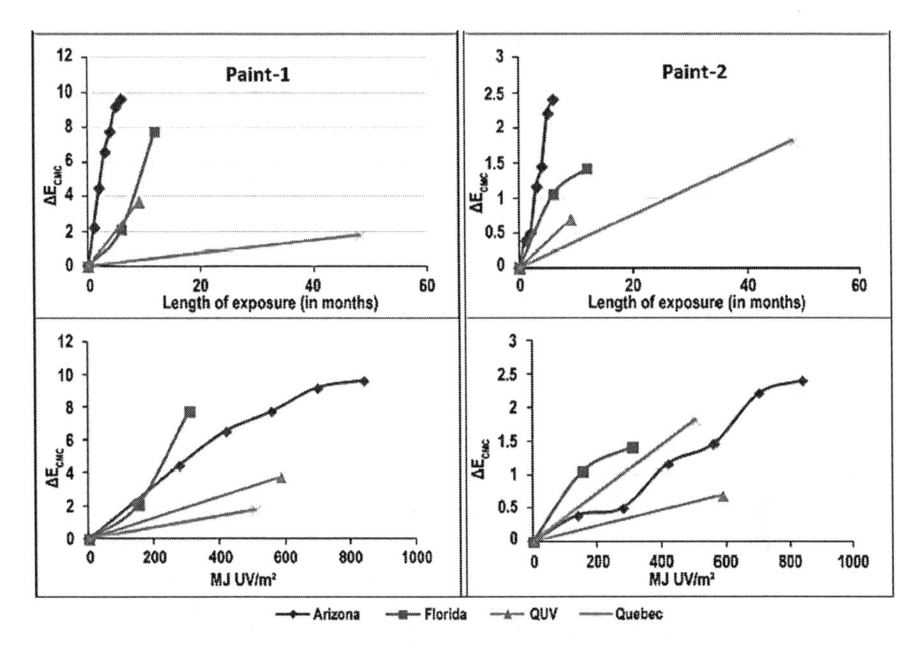

FIGURE 9.23
ΔE_{CMC} (total colour measurement change) parameters for different paints [71].

legitimacy of an accelerated testing method. Various factors such as the effect of spectral power distribution, comparison of a light source, the effect of irradiance or dose rate, and the effect of light/dark cycle were studied. The spectral power distributions of the common artificial UV sources are shown in Figure 9.24 along with a reasonable representation of Florida sunlight [72]. Xenon arcs can be filtered using a variety of materials; common filters are quartz, type S borosilicate, soda-lime glass, and CIRA (quartz with an IR reflecting coating).

Laboratory weatherometer has also been extensively used to simulate UV radiation to terrestrial sunlight [73, 74]. In a systematic way, thermoplastic polyurethane (TPU) material exposed to artificial weathering conditions for the different duration was studied by Boubakri et al. [75]. The UV lamp was characterized by irradiance and power of 3 mW/cm^2 at 340 nm and 125 W, respectively. The retained exposure durations were 3, 6, 12, 72, and 144 hours. Weathering results showed the degradation behaviour of TPU films, which initially have a colourless appearance, become yellower in the first stage of ageing (after 6 h), and then turn to brown (after 72 h) and thereafter remain almost unchanged (Figure 9.25). SEM photographs revealed the formation of microcracks on UV-exposed surfaces, especially at long exposure duration. DSC analysis showed that the glass transition temperature (Tg) decreased in the beginning and then increased with increasing UV-exposure time, revealing an increase in crosslink density.

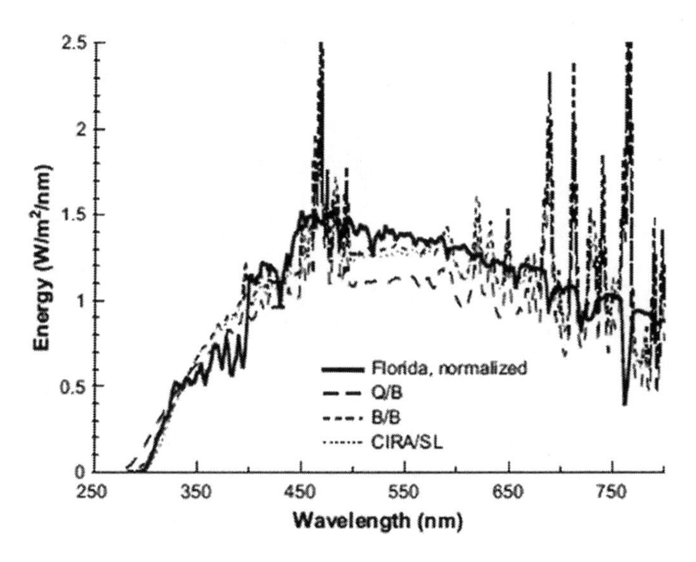

FIGURE 9.24
Spectral power distribution (SPD) of xenon arc lamps equipped with various filter combinations: quartz inner/borosilicate outer (Q/B), borosilicate inner and outer (B/B), IR-reflecting quartz inner/soda-lime outer (CIRA/SL). Adapted from [72].

An approach to predict the in-service weatherability of a model aerostat envelope was proposed by Chatterjee et al. [76] by correlating natural and artificial weathering data. A reliability model with two stress types (UV radiation and temperature) was used to predict the service life of a model aerostat envelope material concerning the gas permeability through the material. Artificial weathering tests were performed in a Xenotest Weather-o-meter. Exposures were carried out at three irradiance levels (i.e., 40, 50, and 60 W/m^2) and three levels of temperature (45°C, 55°C, and 65°C). All other parameters were maintained as per ISO 159 4892 standards. Testing was conducted up to 300 h, and measurements were done after every 100 h. In order to compare the artificial weathering data, natural weathering exposures were also conducted according to the ASTM G7 standard. The predicted mean service life for the loss in gas impermeability of the model coating at the assumed use level and failure criterion was found as 607 h and warranty time as 590 h. The mean life and the 90% warranty time obtained from the natural weathering of the same coating were 642 h and 561 h, respectively. Both experimental results were found to be in good agreement, which validated the model.

9.3.6 Surface Degradation Study in Natural and Artificial Weathering

When any material is exposed to the sun, its properties begin to deteriorate. The continuous exposure in natural or artificial conditions may potentially

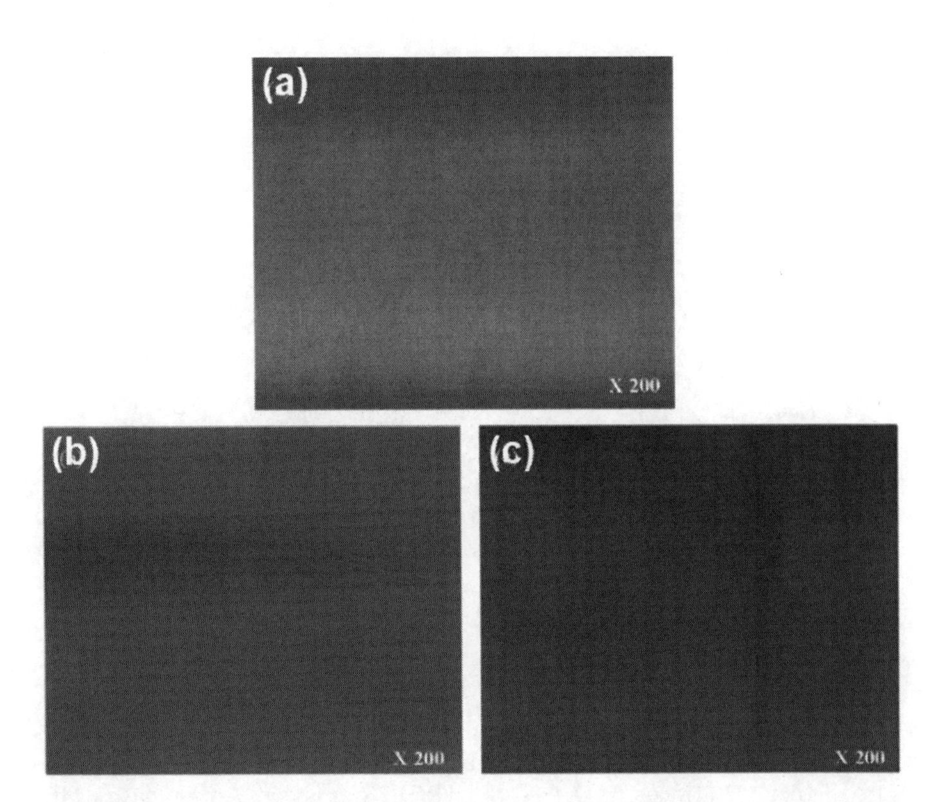

FIGURE 9.25
Colour modifications of UV-aged TPU versus exposure time: (a) without ageing; (b) 6 h; and (c) 72 h [75].

harm the material surface. The surface degradation initiated by harmful ultraviolet (UV) radiation further propagate in the presence of environmental influences (e.g., oxygen, temperature, and humidity), capable of breaking the chemical bonds present in the material; hence, the degradation starts at the surface. With continued exposure, the surface properties of the material continue to degrade. The change in appearance, optical property, chemical structure, and surface microstructure can be assessed primarily by scanning electron microscopy (SEM), atomic force microscopy (AFM) and Fourier transformed infrared spectroscopy (FTIR) during surface degradation.

9.3.6.1 Surface Degradation Study by FTIR

Chemical changes and surface degradation of polyurethane (PU) coatings under laboratory accelerated weathering conditions has been evaluated by FTIR analysis [77]. Tests were performed in the QUV weather-o-meter (from Q-Panel), fluorescent UV lamps (UVA) with a peak wavelength of 340 nm

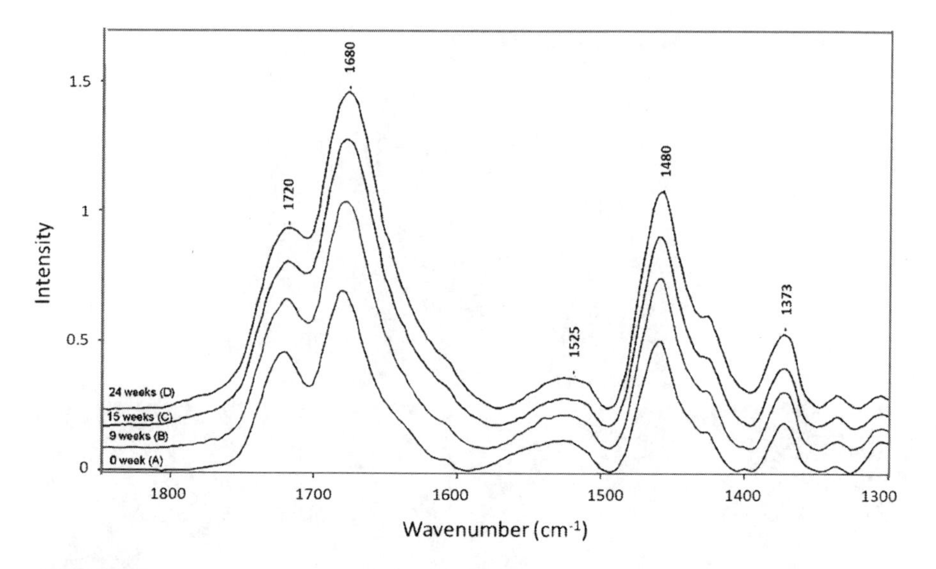

FIGURE 9.26
The diagram illustrates ATR FTIR spectroscopic analysis in the 1850–1300 cm^{-1} regions, which shows the functional group changes at 1680 and 1720 cm^{-1}. Adapted from [77].

were mounted. FTIR analysis was utilized to monitor the chemical changes in high-gloss polyurethane topcoat at 0, 9, 15, and 20 weeks of QUV exposure. Figure 9.26 traces A–D reveals ATR FTIR analysis. An increase of the 1680 cm^{-1} band linked to –NH–(C=O)–NH– of polyurea stretching was observed. The increase of the 1680 cm^{-1} band indicates that the formation of polyurea [78] increased as exposure time was increased. Not only a reduction of polyurethane components as indicated by the decrease in the 1720 cm^{-1} (polyurethane C=O stretching) band but also a reduction of the C–H and C–O stretching at ATR and FTIR results illustrated that coating degradation caused the urea groups to increase on top surfaces, while the proportion of urethane formation increased as depth into the coating increased.

Traces A–D in Figure 9.27 supported the formation of polyurea (N–H) stretching with the increase of the 3370 cm^{-1} band. Figure 9.27 clearly showed that, as QUV exposure time increases, the N–H stretching at 3370 cm^{-1} becomes broader. Another interesting feature was the appearance of the band at 2870 cm^{-1} attributed to C–H_2 stretching. This band appeared to increase after 24 weeks of exposure. As exposure time was increased, the coating appeared to undergo some molecular conformational changes as indicated by the increase of the 1740 and 1728 cm^{-1} bands together with the increase of both the 1680 and 1691 cm^{-1} stretching modes. Upon exposure to QUV, chain scissions occurred not only in the soft segments of polyurethane upon the QUV exposure, but chain scissions and re-association also took place in the hard segments. It is also possible that upon exposure, surface species

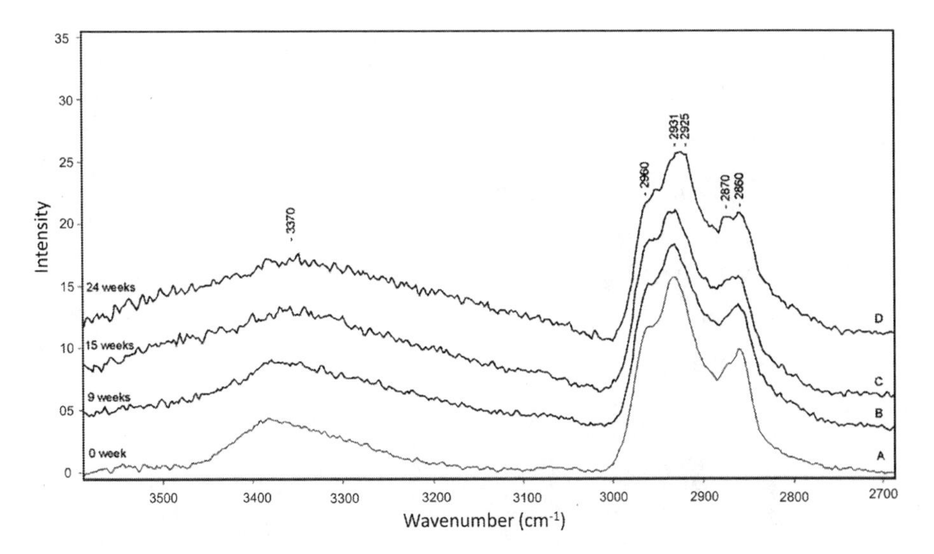

FIGURE 9.27
The diagram illustrates ATR FTIR spectroscopic analysis in the range of 2700–3600 cm^{-1} regions are depicted. The appearance of 2870 cm^{-1} is observed as well as an increased or broader N–H band at 3370 cm^{-1}. Adapted from [77].

were oxidized leading to the formation of carbonyl groups, particularly the formation of polyurea. This particular behaviour was documented by Kim and Urban [78].

Furthermore, the degradation of methylene diphenyl diisocyanate (MDI)-based polyether and polyester PU film upon UV exposure was studied [79]. For this purpose, PU cast films with different thicknesses were prepared and assessed by FTIR analysis to investigate chemical changes during the UV treatment as well as the penetration depth of the UV radiation. FTIR results are shown in Figure 9.28. The sample side labelled as 'front' is the side which was irradiated directly while the side labelled as 'back' was opposite to the radiation. After an exposure of 50 hours, no remarkable changes between the front and backside were found. After 200 h exposure, three regions were observed with noticeable changes. Region I was attributed to changes of the NH– (3325 cm^{-1}) and CH– valence bands. This decrease was attributed by several authors to oxidation of the CH_2 bridge and might result in a formation of quinone structures responsible for the observed yellowing of the samples [80, 81]. The changes in region II (C=O bands, 1680–1730 cm^{-1}) were less pronounced. These bands are attributed to ester structures, which seem to be unaffected by the UV radiation. A more significant difference between the front and backside was visible in region III, which characterizes C–O–C bands (1080–1105 cm^{-1}). However, these changes can be detected only in the spectra of the ether-based materials. Figure 9.28(c–d) shows the FTIR spectra

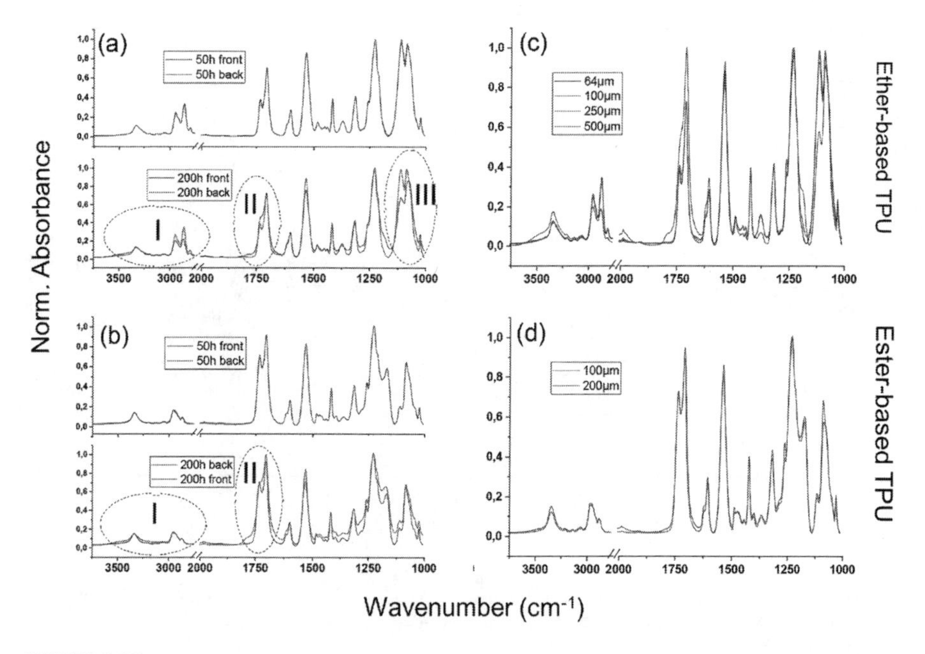

FIGURE 9.28
(Left) ATR-infrared spectra of 250 μm ether- (top) and 200 μm ester-based (bottom) PU cast films after 50 and 200 h UV exposure recorded on the front and backside of cast films; (right) of films with different thicknesses after 200 h (back side only). Spectra were normalized (peak at 1225 cm^{-1}) [79].

of the backside of cast films with different thicknesses after 200 h UV exposure were compared. It was noticed that only the ether-based sample with a thickness of 64 μm showed changes in regions I–III (Figure 9.28I), whereas thicker film samples did not show any variation. These results indicated that the effect of a 200 h UV exposure is limited to thicknesses lower than 100 μm. The results of the UV exposure of the ester-based materials (Figure 9.28(d)) support these findings.

9.3.6.2 Surface Degradation Study by SEM and AFM

Surface morphology studies such as scanning electron microscopy (SEM) and atomic force microscopy (AFM) can be effective tools in analysing the impact of weathering on the envelope materials used in aerostat and airship applications. AFM can give semi-quantitative and qualitative information about the polymer degradation and ageing of the materials under exposure to UV radiation. Polymer films or coating generally prior to degradation have smooth and relatively flat surfaces. Upon being subjected to the harsh conditions such as exposure to UV radiation, atomic oxygen, ozone, and large temperature fluctuations, these polymer films are subjected to degradation, and the surface tends to become rougher, the erosion and dissolution of the degradation products create micro-cracks and pores which tend to increase the surface roughness. Scanning electron microscopy is an effective visualization tool used to study the surface morphology of polymer films and coating. Owing to its large depth of focus, high lateral resolution, and capacity for X-ray microanalysis, it delivers consistent images of polymer surfaces and can be highly effective in studying the degradation of envelope materials subjected to weathering.

The surface degradation of polyurethane coating under different laboratory-accelerated weathering environment was studied [82]. An epoxy primer with a high gloss polyurethane topcoat coating system was exposed either only in a QUV chamber or in a QUV chamber and a Prohesion chamber, alternating exposure. Figure 9.29(a–c) shows the surface morphology of the coating before and after exposure. After 27 weeks of QUV/Prohesion alternating exposure, the uppermost layer of the coating was still coherent, with little evidence of pigment exposure on the surface (Figure 9.29(b)). However, after 24 weeks of QUV exposure, the coating outer layer was significantly

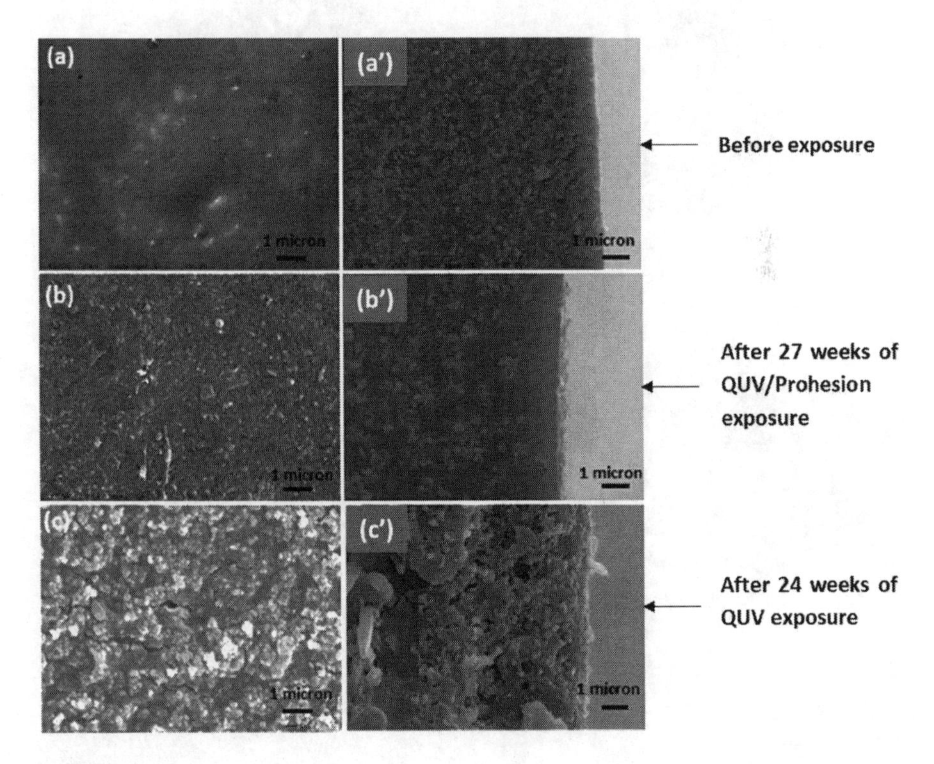

FIGURE 9.29
SEM images of surface and cross-section view of coating before and after exposure [82].

damaged, with pigments exposing on the coating surface (Figure 9.29(c)). It was concluded that QUV exposure damaged the coating surface faster and more seriously.

Furthermore, AFM analysis revealed that blisters formed on the coating surface after two weeks of QUV exposures (Figure 9.30). Under the QUV exposure, these blisters increased in size during the first nine weeks, and pigments in the coating bulk were exposed to the surface after the coating surface was damaged. Therefore, some spikes were shown in Figure 9.31(d) are most likely pigment particles, not blisters. However, under the QUV/ Prohesion alternating exposure, blisters did not significantly change their size over the first 15 weeks of exposure, and sharp features associated with exposed pigment particles were not observed on the coating surface (Figure 9.31). A comparison of Figures 9.30 and 9.31 indicates that the development and the breakage of those blisters caused the damage of the coating surface layer.

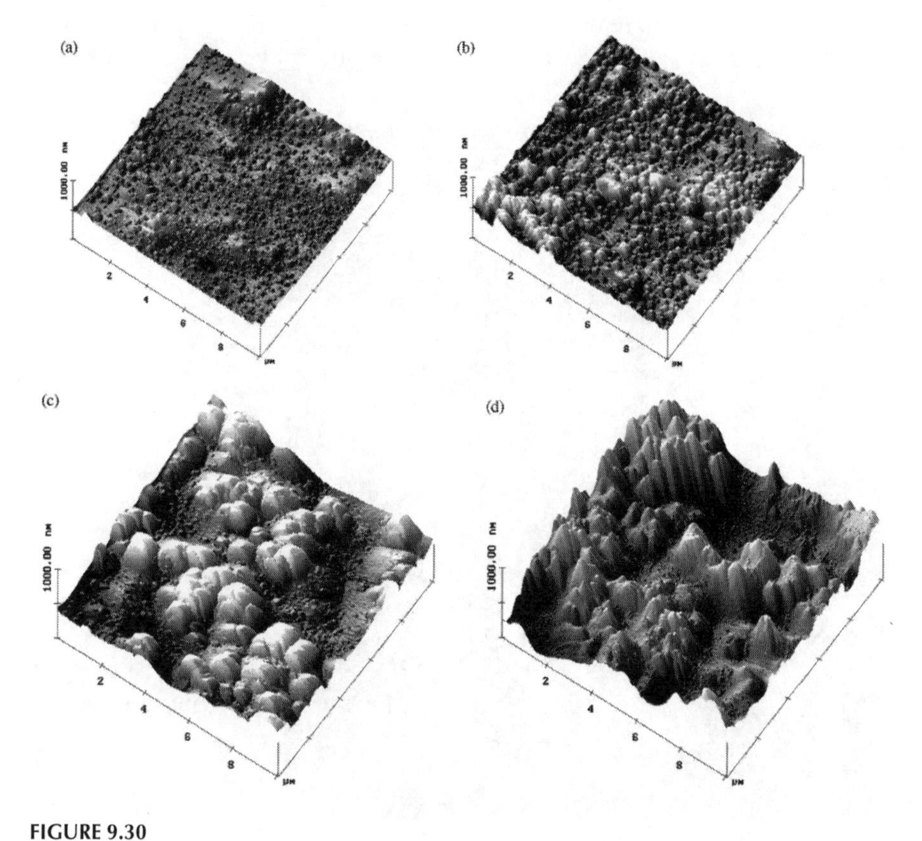

FIGURE 9.30
Ten micron AFM height images showing the topography change of the high gloss polyurethane topcoat under QUV exposure: (a) 2; (b) 5; (c) 9; and (d) 18 weeks' exposure [82].

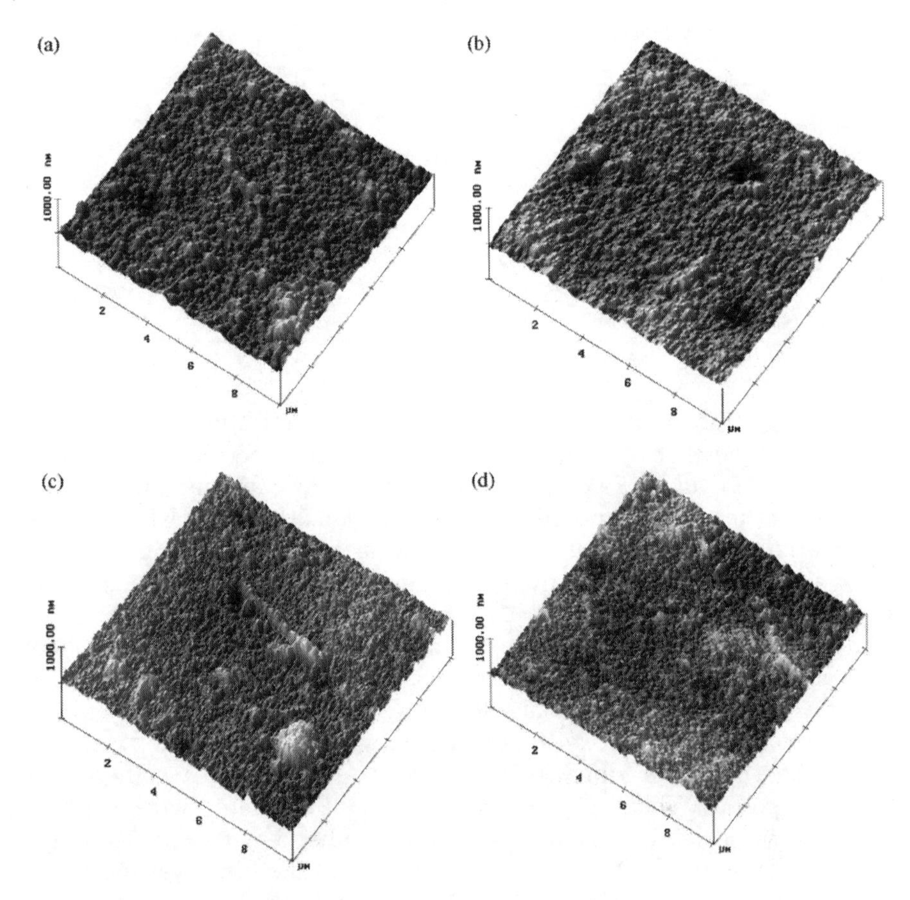

FIGURE 9.31
Ten micron AFM height images showing the topography change of the high gloss polyurethane topcoat under QUV/Prohesion alternating exposure: (a) 6; (b) 9; (c) 12; and (d) 15 weeks' exposure [82].

To analyse the effect of ageing on the surface of TPU films for the application in high-altitude airship envelope materials, artificial weathering was performed using ATLAS Ci3000+ Xenon arc weather-o-meter for 480 hours under the combined influence of irradiance, wind, and temperature [83]. SEM analysis revealed that the unexposed film had a relatively smooth surface without any pores or irregularities; and upon 480 hours of exposure, a considerable amount of micro-cracks, blisters, and voids were detected, indicating the specimen had turned comparatively rough due to photodegradation. Furthermore, the fracture topography (AFM analysis) of the tensile specimens before and after weathering showed the unaged film's fracture surface remained relatively ductile, while the exposed film

TABLE 9.7

Average Surface Roughness (Ra) of Unaged and Aged TPU Film

Irradiation time (h)	R_a (nm)
0	7.72
48	8.91
120	9.97
240	10.81
360	12.43
480	17.01

Adapted from [83].

had a broad fracture path with several tear lines indicating that exposure to artificial weathering reduced the plasticity and increased the brittleness of the TPU films. To quantify the effect of the irradiation, the surface roughness (Ra) of the exposed film was calculated from the AFM images, summarized in Table 9.7. It is noticed that as the time of aging increases the roughness (Ra) and many straight pillars with different height are observed (Table 9.7).

In the last few years, research has been carried out to improve the weather resistance properties of the protective layer of envelope material of LTA systems by using nanomaterials. Recently, Adak et al. [84] explored the possibility of using hydroxyl functionalized graphene to develop polyurethane nanocomposite film with enhanced gas barrier and weather resistance properties. The neat polyurethane and PGN films were exposed under artificial weathering conditions in a weather-o-meter. Change in the surface morphology before and after exposure of the films was evaluated by SEM analysis shown in Figure 9.32, the unexposed film (Figure 9.32(a)) showed a relatively smooth surface with no irregularities whereas the PU film after 300 hours exposure had micro-cracks indicating the occurrence of photo-oxidation under the influence of UV radiation and moisture in the weatherometer (Figure 9.32(c)). Graphene added film even after 300 hours of accelerated artificial weathering showed just a few patches and blisters (Figure 9.32(d–f)), unlike the micro-cracks and pores that were visible in the exposed neat PU film, showing improved weather resistance properties of the functionalized graphene PU nanocomposite films due to the intrinsic UV absorption capability of graphene. Nuraje et al. reported a similar finding by studying the weathering properties of graphene-reinforced PU coatings [85]. The UV chamber in QUV from Q-Panel was used for accelerated weathering tests. It was concluded that the addition of graphene improved protection against environmental factors like UV degradation and corrosion. The PU coating containing graphene nanoflakes also provides hydrophobicity and improved mechanical properties to coated PU.

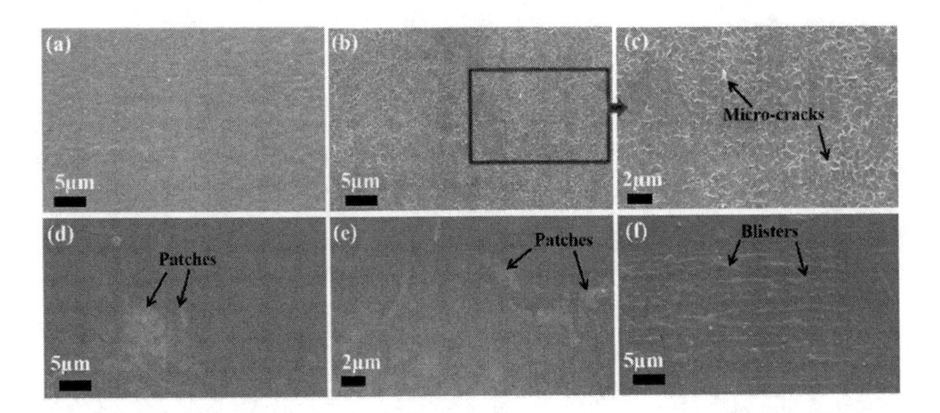

FIGURE 9.32
SEM images of the film surface before and after weathering; (a) unexposed neat PU film; (b, c) artificial weather-exposed film surfaces for neat PU; (d, e) PU 2 wt % graphene; and (f) PG 3 wt % graphene. Adapted from [84].

9.4 Conclusion

The revived interest in LTA systems such as airships and aerostats for applications in broadcasting, telecommunication, weather monitoring, etc., has bought about many advancements in the testing and evaluation of the materials used to construct them. The envelope material must possess properties such as low gas permeability, lightweight, high strength, and UV and ozone resistance. Thus, there exists a challenge to develop materials for LTA systems that can provide the right balance of these diverse properties, and stringent testing and evaluation of these materials is a crucial task to achieve the same. It is also of utmost importance to carry out proper testing in accordance to ASTM/ISO standards to ensure the safety of the payload, be it sophisticated sensors and cameras or human lives on board these LTA systems. This chapter presented an exhaustive review of the basic principle and testing methodology of traditional tests such as gas permeability, uniaxial tensile, tear and seam strength test. In addition to this, advanced testing procedures such as biaxial tensile testing, ozone and atomic oxygen resistance, and absorptivity and emissivity determination studies have also been examined. Special attention has been given to weatherability – both natural and artificial, and surface degradation studies using characterization techniques such as FTIR, SEM, and AFM. The incorporation of state-of-the-art non-destructive testing along with simulation and mathematical model will enable researchers to better evaluate materials leading to the development of more advanced LTA systems.

References

1. Stockbridge, C., A. Ceruti, and P. Marzocca. 2012. "Airship Research and Development in the Areas of Design, Structures, Dynamics and Energy Systems." *International Journal of Aeronautical and Space Sciences* 13 (2): 170–187. doi:10.5139/IJASS.2012.13.2.170.
2. Kumar, A., S. C. Sati, and A. K. Ghosh. 2016. "Design, Testing, and Realisation of a Medium Size Aerostat Envelope." *Defence Science Journal* 66 (2): 93–99. doi:10.14429/dsj.66.9291.
3. Miller, T., F. De, and M. Mandel. 2000. "Airship Envelopes: Requirements, Materials and Test Methods Abstract: Current Airships All Employ the Pressure Envelope Design Principle. Thus the Envelope Must Be Considered as a Main Structural Element of the Airship. This at the Beginning of Devel," 1–5. Available at: https://docplayer.net/15357262-Airship-envelopes-requirements-materials-and-test-methods.html
4. Aglan, H., M. Calhoun, and L. Allie. 2008. "Effect of UV and Hygrothermal Aging on the Mechanical Performance of Polyurethane Elastomers." *Journal of Applied Polymer Science* 108 (1): 558–564. doi:10.1002/app.27049.
5. ASTM. 2008. "Standard Test Method for Breaking Force and Elongation of Textile Fabrics (Strip Method)." ASTM D 5035-06. www.astm.org.
6. Hu, J., Y. Li, W. Chen, T. Zhang, C. Gao, T. Shi, and D. Yang. 2018. "Uniaxial Mechanical Properties of Multi-Layer Thin Films in Use for Scientific Balloons." *Advances in Space Research*, 62 (5): 1165–1176. doi:10.1016/j.asr.2018.06.003.
7. Komatsu, K., M. Sano, and Y. Kakuta. 2003. "Development of High-Specific-Strength Envelope Materials." In *AIAA's 3rd Annual Aviation Technology, Integration, and Operations (ATIO) Forum*. Reston, Virginia: American Institute of Aeronautics and Astronautics. doi:10.2514/6.2003-6765.
8. Kang, W., Y. Suh, K. Woo, and I. Lee. 2006. "Mechanical Property Characterization of Film-Fabric Laminate for Stratospheric Airship Envelope." *Composite Structures* 75 (1–4): 151–155. doi:10.1016/j.compstruct.2006.04.060.
9. Meng, J., M. Lv, D. Tan, and P. Li. 2016. "Mechanical Properties of Woven Fabric Composite for Stratospheric Airship Envelope Based on Stochastic Simulation." *Journal of Reinforced Plastics and Composites* 35 (19): 1434–1443. doi:10.1177/0731684416652947.
10. Chen, J., W. Chen, and D. Zhang. 2014. "Experimental Study on Uniaxial and Biaxial Tensile Properties of Coated Fabric for Airship Envelopes." *Journal of Reinforced Plastics and Composites* 33 (7): 630–647. doi:10.1177/0731684413515540.
11. Liao, L., and I. Pasternak. 2009. "A Review of Airship Structural Research and Development." *Progress in Aerospace Sciences* 45 (4–5): 83–96. doi:10.1016/j.paerosci.2009.03.001.
12. Vasudevan, D., H. Mohan, S. Gaggar, and R. B. Gupta. 2019. "Analysis and Retardation of Helium Permeation for High Altitude Airship Envelope Material." *AIP Conference Proceedings* 2148 (September). doi:10.1063/1.5123949.

13. Pechar, T. W., S. Kim, B. Vaughan, E. Marand, M. Tsapatsis, H. K. Jeong, and C. J. Cornelius. 2006. "Fabrication and Characterization of Polyimide-Zeolite L Mixed Matrix Membranes for Gas Separations." *Journal of Membrane Science* 277 (1–2): 195–202. doi:10.1016/j.memsci.2005.10.029.
14. Gorrasi, G., M. Tortora, V. Vittoria, E. Pollet, B. Lepoittevin, M. Alexandre, and P. Dubois. 2003. "Vapor Barrier Properties of Polycaprolactone Montmorillonite Nanocomposites: Effect of Clay Dispersion." *Polymer* 44 (8): 2271–2279. doi:10.1016/S0032-3861(03)00108-3.
15. George, S. C., and S. Thomas. 2001. "Transport Phenomena through Polymeric Systems." *Progress in Polymer Science (Oxford)* 26 (6): 985–1017. doi:10.1016/S0079-6700(00)00036-8.
16. Joshi, M., A. Bapan, and B. S. Butola. 2018. "Polyurethane Nanocomposite Based Gas Barrier Films, Membranes and Coatings: A Review on Synthesis, Characterization and Potential Applications." *Progress in Materials Science* 97 (May): 230–282. doi:10.1016/j.pmatsci.2018.05.001.
17. Siracusa, V.. 2012. "Food Packaging Permeability Behaviour: A Report." *International Journal of Polymer Science* 2012 (i). doi:10.1155/2012/302029.
18. Duncan, B., J. Urquhart, and S. Roberts. 2005. "Review of Measurement and Modelling of Permeation and Diffusion in Polymers." *Engineering*, (January), p. 73. Available at: http://resource.npl.co.uk/materials/polyproc/iag/october2005/depc_mpr_012.pdf
19. American Society for Testing and Material–ASTM. 2009. *ASTM D 1434-82: Standard Test Method for Determining Gas Permeability Characteristics of Plastic Film and Sheeting*. West Conshohocken: ASTM. www.astm.org.
20. Adak, B., M. Joshi, and B. S. Butola. 2018. "Polyurethane/Clay Nanocomposites with Improved Helium Gas Barrier and Mechanical Properties: Direct versus Master-Batch Melt Mixing Route." *Journal of Applied Polymer Science*, 135: 46422. doi: 10.1002/app.46422.
21. Islam, S., and P. Bradley. 2012. "Materials." In *Airship Technology*, pp. 113–148. 10.1016/B978-0-08-044951-7.50029-4.
22. Zhai, H., and A. Euler. 2005. "*Material Challenges for Lighter-than-Air Systems in High Altitude Applications*." In *AIAA 5th Aviation, Technology, Integration, and Operations Conference (ATIO)*. pp. 1–12. Available at: http://arc.aiaa.org/doi/pdf/10.2514/6.2005-7488.
23. Mokwena, K.K., and J. Tang. 2012. "Ethylene Vinyl Alcohol: A Review of Barrier Properties for Packaging Shelf Stable Foods." *Critical Reviews in Food Science and Nutrition*, 52 (7): 640–650. Available at: http://www.ncbi.nlm.nih.gov/pubmed/22530715.
24. Maekawa, S., M. Nakadate, and A. Takegaki. 2007. "Structures of the Low-Altitude Stationary Flight Test Vehicle." *Journal of Aircraft*, 44(2): 662–666.
25. Williams, J. G., and P. D. Ewing. 1972. "Fracture under Complex Stress – the Angled Crack Problem." *The International Journal of Fracture*, 8: 441–446.
26. Ueda, Y. 1983. "Characteristics of Brittle Fracture under General Combined Modes Including Those under Bi-Axial Tensile Loads." *Engineering Fracture Mechanics*, 6: 1131–1158.

27. Chu, C. Y., and F. Chen. 1992. "Tearing Failure Mechanism of Woven Fabrics and Comparison Between Tear Test Methods Text." *Research Journal*, 13: 196–200.
28. Minami, H. 1978. "Strength of Coated Fabrics with Crack." *Journal of Industrial Textiles*, 7: 269–292.
29. Chen, J. W., and W. J. Chen. 2016. "Central Crack Tearing Testing of Laminated Fabrics Uretek 3216LV under Uniaxial and Biaxial Static Tensile Loads." *The Journal of Materials in Civil Engineering*, 28 (7): 04016028. doi:10.1061/(ASCE) MT.1943-5533.0001537.
30. Maekawa, S., K. Shibasaki, T. Kurose, T. Maeda, Y. Sasaki, and T. Yoshino. 2008. "Tear Propagation of a High-Performance Airship Envelope Material." *Journal of Aircraft*, 45 (5): 1546–1553. doi:10.2514/1.32264.
31. Krook, C. M., & Fox, K. R. 1945. "Study of the Tongue-Tear Test." *Textile Research*, 15(11): 389–396. doi:10.1177/004051754501501102.
32. Hager, O. B., D. D. Gagliardi, and H. B. Walker. 1947. "Analysis of Tear Strength." *Textile Research Journal*, 17(7): 376–381. doi:10.1177/004051754701700704.
33. Singh, G., Bairwa, D. K., Chattopadhyay, R., Joshi, M., & Setua, D. K. (n.d.). *Investigation on influence of time on tear behaviour of airship envelope*. https://doi.org/10.1177/1528083720907704.
34. Topping, A. D. 1973. The Critical Slit Length of Pressurized Coated Fabric Cylinders. *Journal of Coated Fabrics*, 3(2), 96–110. https://doi.org/10.1177/152808377300300203.
35. Wang, F. X., W. Xu, Y. L. Chen, and G. Y. Fu. 2016 "Tearing Analysis of a New Airship Envelope Material under Uniaxial Tensile Load." *IOP Conf. Ser.: Mater. Sci. Eng*. 137 (1): 1–12.
36. Chen, J., W. Chen, H. Zhou, B. Zhao, M. Wang, W. Sun, and S. He. 2017. "Fracture Failure Analysis and Bias Tearing Strength Criterion for a Laminated Fabric." *Journal of Industrial Textiles*, 47 (7): 1496–1527.
37. Yang, Y., P. Zeng, and L. Lei. 2015. "Numerical Simulation Study of the T-Peel Behavior of Coated Fabric Films Used in Inflatable Structures." *Journal of Applied Polymer Science*, 132 (3): 1–9. doi:10.1002/app.41299.
38. Shi, T., W. Chen, C. Gao, J. Hu, B. Zhao, Xu. Wang, Xi. Wang, and G. Lu. 2018. "Investigation of Mechanical Behavior of Weld Seams of Composite Envelopes in Airship Structures." *Composite Structures*, 2018 (May): 1–12. doi:10.1016/j.compstruct.2018.06.019.
39. ASTM D5035-11. 2019. *Standard Test Method for Breaking Force and Elongation of Textile Fabrics (Strip Method)*."ASTM International, West Conshohocken, PA, www.astm.org
40. ASTM D4851-07 (2019)e1. 2015. "Standard Test Methods for Coated and Laminated Fabrics for Architectural Use." ASTM International, West Conshohocken, PA, www.astm.org
41. ASTM D1876-08(2015)e1. 2015. "Standard Test Method for Peel Resistance of Adhesives (T-Peel Test)." ASTM International, West Conshohocken, PA, www.astm.org
42. Petković, G., M. Vukoje, J. Bota, and S. P. Preprotić. 2019. "Enhancement of Polyvinyl Acetate (PVAc) Adhesion Performance by SiO2 and TiO2 Nanoparticles." *Coatings*, 9 (11): 1–17. doi:10.3390/coatings9110707.

43. Tang, W., and M. Keefe. 2003. "Stress Analysis and Structural Modifications of a Polyester Double-Tape Seam." *Proceedings of the Institution of Mechanical Engineers Part L: Journal of Materials: Design and Applications*, 217 (2): 101–112. doi:10.1243/146442003321673608.
44. Theller, H. W. 1989. "Heatsealability of Flexible Web Materials in Hot-Bar Sealing Applications." *Journal of Plastic Film & Sheeting*, 5 (1): 66–93.
45. Hollande, S., L. Jean-Louis and L. Thierry. 1998. "High-Frequency Welding of an Industrial Thermoplastic Polyurethane Elastomer-Coated Fabric. *Polymer*, 39: 5343–5349. doi:10.1016/j.ijadhadh.2013.09.019
46. Johnson, E. C., D. N. Patel, C. J. Panetta, O. Esquivel, and Y. M. Kim. 2013 "*NDE of Lap Seams in Composite Laminate Fabric Structures.*" In *Proceedings of the 13th International Symposium on Nondestructive Characterization of Materials*, Le Mans, France, pp. 20–24. https://www.ndt.net/article/ndcm2013/content/papers/45_Johnson_Rev1.pdf
47. ASTM D1149-18. 2018. "Standard Test Methods for Rubber Deterioration—Cracking in an Ozone Controlled Environment." ASTM International, West Conshohocken, PA, www.astm.org.
48. Maekawa, S., T. Maeda, Y. Sasaki, and T. Kitada. 2005, "*Development of Advanced Lightweight Envelope Materials for Stratospheric Platform Airship," 43rd Aircraft Symposium, Aeronautical and Space Sciences Japan, Japan Society for Aeronautical and Space Sciences Paper 2005-1B12*, pp. 120–124.
49. Wang, Y., H. Wang, X. Li, D. Liu, Y. Jiang, and Z. Sun. 2013. "O_3/UV Synergistic Aging of Polyester Polyurethane Film Modified by Composite UV Absorber." *Journal of Nanomaterials*, 2013, doi:10.1155/2013/169405.
50. Wang, H., Y. Wang, D. Liu, Z. Sun, and Wang, H. 2014. "Effects of Additives on Weather-Resistance Properties of Polyurethane Films Exposed to Ultraviolet Radiation and Ozone Atmosphere. *Journal of Nanomaterials*, 2014, doi:10.1155/2014/487343.
51. Xia, X. L., D. F. Li, C. Sun, and L. M. Ruan. 2010. "Transient Thermal Behavior of Stratospheric Balloons at Float Conditions." *Advances in Space Research*, 46 (9): 1184–1190. doi:10.1016/j.asr.2010.06.016.
52. Yao, W., X. Lu, C. Wang, and R. Ma. 2014. "A Heat Transient Model for the Thermal Behavior Prediction of Stratospheric Airships." *Applied Thermal Engineering*, 70 (1): 380–387. doi:10.1016/j.applthermaleng.2014.05.050.
53. ASTM E903-12. "Standard Test Method for Solar Absorptance, Reflectance, and Transmittance of Materials Using Integrating Spheres." ASTM International, West Conshohocken, PA, 2012, www.astm.org
54. ASTM E490-00a. "Standard Solar Constant and Zero Air Mass Solar Spectral Irradiance Tables." ASTM International, West Conshohocken, PA, 2019, www.astm.org.
55. ASTMe E408-13. 2019. "Standard Test Methods for Total Normal Emittance of Surfaces Using Inspection-Meter Techniques." ASTM International, West Conshohocken, PA, www.astm.org.
56. Fulton, M. L., M. A. Zimmerman, and R. S. Dummer. 2013. "*Volume Roll-to-Roll Thermal Control Coatings for High Altitude Airship Applications." Optics InfoBase Conference Papers*, 5–7.

57. ATLAS Inc, USA. 2001. *Weathering Testing Guidebook*. ATLAS, Weathering Services Group.
58. Hamid, S. H., M. B. Amin, and A. G. Maadhah. 1992. "*Handbook of Polymer Degradation*." https://www.bookdepository.com/Handbook-Polymer-Degradation-S-Halim-Hamid/9780824786717
59. CIE. 2020. "CIE 85- Solar Spectral Irradiance (1st Edition)." (CIE), International Commission on Illumination. Accessed February 28. https://standards.globalspec.com/std/443758/CIE 85
60. Lu, T., E. Solis-Ramos, Y. Yi, and M. Kumosa. 2018. "UV Degradation Model for Polymers and Polymer Matrix Composites." *Polymer Degradation and Stability* 154: 203–210. doi:10.1016/j.polymdegradstab.2018.06.004.
61. Yousif, E., and R. Haddad. 2013. "Photodegradation and Photostabilization of Polymers, Especially Polystyrene: Review." *SpringerPlus*, 2 (1): 1–32. doi:10.1186/2193-1801-2-398.
62. Xie, F., T. Zhang, P. Bryant, V. Kurusingal, J. M. Colwell, and B. Laycock. 2019. "Degradation and Stabilization of Polyurethane Elastomers." *Progress in Polymer Science*, 90: 211–268. https://doi.org/10.1016/j.progpolymsci.2018.12.003.
63. Raza, W., G. Singh, S. B. Kumar, and V. B. Thakare. 2016. "Challenges in Design and Development of Envelope." *International Journal of Textile and Fashion Technology (IJTFT)*, 6 (2): 27–40.
64. Pal, S. K., V. B. Thakare, G. Singh, and M. K. Verma. 2011. "Effect of Outdoor Exposure and Accelerated Ageing on Textile Materials Used in Aerostat and Aircraft Arrester Barrier Nets." *Indian Journal of Fibre & Textile Research*, 36.
65. Shi, Y., J. Qin, Y. Tao, G. Jie, and J. Wang. 2019. "Natural Weathering Severity of Typical Coastal Environment on Polystyrene: Experiment and Modeling." *Polymer Testing*, 76 (July): 138–145. doi:10.1016/j.polymertesting.2019.03.018.
66. Quill, J. 2001. "*The Essentials of Laboratory Weathering Q-Lab Corporation*." Q-Lab Corporation. Westlake, USA.
67. ISO. 2008. "Methods of Exposure to Solar Radiation – Part 3: Intensified Weathering Using Concentrated Solar Radiation." ISO 877-3. https://www.iso.org/standard/67566.html.
68. ASTM. 2016. "Standard Practice for Performing Accelerated Outdoor Weathering of Nonmetallic Materials Using Concentrated Natural Sunlight." ASTM G90. https://www.astm.org/DATABASE.CART/HISTORICAL/G90-10.htm.
69. Andrady, A. L., ed. 2003. *Plastics and the Environment*. Hoboken, NJ, USA: John Wiley & Sons, Inc. doi:10.1002/0471721557.
70. Ribeiro, M. C. S., A. J. M. Ferreira, and A. T. Marques. 2009. "Effect of Natural and Artificial Weathering on the Long-Term Flexural Performance of Polymer Mortars." *Mechanics of Composite Materials*, 45 (5): 515–526. doi:10.1007/s11029-009-9104-7.
71. Cogulet, Antoine, Pierre Blanchet, and Véronic Landry. 2019. "Evaluation of the Impacts of Four Weathering Methods on Two Acrylic Paints: Showcasing Distinctions and Particularities." *Coatings*, 9 (2): 121. doi:10.3390/COATINGS9020121.
72. Pickett, J. E., D. A. Gibson, and M. M. Gardner. 2008. "Effects of Irradiation Conditions on the Weathering of Engineering Thermoplastics." *Polymer Degradation and Stability*, 93 (8), 1597–1606. doi:10.1016/j.polymdegradstab.2008.02.009.

73. Makki, H., K. N. S. Adema, E. A. J. F. Peters, J. Laven, L. G. J. Van Der Ven, R. A. T. M. Van Benthem, and G. De With. 2014. "A Simulation Approach to Study Photo-Degradation Processes of Polymeric Coatings." *Polymer Degradation and Stability,* 105 (1): 68–79. doi:10.1016/j.polymdegradstab.2014.03.040.
74. Chatterjee, U., B. S. Butola, and M. Joshi. 2016. "Optimal Designing of Polyurethane-Based Nanocomposite System for Aerostat Envelope." *Journal of Applied Polymer Science,* 133 (24): 1–9. doi:10.1002/app.43529.
75. Boubakri, A., N. Guermazi, K. Elleuch, and H. F. Ayedi. 2010. "Study of UV-Aging of Thermoplastic Polyurethane Material." *Materials Science and Engineering A,* 527 (7–8): 1649–1654. doi:10.1016/j.msea.2010.01.014.
76. Chatterjee, U., S. Patra, S. B. Bhupendra, and M. Joshi. 2017. "A Systematic Approach on Service Life Prediction of a Model Aerostat Envelope." *Polymer Testing,* 60 (July), 18–29. doi:10.1016/j.polymertesting.2016.10.004.
77. Yang, X. F., C. Vang, D. E. Tallman, G. P. Bierwagen, S. G. Croll, and S. Rohlik. 2001. "Weathering Degradation of a Polyurethane Coating." *Polymer Degradation and Stability,* 74 (2): 341–351. doi:10.1016/S0141-3910(01)00166-5.
78. Kim, H., and M. W. Urban. 2000. "Molecular Level Chain Scission Mechanisms of Epoxy and Urethane Polymeric Films Exposed to UV/H2O. Multidimensional Spectroscopic Studies." *Langmuir,* 16 (12): 5382–5390. doi:10.1021/la990619i.
79. Scholz, P., V. Wachtendorf, U. Panne, and S. M. Weidner. 2019. "Degradation of MDI-Based Polyether and Polyester-Polyurethanes in Various Environments - Effects on Molecular Mass and Crosslinking." *Polymer Testing,* 77 (May): 105881. doi:10.1016/j.polymertesting.2019.04.028.
80. Theiler, G., V. Wachtendorf, A. Elert, and S. Weidner. 2018. "Effects of UV Radiation on the Friction Behavior of Thermoplastic Polyurethanes." *Polymer Testing,* 70 (September): 467–473. doi:10.1016/j.polymertesting.2018.08.006.
81. Schollenberger, C. S., and K. Dinbergs. 1961. "A Study of the Weathering of an Elastomeric Polyurethane." *Polymer Engineering and Science,* 1 (1): 31–39. doi:10.1002/pen.760010108.
82. Yang, X. F., D. E. Tallman, G. P. Bierwagen, S. G. Croll, and S. Rohlik. 2002. "Blistering and Degradation of Polyurethane Coatings under Different Accelerated Weathering Tests." *Polymer Degradation and Stability,* 77 (1): 103–109. doi:10.1016/S0141-3910(02)00085-X.
83. Liu, Yuyan, Yuxi Liu, S. Liu, and H. Tan. 2014 "Effect of Accelerated Xenon Lamp Aging on the Mechanical Properties and Structure of Thermoplastic Polyurethane for Stratospheric Airship Envelope." *Science and Technology on Advanced Composites in Special.* doi:10.1007/s11595-014-1080-7.
84. Adak, B., M. Joshi, and B. S. Butola. 2019. "Polyurethane/Functionalized-Graphene Nanocomposite Films with Enhanced Weather Resistance and Gas Barrier Properties." *Composites Part B: Engineering,* 176 (November): 107303. doi:10.1016/j.compositesb.2019.107303.
85. Nuraje, N., S. I. Khan, H. Misak, R. Asmatulu, H. Jafari, and J. I. Velasco. 2013. "The Addition of Graphene to Polymer Coatings for Improved Weathering." *ISRN Polymer Science.* doi:10.1155/2013/514617.

10

Modelling for Performance Analysis of Aerostats/Airships

Bapan Adak
Kusumgar Corporates Pvt Ltd, Gujarat, India

S. Parasuram, Mangala Joshi and B. S. Butola
Indian Institute of Technology, New Delhi, India

CONTENTS

10.1 Introduction

In recent years, there has been a renewed interest in lighter-than-air (LTA) vehicles, primarily aerostats and airships, which are aerodynamic systems

DOI: 10.1201/9780429432996-10

that attain their static lift by using a lighter-than-air or buoyant gas such as hydrogen or helium. Aerostats/airships have a broad spectrum of potential applications. In addition to their conventional usage in military surveillance, homeland security, etc., these LTA systems are finding enormous applications in scientific and civilian applications such as traffic management, telecommunication relay, disaster relief, weather forecasting, and monitoring of air pollution and composition over cities. The development of materials for LTA systems, especially airships, which operate in the stratosphere, is a complex task, as it requires a combination of properties that one single material cannot deliver. The material system needs to be lightweight, capable of maintaining a high gas barrier property and flexibility at low temperatures, and be resistant to environmental degradation (mainly intense UV radiation, temperature, humidity, etc.). Generally, composite fabric structures are used to accomplish this, where the fabric acts as the strength layer, and the multilayered coating or lamination set up accounts for the other properties [1]. In general, evaluation of performance of materials used in aerostats and airships through experimental studies is a cost-intensive and time-consuming process which also requires a wide variety of sophisticated instruments. Thus, there exists a need to incorporate more simulations and mathematical models in studies to evaluate the materials used for applications in aerostats/airships. The polymeric materials used in the weather protection and gas retention layers are seldom single materials, and in most cases, additives such as UV absorbers, UV reflectors, light stabilizers, and nanomaterials are incorporated in the polymer to enhance the weather resistance and gas barrier properties, and models are used in deciding optimal compositions of polymer-additive-based films and coating. Evaluation of the service life of these materials through field studies takes a prolonged time, and natural weathering consists of many uncontrolled parameters which make the task rather complicated. Thus, researchers opt for artificial weathering, where the exposure condition can be accurately monitored, and develop models to correlate the results from artificial weathering to predict the service life of these LTA vehicles [2, 3]. In this chapter, a brief summary of (i) challenges in designing models for LTA systems, (ii) models used for predicting thermal performance of aerostats and airships, (iii) simulations and models for predicting the mechanical performance of an aerostat or airship envelope, and (iv) conventional, relativistic models used to predict the service life of aerostat and airship envelopes is presented.

10.2 Challenges in Designing Models for LTA Systems

Although the modelling is very important for analysing the performance of LTA systems, there are many challenges in designing respective models. It is

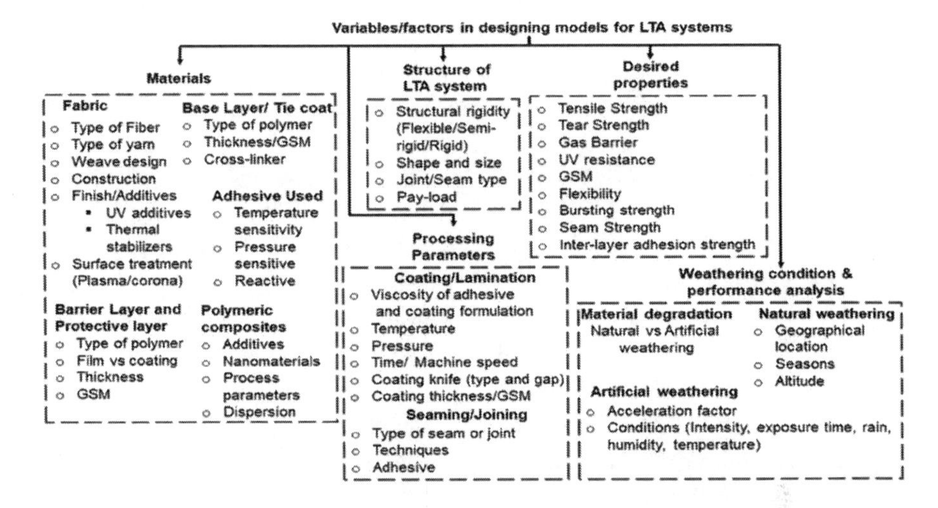

FIGURE 10.1
Different factors which are to be considered in designing models for analysing the performance of LTA system.

because of the large number of factors/variables which are to be considered for designing the models, as summarised in Figure 10.1. Among these, the six main broad factors are related to (i) materials, (ii) processing parameters, (iii) desired properties, (iv) structure of LTA systems, and (v) weathering conditions and performance analysis.

10.2.1 Variables in Materials

There is huge variability in the raw material itself, which introduces difficulty in developing models for LTA systems. The main raw materials for the envelope of an LTA system are fabric, polymer (in the form of film or coat), and adhesive (in case of lamination). The properties of the fabric, as well as the final structure (coated/laminated), depends on many factors, such as weave, type of yarn, yarn count, yarn strength, thread density, cover factor, and surface treatment. Polymers are used for coating or lamination to improve certain functionality of the fabric or to retain the properties of the whole structure throughout its service life. The morphology and properties (physical as well as chemical) of polymers such as density, glass transition temperature, crystallinity, reactivity (based on functional groups), and hardness control the functionality improvement of fabric and properties of the coated/laminated structure. In case of lamination, adhesives which are also of different types (mainly thermoplastic polyurethane, reactive polyurethane, and acrylate type) play the main role in controlling the bonding of different layers in laminates and maintain the integrity of the system.

10.2.2 Variables in Processing Parameters

There are various processing parameters at different stages of processing of LTA envelope materials, which are summarized as follows:

1. Wet processing of fabric (variables: washing and scouring recipe, process time, temperature, pH)
2. Surface treatment of fabric (variables: intensity of corona or plasma treatment, time)
3. Mixing of polymer and additives (variables: screw speed, temperature, residence time, additive type and concentration, polymer)
4. Coating (variables: technique, machine speed, coating thickness/weight, viscosity of coating formulation, polymer, fabric)
5. Lamination (variables: technique, adhesive type, viscosity of adhesive, film, fabric)
6. Sealing or joining (variables: type of joint, technique, RF, temperature)

10.2.3 Desired Properties for the Envelope of an LTA System

The desired properties for envelope materials for different LTA systems have been discussed in detail in Chapter 1. The performance of these LTA systems depends on these properties (tensile strength, tear strength, weight, flexibility, weather resistance, gas barrier, creep, seam strength, and many more). Because of the large number of property requirements, it is very difficult to consider all these properties during designing of models for performance analysis of LTA systems. However, researchers have developed many models which describe performance analysis of LTA system in terms of a specific set of properties such as tensile properties [4–7], tearing properties [8, 9], etc.

10.2.4 Variables Regarding the Structure of LTA Systems

There are again a large number of structural parameters which are also needed to be considered during modelling for LTA systems. These are mainly: (i) flexibility and rigidity of the structure (rigid, semi-rigid and flexible system), (ii) size, (iii) shape, (iv) joint type (lap joint, single butt joint, double butt joint), and (v) payload.

10.2.5 Variation in Weathering Conditions

A long service life is always desired for any LTA system. However, the service life analysis by exposing under natural weathering is a very time-consuming process. Therefore, researchers have adopted some numerical approaches for estimation of service life of an LTA system. Here, the materials are exposed under accelerated artificial weathering conditions and the data of short-term exposure are correlated with long-term natural exposure by some mathematical models. However, these models are empirical, and a particular model is applicable to only a specific material. Moreover, the model may not be valid for repetition of natural weathering. The reason is that, though the weathering conditions remain unchanged for artificial weathering, the conditions in natural weathering vary depending on geographical location, seasons, and altitude. Hence, service life prediction of LTA systems using the data generated by artificial weathering and correlating it with the natural or actual weathering is a very challenging task.

10.3 Different Models Related to LTA Systems

Although developing models for performance analysis of LTA systems is a challenging task considering all these factors, researchers are putting continuous efforts for establishing new models.

10.3.1 Models for Predicting Thermal Performance of Aerostats and Airships

In recent decades, many researchers have focused on prediction of thermal performance of LTA systems such as balloons, aerostats, and airships mainly by using the lumped method. For example, Stefan [10] developed an analytical model for describing the internal and external heat transfer processes for a hot air balloon with respect to fuel consumption. Later, the study was extended to the high-altitude airship by dividing it in two parts (top and bottom), and the thermal parameters were calculated by taking an average [11]. In 1983, Carlson and Horn [12] proposed a new model for estimating the thermal and trajectory behaviour of high-altitude airships. In this model, the radiative emission and absorption properties of lifting gas and also the daytime gas temperature above the balloon film were considered. The model predicted data was compared with flight data. Shi et al. [13] used the Runge–Kutta method for developing a thermal model for predicting thermal performance of stratospheric airships considering the airship control style, flight state, and working environment.

Anderson et al. [14] and Colonius et al. [15] developed some computational models for analysing the natural buoyant convections inside and around the high-altitude balloon caused by solar heating. However, the radiative load of the actual environment was not considered in these studies. On the contrary, Franco and Cathey [16] analysed the thermal performance of NASA's (National Aeronautics and Space Administration, USA) spherical scientific balloon in floating condition without considering the convection effect.

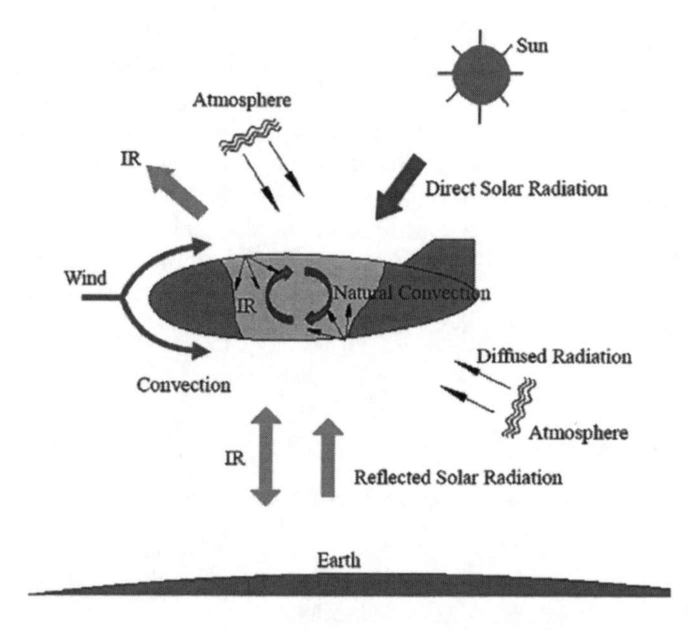

FIGURE 10.2
Thermal environment of an aerostat or airship [19]. Reprinted with permission from Elsevier.

Xia et al. [17] numerically investigated the temperature distribution around a stratospheric balloon based on a three-dimensional transient thermal model considering the effect of different radiations (direct and reflected solar radiation, IR and diffuse radiation, surface radiation) and convective heat transfer coefficient. Wang and Yang [18] proposed a new approach (steady simulation and unsteady simulation) for analysing the thermal distribution around a stratospheric airship considering natural convection, solar radiative flux, and earth inferred radiative flux. The simulation result was in good accord with the experimental results. In another study, they improved the model further by using the control volume method for analysing the transient thermal performance of airships under different environmental conditions (Figure 10.2). The location-wise infrared radiation variation, the solar irradiative heat flux, the convection both inside and outside the airship, and also the convective velocity were considered in this simulation and there was a good agreement with experimental results [19].

10.3.2 Models for Deciding Optimal Composition of a Polymer/Additive-Based Film or Coating

In recent days, researchers are incorporating different fillers (micro/nano) and additives to enhance some specific properties (gas barrier, weather resistance, mechanical, etc.) of a polymer before using it in any particular layer of the LTA envelope. Sometimes, these additives/fillers are used in combination in the same polymer matrix and other times in different layers of polymeric materials.

However, there is scant literature on optimization of the composition of different fillers/additives to obtain best performance. Recently, Chatterjee et al. [20] studied the effect of two different nanofillers (nanoclay and graphene) as a gas barrier property enhancer and also organic UV additives (Tinuvin B75 which is a mixture of UV absorber, HALS, and antioxidant in a ratio of 2:2:1) for improving the weather resistance property of thermoplastic polyurethane (TPU). These fillers/additives were incorporated separately in the TPU matrix in different concentrations, and these resulting formulations were coated on a polyester fabric in varying combinations. It was observed that nanoclay and graphene increased the helium gas barrier property, and the loss in properties after 100 h exposure to accelerated weathering was less in the presence of UV stabilizers. Design Expert software was used to optimise the concentration of different fillers/additives by the desirability function approach. A '9-run' design mixture was used for this purpose, as shown in Figure 10.3. An optimum result (minimum loss in UPF value and gas barrier property) was obtained with the formulation containing 1.36 wt % nanoclay (Cloisite 30B), 3.03 wt % graphene, and 0.61 wt % organic UV stabilizer, showing minimal degradation of surface coating and indicating a good potential to be used in the aerostat hull.

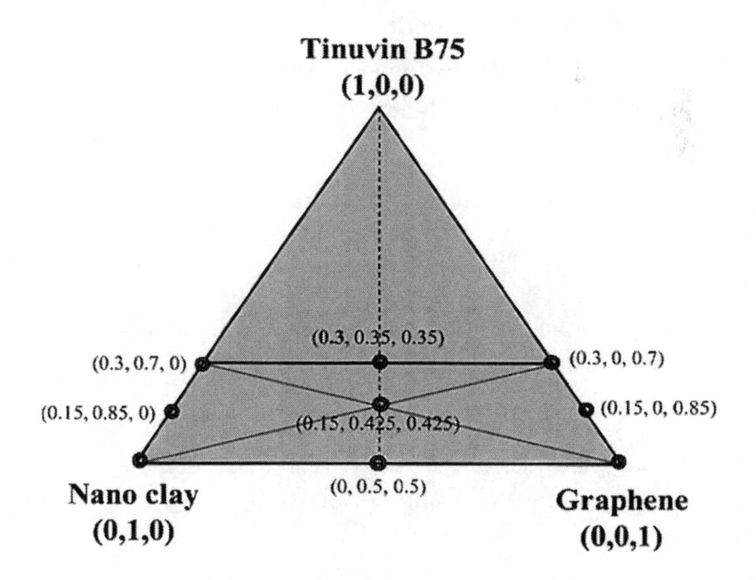

FIGURE 10.3
Layout of the mixture design with an upper bound constraint for one component (values in the bracket define the weight proportion of each component at that specific point) [20]. Reprinted with permission from Wiley.

10.3.3 Models Regarding Gas Barrier Property of Films and Coated/Laminated Structures

There are many models which can predict the permeability of gas through films and coated and laminated structures. The permeability of gas through any polymeric material depends on the free volume created by the segmental motion of the polymeric chains. Moreover, most of the polymers are semi-crystalline in nature, where crystallinity, size of crystallite, and orientation of polymeric chains play a crucial role in controlling permeability through the polymeric material.

Recently, polymer nanocomposites are increasingly being used for different gas barrier applications. Extensive ongoing research is focused on improvement of the gas barrier property of polymeric films and coated/laminated fabric by incorporating different-layered structured nanomaterials (nanoclay, graphene) in the polymer matrix [21–24]. These gas barrier polymer nanocomposites are potential material for the gas barrier layer of envelope for LTA systems. Because of their plate-like structure, these nanomaterials have huge potential in increasing gas barrier properties of polymers by increasing the 'tortuous path' for gas diffusion, when dispersed properly in polymer matrix. However, improvement in the gas barrier property depends on several factors, such as:

1. Geometric shape and size of the nanomaterials
2. Aspect ratio of the nanomaterial
3. Interaction between polymer and nanomaterials
4. Mixing, dispersion, and exfoliation of nanoparticles in the polymer matrix

Incorporation of these nanomaterials in the polymer matrix modifies the intrinsic properties of the material such as crystallinity, morphology, Tg, density, orientation, and flexibility, thereby helping to improve the gas barrier properties.

The mechanism of gas permeation through a polymeric film or coating is very complex and consists of three distinct steps as follows [25]:

1. Adsorption of gas molecules on the surface of film or coating
2. Diffusion of gas molecules through the film or coating
3. Desorption of the gas molecules from the other surface of the film or coating

The transmission of gas through polymeric films or membranes follows diffusion models related to Henry's and Fick's law. If the solubility or sorption coefficient 'S' (in $mol.cm^{-3}.Pa^{-1}$) is independent of the concentration, the gas permeability coefficient, P (in mol.Pa^{-1}.cm^{-1}.s^{-1}), can be expressed as follows:

$$P = D.S \tag{10.1}$$

where D is the diffusion constant (in cm^2.s^{-1}) [24]. There are several models for predicting the gas barrier property of polymer nanocomposites in terms of relative permeability (R_p), expressed by the ratio of P/P_0 where P_0 is the gas permeability of pure polymer matrix. Some of these popular models are summarized in Table 10.1.

TABLE 10.1

Different Models Related to Gas Permeability through Polymer Nanocomposites

Model name	Filler type	Orientation of nanoplatelets	Filler geometry	Formula
Nielsen model [26]	Ribbon	Regular planar arrangement	2D rectangular (L, W)	$R_p = \dfrac{1-\phi}{1+\dfrac{\alpha\varphi}{2}}$
Cussler-regular array [27]	Ribbon	Regular arrangement	2D rectangular (L, W)	$R_p = \dfrac{1-\phi}{1-\phi+\alpha^2\varphi^2}$
Maity and Bhowmik	Ribbon	Random arrangement	2D rectangular (L, W)	$R_p = \dfrac{1-\phi}{\left(1+\dfrac{\alpha\varphi}{2}\right)^2}$
Cussler-random array [27]	Ribbon	Random arrangement	2D rectangular (L, W)	$R_p = \dfrac{1-\phi}{\left(1+\dfrac{2\alpha\varphi}{3}\right)^2}$
Bhardwaj model [28]	Ribbon	Planar or random or orthogonal orientation	2D rectangular (L, W)	$R_p = \dfrac{1-\phi}{1+\dfrac{\alpha}{3}\left(S+\dfrac{1}{2}\right)\phi}$ $S = 1$ means planar arrangement, S = 0 means random arrangement, and $S = -1/2$ means orthogonal arrangement of nanoplatelets
Gusev-Lusti [29]	disk shape	Random dispersion and perfectly aligned in an isotropic matrix	3D round	$R_p = \exp\left[\left(-\dfrac{\alpha\varphi}{3.47}\right)^{0.71}\right]$

Note: R_p = Relative permeability, Φ = volume fraction of nanoplatelets, S = orientation or order parameter, L = length of the nanoplatelets, W = width of the nanoplatelets, α = L/W = aspect ratio.

10.3.4 Models for Predicting Mechanical Performance of an Aerostat or Airship Envelope

There has been a consistent effort by many researchers in simulation and analysis of the mechanical performance of aerostat/airship envelopes using different analytical models. The primary objective is to negate the dependence on time-consuming and cumbersome experimental studies to analyse the performance of envelope materials and incorporate simulations or mathematical models to carry out the task more effectively. Some of the parameters/factors that are generally modelled are ultimate tensile strength, tear strength and weld seam strength. In the coming paragraphs each of these factors are briefly introduced, and literature available on the same are discussed.

The uniaxial tensile test is one of the most preliminary and quintessential testing methodologies used to analyse the mechanical properties of material systems developed for airships and aerostats. Researchers have favoured this test method compared to biaxial testing owing to its simplicity and ease with which the essential material parameters can be extracted for simulating structural behaviour. Hu et al. [4] have studied the structural characteristics of a plain-woven fabric, URETEK 5893, in uniaxial monotonic and cyclic loading under off-axis and on-axis tensions coupled with the Digital Image Correlation (DIC) technique for better understanding of the failure mechanism. The Elastic modulus obtained from experimental studies in the warp and weft direction were in close proximity to the theoretical results, while the values deviated slightly for 45° loading condition owing to the nonlinearity of the material.

Kang et al. [5] analysed the tensile performance of the Tedlar-Vectran-TPU laminate. The tensile testing was carried out inside a thermal chamber at very low to high temperatures (–75°C, –50°C, –25°C, 25°C, 50°C, 65°C) to study the temperature dependency. The results obtained from experiments were compared with the tensile properties predicted by geometrically nonlinear finite element analysis, as shown in Figure 10.4. The stress-strain response showed significant nonlinearity for all the material systems studied at different temperatures, indicating strong dependence of stiffness on temperature. As stratospheric airships are exposed to a diverse range of temperatures, this analysis can be considered quite useful in analysing their performance.

In a recent study, Meng et al. [6] studied the strength criteria and mechanical properties of laminated fabric for applications in stratospheric airships, by testing the fabric with or without initial notch and applying uniaxial or biaxial tension. The simplified maximum stress criterion was used to forecast the failure of the laminated fabric under on-axial tension, while a new modified Tsai–Hill criterion was used for off-axial tension which takes into account the interaction between normal stress and shear stress components [6].

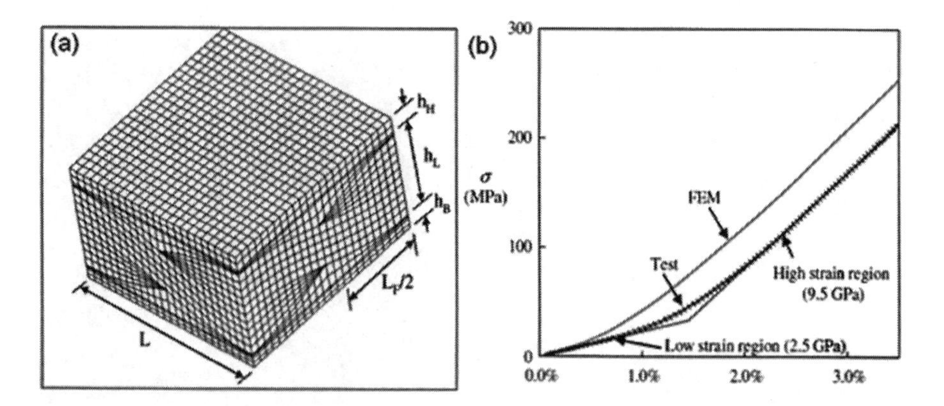

FIGURE 10.4
(a) Finite element mesh for quarter unit cell of film-fabric laminate. (b) Stress-strain curves of the film-fabric laminate at room temperature (25°C) [5]. Reprinted with permission from Elsevier.

$$\frac{\sigma_1^2}{X^2}+\frac{\sigma_2^2}{Y^2}-\frac{\sigma_1\sigma_2}{X^2}+\frac{\tau^2}{S^2}=1 \tag{10.2}$$

$$\frac{\sigma_1^2}{X_t^2}+F_{12}\frac{\sigma_1\sigma_2}{X_tY_t}+\frac{\sigma_2^2}{Y_t^2}+\frac{\tau^2}{S^2}+F_{16}\frac{\sigma_1\tau}{X_tS}=1 \tag{10.3}$$

where, σ_1 and $\sigma_2$represent the normal biaxial stresses in the x and y direction, and τ represents the shear stress in the xy plane, respectively. Similarly, X and Y represent the strength or maximum allowable stress in the x and y direction and *S* represents the shear strength in the xy plane, respectively. F_{12} is the factor introduced to take into consideration the interaction between warp and weft stress, and F_{16} represents the interactions between normal and shear stresses.

Despite the benefits of using the uniaxial tensile test method, to ideally recreate the volumetric stress condition the airship and aerostats are subjected to, researchers have explored the use of biaxial testing.

In a recent study, Shi et al. [7] carried out biaxial tensile tests under various stress ratios using self-designed double-layer cruciform specimens to develop a valid strength criterion (Figure 10.5) and formulate biaxial constitutive relationships. The existing strength criterions such as Tsai–Hill, Norris, and Yeh–Stratton criterions showed deviation from the experimental results especially in the normal biaxial stress range of the composite which prompted the researchers to develop a new biaxial strength criterion, the 'Chen-Shi failure criterion' (Figure 10.6) [7].

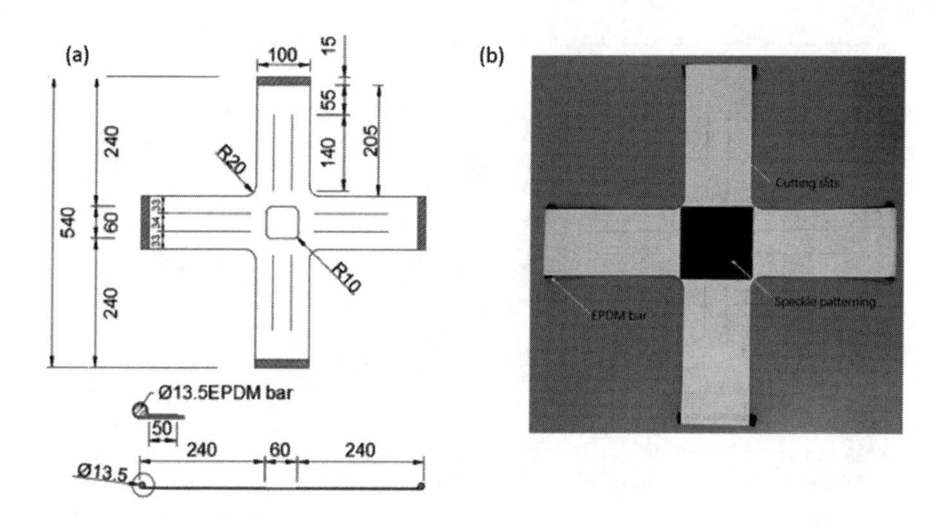

FIGURE 10.5
(a) Configuration of double-layer cruciform specimens, and (b) cruciform specimen product [7]. Reprinted with permission from Elsevier.

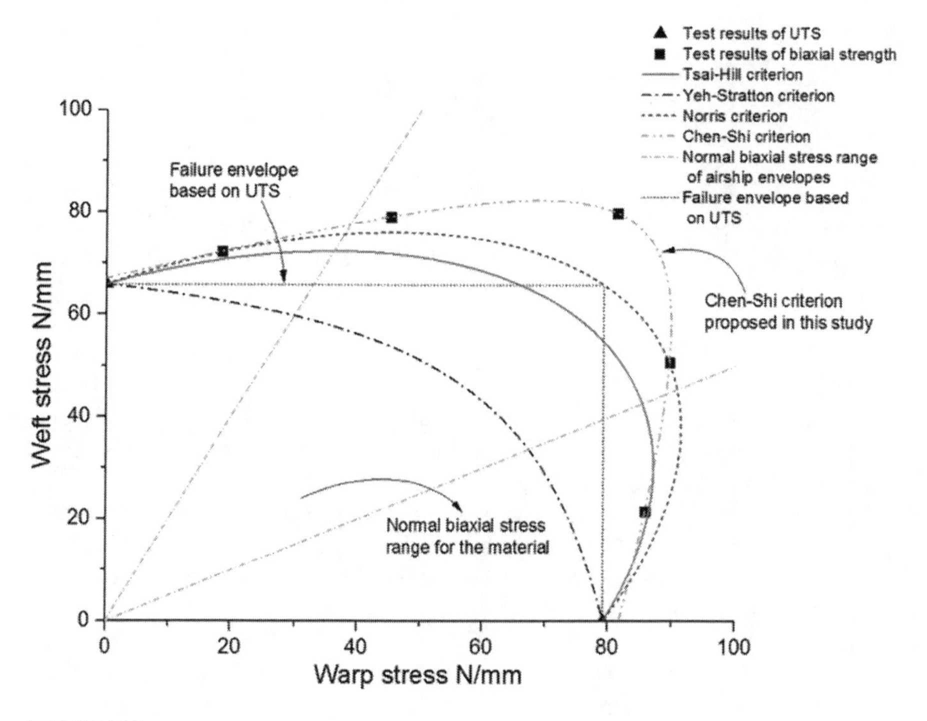

FIGURE 10.6
Failure of envelopes defined by different strength criteria [7]. Reprinted with permission from Elsevier.

$$\frac{\delta_x^2}{X^2}+\frac{\delta_y^2}{Y^2}-\frac{\delta_x\delta_y}{X^2}+\frac{\tau_{xy}^2}{S^2}=1 \quad \text{Tsai Hill failure criterion} \tag{10.4}$$

$$\frac{\delta_x}{X}+\frac{\delta_y}{Y}-\frac{\delta_x\delta_y}{X^2}+\frac{\tau_{xy}^2}{S^2}=1 \quad \text{Yeh Stratton failure criterion} \tag{10.5}$$

$$\frac{\delta_x^2}{X^2}+\frac{\delta_y^2}{Y^2}-\frac{\delta_x\delta_y}{XY}+\frac{\tau_{xy}^2}{S^2}=1 \quad \text{Norris failure criterion} \tag{10.6}$$

$$\frac{\delta_x^2}{C_1}+\frac{\delta_y^2}{C_2}+\frac{\delta_x\delta_y}{C_3}+\frac{\delta_x}{C_4}+\frac{\delta_y}{C_5}+\frac{\tau_{xy}^2}{C_6}=1 \quad \text{Chen}-\text{Shi biaxial failure criterion} \tag{10.7}$$

where, δ_x and δ_y represent the normal biaxial stresses in the x and y direction, and τ_{xy} represents the shear stress in the xy plane, respectively. Similarly, X and Y represent the strength or maximum allowable stress in the x and y directions, and *S* represents the shear strength in the xy plane, respectively. The parameters C_1 to C_6 in the Chen–Shi biaxial failure criterion are introduced to take into account the yarn interactions and material characteristics which enable it to accurately predict the ultimate tensile strength (UTS) of composite fabric with a deviation as low as 1.79–3.17% from the experimental results.

The tear strength has no definite relation with the actual tear propagation characteristics of an airship envelope. Therefore, many works of literature have tried to establish a relationship by applying uniaxial or biaxial stress. Maekawa et al. [30] took a unique approach to simulate the actual stress field of an airship envelope and also to find the appropriate formula for estimating the tear propagation stress, applying the stress of both types. The data were fitted with Thiele's empirical equation, and an excellent correlation was found with test results.

Meng et al. [8] reported the tearing behaviour of a flexible multi-layered film laminate, reinforced by a fabric composed of a high-performance fibre, which was developed for stratospheric airship envelope. A MATLAB programme was developed for analysing the tearing behaviour of the laminated fabric considering different factors such as interaction forces between warp and weft yarns, frictional coefficient, and the hinge between deformed and non-deformed regions. The simulation result showed a high degree of precision in the estimation of tear propagation strength [8]. In a more recent study, Zhang et al. [9] analysed the influence of initial notch on the tear behaviour of PVC-coated fabric. They also studied the effect of notch size, shape, specimen dimension, and the rate of loading on the same. A tearing model was proposed based on exponential stress field theory, with a notch size function that takes into account the specimen and notch size at once. The proposed model, when compared to classical models such as the stress intensity factor

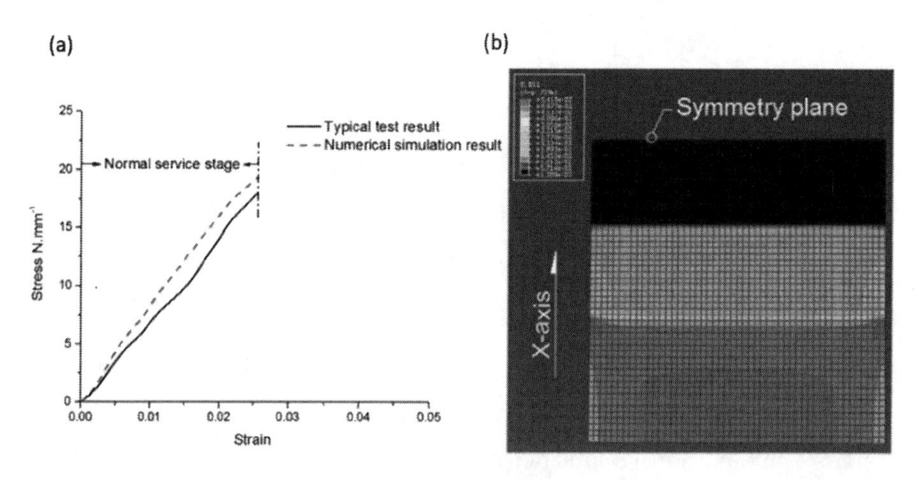

FIGURE 10.7
Numerical simulation results for (a) stress-strain curve and (b) global strain field [31]. Reprinted with permission from Elsevier.

theory, Theile's empirical theory, and the critical stress theory, proved to be a better fit as it takes into account in greater detail the effect of notch and specimen size which the classical model tended to neglect. Weld seam performance characteristics modelling, despite being a critical factor, has been generally not addressed by researchers while studying the mechanical performance of airships, as most studies consider them to be ideal continuous capsules. In a recent study, Taibai Shi et al. [31] developed a numerical model using ABAQUS software to study the tensile behaviour of weld seams in composite structures used as airship envelope materials. Variable parameters such as weld temperature and welding duration were not considered in this study. The numeric simulation results showed a close similarity with the actual test data in terms of strain simulation and strain field reappearance (Figure 10.7).

10.3.5 Models for Predicting Service Life of Aerostats and Airships

The development of accurate models to predict the life of aerostats and airships is a highly tedious task given the number of variables. In an attempt to accelerate the process of service life prediction, a 'speeding up the clock' approach is generally adopted in laboratories where the critical experimental factors such as temperature, relative humidity, and UV radiation irradiance are increased manifold compared to what the materials are generally subjected to during their use. The problem with this approach is that by increasing the intensity of multiple experimental factors at once, there is a high likelihood of new failure modes being generated in the material system which do not occur in actual use. This also explains the significant deviation that is noticed between the results obtained from accelerated and natural exposure testing. Thus, to predict the life cycle of material systems used in airships and aerostats accurately, a combination of natural exposure and accelerated testing needs to be adopted with correction factors to take into account the interdependency of multiple experimental factors. The two generally adopted routes to developing models to predict life cycle are statistical and mechanistic modelling; the former takes a reliability theory-based approach, and the latter considers the photodegradation caused by individual factors and also combines influence of the multiple factors in detail [2, 3].

In one of the earlier works, Shallow et al. [32] attempted to correlate and derive a relationship between natural weathering and artificial weathering testing in a Weather-o-meter. They observed a substantial dissimilarity in the results obtained for exposure under sunlight and xenon arc lamps. An acceptable correlation can be achieved between outdoor-actual field tests and artificial weathering by incorporating suitable assumptions and maintaining accurate control over the accelerated atmosphere. The fact remains that despite the best efforts to recreate the actual working conditions through a weather-o-meter, correlation to actual natural weathering is a complex process, and thus the literature available with this regard is also scarce. Vaca-Trigo et al. [33] developed a simple life prediction model for an amine-based cross-linked coating and correlated outdoor exposure to laboratory-accelerated weathering results. The environmental degradation factors that were taken into account for the indoor studies include UV radiation and spectrum, temperature, and humidity. Individual models for each factor were developed and a combined model was also established. The model developed based on laboratory exposure conditions proved to be in line with the outdoor exposure data when considering the variability in the outdoor exposure parameters. In a more recent study, Chatterjee et al. [34] theorized an approach to systematically assess the service life of TPU-coated polyester fabric-based aerostat systems. For this purpose, a reliability model was developed considering two principal stresses, namely UV radiation and temperature. Accelerated ageing using the Weather-o-meter was performed to establish a life stress model through which the service life of the material could be predicted. Validation of the model with actual field test data was also performed, and it was found to be in good agreement.

10.4 Conclusion

The primary requirements for the envelope of an aerostat or airship are excellent flexibility, high specific strength, gas barrier, and weather resistance, all of which ensure their long service life. However, service life analysis is a very challenging and long-term procedure without any analytical model. Therefore, different analytical models have huge significance in predicting the performance and service life of different LTA systems. However, the task is very challenging mainly because of the large number of variables related to materials, processing conditions, properties, and weathering conditions. In spite of these challenges, many researchers have tried to predict the overall service life or retention in specific properties (gas barrier, tensile, tear, flexural, inter-layer adhesion, joint strength, etc.) of LTA systems after exposure under natural/artificial weathering. These models are very helpful in anticipating the serviceable life of LTA systems. However, the existing models are few, and most of these models are specific for a particular LTA system and set of weather conditions. Therefore, extensive research work is going on to establish better models covering a wide spectrum of LTA systems in terms of shape/size, properties, location, and application.

References

1. H. Zhai, A. Euler, *Material challenges for lighter-than-air systems in high altitude applications*, in: *AIAA 5th ATIO and 16th Light. Sys Tech. Balloon Syst. Conf.*, American Institute of Aeronautics and Astronautics, Reston, Virginia, 2005. doi:10.2514/6.2005-7488.
2. R. Martin, *Ageing of Composites*, Woodhead Publishing, 2008. doi:10.1533/9781845694937.
3. D.R. Bauer, W.M. Jonathan, *Service Life Prediction of Organic Coatings*, American Chemical Society, 1999.
4. J. Hu, C. Gao, S. He, W. Chen, Y. Li, B. Zhao, T. Shi, D. Yang, Effects of on-axis and off-axis tension on uniaxial mechanical properties of plain woven fabrics for inflated structures, *Compos. Struct.* (2017). doi:10.1016/j.compstruct.2017.02.009.
5. W. Kang, Y. Suh, K. Woo, I. Lee, Mechanical property characterisation of film-fabric laminate for stratospheric airship envelope, *Compos. Struct.* (2006). doi:10.1016/j.compstruct.2006.04.060.
6. J. Meng, M. Lv, Z. Qu, P. Li, Mechanical properties and strength criteria of fabric membrane for the stratospheric airship envelope, *Appl. Compos. Mater.* (2017). doi:10.1007/s10443-016-9515-2.
7. T. Shi, W. Chen, P. Wang, & M. Wang. Biaxial constitutive relationship and strength criterion of composite fabric for airship structures. *Composite Structures*, 214 (2017), 379–389. doi:10.1016/j.compstruct.2019.02.028.
8. J. Meng, P. Li, G. Ma, H. Du, M. Lv, Tearing behaviors of flexible fiber-reinforced composites for the stratospheric airship envelope, *Appl. Compos. Mater.* (2017). doi:10.1007/s10443-016-9539-7.
9. Y. Zhang, J. Xu, Y. Zhou, Q. Zhang, F. Wu, Central tearing behaviors of PVC coated fabrics with initial notch, *Compos. Struct.* (2019). doi:10.1016/j.compstruct.2018.09.104.

10. K. Stefan, Performance theory for hot air balloons, *J. Aircr.* 16 (1979) 539–542.
11. K. Stefan, *Thermal effects on a high altitude airship*, in: *Lighter-Than-Air Syst. Conf.* AIAA, 1983.
12. L.A. Carlson, W.J. Horn, New thermal and trajectory model for high-altitude balloons, *J. Aircr.* (1983). doi:10.2514/3.44900.
13. H. Shi, B. Song, Q. Yao, X. Cao, Thermal performance of stratospheric airships during ascent and descent, *J. Thermophys. Heat Transf.* (2009). doi:10.2514/1.42634.
14. W. Anderson, P. Jungsun, D. Michael, *Numerical analysis concepts for balloon analysis*, in: *32nd Aerosp. Sci. Meet. Exhib.*, 1994: p. 511.
15. A. Samanta, D. Appelö, T. Colonius, J. Nott, J. Hall, Computational modeling and experiments of natural convection for a Titan Montgolfiere, *AIAA J.* (2010). doi:10.2514/1.45854.
16. H. Franco, H.M. Cathey, Thermal performance modeling of NASA's scientific balloons, *Adv. Sp. Res.* (2004). doi:10.1016/j.asr.2003.07.043.
17. X.L. Xia, D.F. Li, C. Sun, L.M. Ruan, Transient thermal behavior of stratospheric balloons at float conditions, *Adv. Sp. Res.* (2010). doi:10.1016/j.asr.2010.06.016.
18. Y.W. Wang, C.X. Yang, *Thermal analysis of stratopheric airship in working process*, in: *AIAA Balloon Syst. Conf.*, 2009.
19. Y.W. Wang, C.X. Yang, A comprehensive numerical model examining the thermal performance of airships, *Adv. Sp. Res.* (2011). doi:10.1016/j.asr.2011.07.013.
20. U. Chatterjee, B.S. Butola, M. Joshi, Optimal designing of polyurethane-based nanocomposite system for aerostat envelope, *J. Appl. Polym. Sci.* 133 (2016). doi:10.1002/app.43529.
21. B. Adak, B.S. Butola, M. Joshi, Effect of organoclay-type and clay-polyurethane interaction chemistry for tuning the morphology, gas barrier and mechanical properties of clay/polyurethane nanocomposites, *Appl. Clay Sci.* 161 (2018) 343–353. doi:10.1016/j.clay.2018.04.030.
22. B. Adak, M. Joshi, B.S. Butola, Polyurethane/functionalised-graphene nanocomposite films with enhanced weather resistance and gas barrier properties, *Compos. Part B Eng.* 176 (2019) 107303. doi:10.1016/j.compositesb.2019.107303.
23. B. Adak, M. Joshi, B.S. Butola, Polyurethane/clay nanocomposites with improved helium gas barrier and mechanical properties: Direct versus masterbatch melt mixing route, *J. Appl. Polym. Sci.* (2018). doi:10.1002/app.46422.
24. M. Joshi, B. Adak, B.S. Butola, Polyurethane nanocomposite based gas barrier films, membranes and coatings: A review on synthesis, characterisation and potential applications, *Prog. Mater. Sci.* (2018). doi:10.1016/j.pmatsci.2018.05.001.
25. G. Choudalakis, A.D. Gotsis, Permeability of polymer/clay nanocomposites: A review, *Eur. Polym. J.* 45 (2009) 967–984. doi:10.1016/j.eurpolymj.2009.01.027.
26. L.E. Nielsen, Models for the Permeability of Filled Polymer Systems, *J. Macromol. Sci. Part A-Chem.* 1 (1967) 929–942. doi:10.1080/10601326708053745.
27. N.K. Lape, E.E. Nuxoll, E.L. Cussler, Polydisperse flakes in barrier films, *J. Memb. Sci.* (2004). doi:10.1016/j.memsci.2003.12.026.
28. R.K. Bharadwaj, Modeling the barrier properties of polymer-layered silicate nanocomposites, *Macromolecules.* (2001). doi:10.1021/ma010780b.
29. A.A. Gusev, H.R. Lusti, Rational design of nanocomposites for barrier applications, *Adv. Mater.* (2001). doi:10.1002/1521-4095(200111)13:21<1641::AID-ADMA1641>3.0.CO;2-P.

30. S. Maekawa, K. Shibasaki, T. Kurose, T. Maeda, Y. Sasaki, T. Yoshino, Tear propagation of a high-performance airship envelope material, *J. Aircr.* (2008). doi:10.2514/1.32264.
31. T. Shi, W. Chen, C. Gao, J. Hu, B. Zhao, X. Wang, X. Wang, G. Lu, Investigation of mechanical behavior of weld seams of composite envelopes in airship structures, *Compos. Struct.* (2018). doi:10.1016/j.compstruct.2018.06.019.
32. J.E. Shallow, Effects of the dyes and finishes on the weathering of nylon textiles, RAE-TR-74179, 1975.
33. I. Vaca-Trigo, W.Q. Meeker, *A statistical model for linking field and laboratory exposure results for a model coating*, in: *Serv. Life Predict. Polym. Mater. Glob. Perspect.*, 2009. doi:10.1007/978-0-387-84876-1_2.
34. U. Chatterjee, S. Patra, B.S. Butola, M. Joshi, A systematic approach on service life prediction of a model aerostat envelope, *Polym. Test.* (2017). doi:10.1016/j.polymertesting.2016.10.004.

11

Future Trends in the Area of Material Developments for Aerostats/Airships: Future Trends in Material Developments

Subhash Mandal, Debmalya Roy, Kingsuk Mukhopadhyay, and N. Eswara Prasad

Defence Materials and Stores Research and Development Establishment (DMSRDE), DRDO, Kanpur, India

Mangala Joshi

Indian Institute of Technology, New Delhi, India

CONTENTS

11.1 Introduction

A typical lighter-than-air (LTA) system is based on the 'balloon-within-a-balloon' concept. An outer balloon (or hull) is followed by one (or more) internal balloons or ballonets (Figure 11.1). The external envelope is typically filled with lifting gas and the internal ballonet(s) is filled with air. To maintain the internal pressure and the hull shape during ascent, lifting gas is allowed to expand inside the hull while the air is pressed to escape out from the ballonet(s). The reversed situation is usually adopted during descent; lifting gas is allowed to contract, and the air is forced back into the ballonets to maintain pressure and the hull shape.

Based on operational need, the LTA platforms are categorized into three different classes – aerostats, heavy-lift airships, and high-altitude airships. The operational need for aerostats and their capability to deliver the need

DOI: 10.1201/9780429432996-11

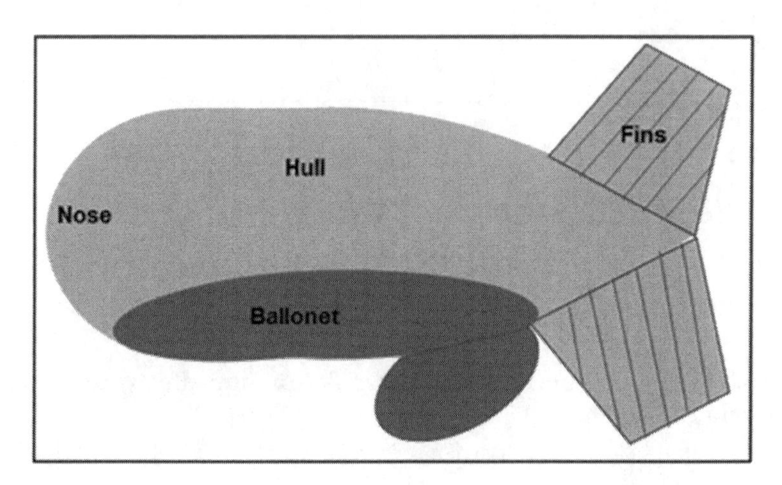

FIGURE 11.1
"Balloon-Within-A-Balloon" Design. Schematic design of typical lighter-than-air system.

appears to be the most promising of the three categories of LTA platforms. The aerostat systems are presently deployed and their limitations as well as capabilities are thoroughly studied and well-documented. Their role of persistent surveillance is the best suited. The aerostat's primary advantage over the other airborne platforms is the capability to provide elevated, persistent surveillance appears to be long life cycle time and low operational cost [1]. The primary operational concern with deploying aerostats is the enemy fire and vulnerability to weather. Aerostats tend not to collapse in the benign weather, while UAVs and aircrafts are more prone to accidents caused by a number of factors, viz. mechanical failure and human errors [2].

The aerostat has prodigious importance in defence for numerous applications such as imaging camera, radar, and surveillance system in patrolling and remote monitoring [3, 4]. Aerostats are categorized as tethered or untethered types based on their applications. Generally, sensors which are placed at the ground have a minimal line of sight (LOS) range due to the curvature of the earth. To get a wide area of LOS, sensors have to be positioned at a higher elevation such as tower, airborne platform, etc. The aerostat is generally preferred, as towers have some limitations as can be seen in Figure 11.2.

The operational need and utility of high-altitude airships (HAA) is far less acceptable than it is for aerostats. US security agencies and other organizations are likely to invest considerable time to study and determine exactly what these types of platforms can potentially offer and how their capabilities could be exploited for robust use of this airborne platform. Some of the potential capabilities of HAA are communications relay, aerial surveillance (long range), relay for internet services, warning for forest fire, and relay of laser weapon for missile defence. The HAA's long endurance goals and

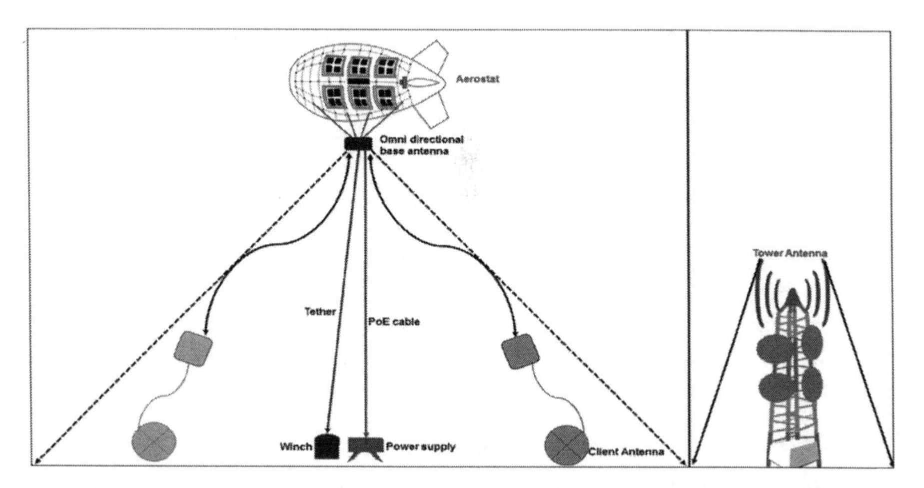

FIGURE 11.2
Importance of the aerostat over the tower.

operational environment present the HAA's technological challenges which appear much more critical than those actually experienced by the aerostats. A huge volume of helium is required to carry even modest payloads in HAA due to very thin atmosphere at the altitude of 70,000 ft. It is estimated that the HAA advanced concept technology demonstrations goal of a 500 lb payload will need an over 500 ft. long airship and capability of holding more than 5 million cubic ft. of helium gas. There are some issues related to the feasibility and the need of heavy-lift airships. The heavy-lift airships advocate the concept that these types of platforms are capable of filling the void between the sea lifting ships which slowly carry the large payloads and the aircraft that can carry smaller loads quickly. The key advantage of the heavy-lift airships would require much less infrastructure, inexpensive components, and long runways than conventional aircraft. The perennial issue in most countries' defence policies is whether future defence budgets would be sufficient to meet the entire defence requirement, and in this context, LTA programs have the potential future surveillance compared to conventional aircraft [5].

11.2 The History of Materials for LTA Systems

The advancement in the paradigm of new materials offers scopes for wide ranges of developments in aerostat technology. The new materials provide the further platform to design challenging airborne platform at harsh environmental conditions. The general requirements for aerostat material are high surface pressure resistance, flexibility, weather resistance, impermeability, and ability to withstand harsh weather conditions at higher altitude. The tethered and untethered helium aerostat is a complex system which requires development and thorough understanding in the area of self-weight, inflated pressure, payload, buoyancy effects, and aerodynamic effects.

In recent years, scientists have worked extensively on the LTA system, which focused mainly on gas barrier/environmental protection layers [6, 7] for aerostat and high-altitude airships. Neoprene, polyurethane, and poly vinyl fluoride are the common materials used for gas barrier properties. Polyester film has excellent properties of gas barrier, good shear stiffness, and high tensile modulus, which make it very suitable for aerostat applications [8–10]. The polyimide film has been found useful as the gas barrier layer for the construction of laminated LTA systems. It has the excellent barrier property of holding the gas inside for inflating the hull structure [11]. One of the main advantages of polyimide film is that it is primarily used for high-temperature applications with the additional advantage of diffusing the enemy targets like laser light by virtue of the high dielectric constant of polyimide films. The polyimide polymer, however, degrades by the exposure of UV light so it requires a protective coated with a UV-stable polymeric layer [11].

Polyvinyl copolymers also have effective gas barrier properties but are rarely used for aerostats because of their poor flexibility at low temperature [8, 12]. Polyurethane (PU) and its derivatives are the popular choice for gas barrier material as its properties conform to the fabrication needs of LTA vehicles [8, 12, 13]. The advantages of using a PU coat are many [14]; PU possesses superior mechanical properties (good impact resistance, very high abrasion resistance, good tear and tensile strength), high toughness, flexibility at low temperatures, good bonding with the textile substrates, and good weather-ability. The incorporation of nanomaterials (with high aspect ratio) into the polymer matrix leads to the superior properties of nanocomposite compared to conventional materials due to the large interfacial area between the base polymer matrix and the dispersed nanofillers [15].

The general configuration of an aerostat is the helium carrier with a coating of polyurethane (PU) on laminated nylon-based fabric [7]. A large number of studies are directed towards the development of polyurethane nanocomposite films as it exhibits a high strength-to-weight ratio [16, 17]. A plethora of comprehensive reviews cover polyurethane nanocomposite, discussing the insight into the gas barrier property with a coating that contains platelet-shaped fillers, viz. graphene and clays in the PU matrix [6, 18]. M. Mondal et al. [19] performed an extensive analysis of thermoplastic polyurethane-clay nanocomposites. It has been reported that laponite and montmorillonite types of clays significantly influence the morphological changes and thermal decomposition of the polymer matrix. Nanoclay has been widely used to enhance the gas barrier properties of polymers by plugging the voids in the polymer matrix. The clay particles agglomerate easily due to its hydrophilic nature and deteriorate intrinsic properties of base materials [15]. The problem of agglomeration of clay can be resolved by modifying the clay with the transition metal ions (Fe or Cu) to improve its dispersion and barrier properties [20]. Prof. Joshi et al. [15] reported in detail the various preparation methods of the clay-based PNCs. The compatibilization, exfoliation, re-aggregation, and orientation are the main four important factors required to be monitored while preparing the clay-based PNCs to achieve the maximum barrier properties. It has been reported by varying the clay weight percentages that the increasing clay content improved overall properties such as breaking strength, gas permeability, and tear strength of coatings. It is also been found that the clay content beyond the threshold limit deteriorates the desired properties because of agglomeration and poor dispersion of clay particles [15].

Recently the graphene and its derivatives have been widely studied for the gas barrier properties because of their excellent reinforcement ability to the polymer matrix by atomically thick two-dimensional layered structures. The organic graphene possesses many superior properties when compared with inorganic clay particles. The graphene/polyurethane nanocomposite demonstrated the improved barrier properties with a remarkable enhancement in electrical conductivity [21]. This can be attributed to the flexible, planar, and hexagonal arrangement of sp^2 hybridized carbon atoms in the graphene structure [22, 23]. Graphene and its derivatives are generally prepared by physical and chemical methods. The graphene polymer nanocomposite architectures consist of ultra-thin graphene sheets on the top of the polymer film or the incorporation of the graphene sheets into the polymer matrix [23]. Pierleoni et al. [24, 25] reported the properties of gas barrier of the polymer films using a thin layer of EGO (electrochemically exfoliated graphene oxide) sheets on the top of the polymer film. It has been found that the gas transmission rate of polymer film decreased significantly with the amount of EGO deposited on it. The drastic reduction in permeability of the polymer layer is due to close packing of 2D EGO sheets that suppresses both the vertical and horizontal movement of gas molecules across the deposited EGO layer [25]. Cui et al. [26] reviewed the preparation methods of the graphene and its derivatives followed by various processes of preparing graphene PNCs and the parameters which influence the superior gas barrier properties.

The derivatives of the fluorocarbon-based polymer have been utilized for protection against UV radiations. One such polymer is polyvinyl fluoride (PVF), which is available commercially under the trademark name Tedlar from DuPont [13]. Tedlar retains its properties over the wide temperature range (−72°C to 107°C), making it a suitable candidate for high-altitude applications [7, 15]. The outermost layer is pigmented preferably white to decrease the solar absorption and increase the life of the polymer [13]. The polyvinylidene fluoride or polyvinylidene difluoride (PVDF) is another derivative of the fluorocarbon-based polymer which provides excellent protection against the UV and solar degradation [11]. They can provide excellent environmental protection with little or no maintenance for up to 20 years [9].

The thin metallic layers over the polymeric films were also tried in order to control the diffusion of gas. The thickness of these metal layers varied from 800 to 1200 Å and the layers were applied mostly by vacuum sputtering and metallization techniques [27]. The metallic coating provides more environmental protection at higher altitude compared to rubber/polymer surfaces [12]. The use of metallic coating also aids in the dissipation of static charge and prevents damage caused by lightning strikes [24, 27]. However, the flexibility of metallic films is far behind the rubber/polymeric materials, and hence for dynamic pressure adjustment at high altitude, the metallic coatings may develop microcracks.

PU and its derivatives, viz. polycarbonate polyurethane, can also be used for environmental protection because of their excellent UV stability [8]. The additional protection against chemicals and microbes can also be achieved by infusing certain chemicals as additives in the protection layer. The degradation behaviour of PU-coated nylon fabrics by exposure to the outdoor environment has been studied [28]. It has been observed that maximum strength loss occurs in peak winter, which suggests that moisture plays a vital role in the degradation of the material compared to UV in the sunlight. It is also been reported that after a certain time period of exposure, the decrease of strength is proportional to the duration of exposure. It has been concluded that even though the strength loss is linearly proportional with time, hydrogen gas permeability drastically increased beyond a critical duration of exposure. Thus, the gas permeability is the critical factor for determining the life of an aerostat system [28]. Nuraje et al. [29] reported the properties of weathering of PU coatings reinforced with the graphene. It has been illustrated that the incorporation of the graphene improved the protection against environmental factors, viz. corrosion and UV degradation. The graphene-incorporated PU coating provides the hydrophobicity and improved mechanical properties compared to pristine PU coating. The improvements can be attributed to the high surface area-to-volume ratio and the unique chemical structure of graphene [30]. A number of studies also revealed that the introduction of metal nanoparticles, viz. zinc oxide (ZnO), titanium oxide (TiO_2), and cerium oxide (CeO_2), in polymer coatings inhibits the photo-oxidation by absorbing UV radiations, thereby protecting the base fabric from UV degradation [31–34]. Sinha et al. [35] reviewed the different UV absorbers which have been used in the textile and also the mechanism of the UV absorption by metal oxide particles [29, 35]. The UV-stabilizing nanoparticles have also been used along with graphene and clay materials to improve UV protection and gas barrier properties. The in-depth studies on these hybrid organic-inorganic composite fillers established that the combination is advantageous in retaining both gas barrier properties and UV protection after exposure to weather conditions at high altitude [36].

The aerostat is generally subjected to a pressure of 6 mbar, and the buoyancy effects can be visualized by the Archimedes principle of differences between air density and helium density. The other factors which have significant influence on aerostat are mechanical properties and the properties of material for defining the morphological changes by the environmental impact. Y. Wang et al. [37] have reported the use of inorganic and organic nano absorber in polyester polyurethane film by a spin-coating technique for resisting the synergistic effect of O_3/UV on ageing. The PU resin-coated surface has also been investigated, incorporating the OH-silicone modified polyacrylate additive against UV exposure [38]. The thorough understanding of these states can ultimately solve the issues encountered during aerostat operation. The challenges are also on real-time mechanical properties evaluation due to nonlinearity and large deformation, and the use of advanced material is intended to cater the many difficulties encountered by aerostat film.

11.3 Simulation of Aerostat Design and Properties

Nowadays, it is of topmost priority to have a well-versed analysis of any manufactured product. The evaluation is generally carried out in all climatic conditions and test parameters in which the product is supposed to perform. The Finite Element Analysis (FEA) can be used to have an insight into the performance objectives of the product before the actual standard operating procedures. Various software is available in the market to perform finite element analysis, such as ABAQUS which is aimed to design a product with the complex geometries. It has also a unique feature available to import required geometries from other software packages. This commercially available software can be used to study structural response, thermal response, and effects due to add-on masses. The FEA on the balloon involves a careful judgement of input parameters, the most important of which are the types of element based on localized stresses on the aerostat. The different types of elementary parameters available are depicted in Figure 11.3; the building blocks have to be decided depending on the desired geometries of the aerostat such as hull, fins, and ropes. The meshing of the virtual aerostat model is carried out by increasing nodes where stress is mostly found to be concentrated on the region of interest. The choice of parameters to fit a proper constitutive model (nonlinear analysis) to comprehend the developed material behaviour is critical to receive the desired results.

The simulation by FEA determines the behaviour of a real component with idealized mathematical modelling which includes every detail of the physical conditions. The FEA model is analysed by a solver that calculates the data reflecting design behaviour to the applied boundary conditions and can help

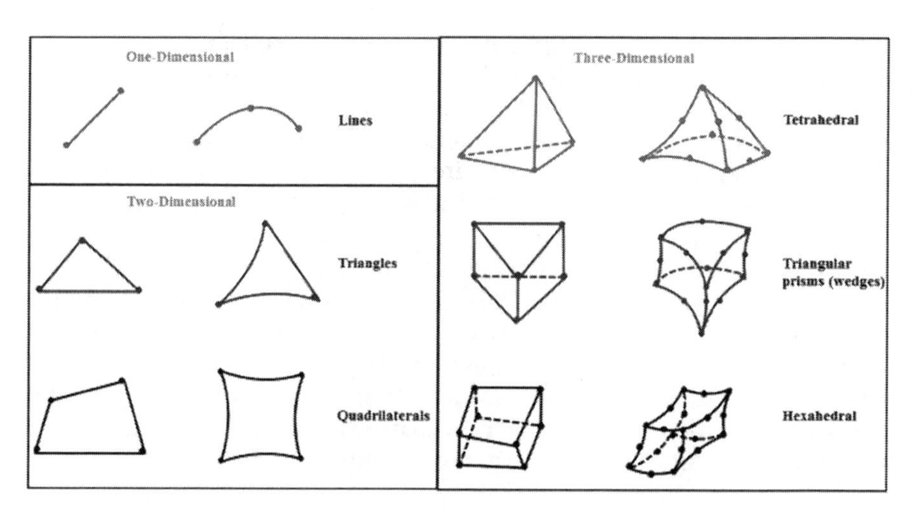

FIGURE 11.3
Different types of Element in Finite element analysis.

to identify the weaknesses or predict potential failure mechanism in design. The FEA of the elastomer components is most complex when compared to the metallic or thermoplastic parts. A three-dimensional FEA was performed to consider the irregularities in geometry using ABAQUS. F. Baginski and colleagues successfully carried out analysis to resolve the issues faced for equilibrium shape determination of the strained balloon, harnessing the minimum total potential energy principle [39–41]. The total potential energy ε of a strained and inflated balloon in a configuration ϕ is given by

$$\varepsilon(\phi) = \varepsilon_p + \varepsilon_f + \varepsilon_t + \varepsilon_{top} + S_t^* + S_f^*$$

where ε_p is hydrostatic pressure potential of lifting gas, ε_f is gravitational potential energy of film, ε_t is gravitational potential energy of load tendons, ε_{top} is gravitational potential energy of apex fitting, S_t^* is relaxed strain energy of the tendons, and S_f^* is relaxed strain energy of film.

FEM analysis was also employed to envisage the deflation and inflation of the tethered aerostat radar system (TARS), W. Anderson and M. Dungan [42] gave an ample numerical analysis in structural, thermal, and acoustic behaviour. It is reported that the aforementioned analytical technique is useful in studying the sustainability of an aerostat at higher altitude. The thermal characteristic is the most important parameter as an aerostat experiences an abrupt change in temperature at higher altitude. The movement of the aerostat also results in the change in weight due to displacement of air. This acoustic behaviour was thoroughly analysed to understand the nature of buoyancy in aerostat at the higher altitude. X. Deng and S. Pellegrino [43] executed the 3D FEA to study the relationship between height, volume, and stress distribution for ultra-long duration on balloon. It has been illustrated that the tensile hoop stress remains on the crown region of the balloon till the pressure in the bottom region reaches negative and produces axisymmetric geometry. The tethered aerostat requires a proper investigation by FEA for stresses, and potential for buckling of the hull, empennage, and suspension system under various configuration and loading conditions to accurately predict its performance at higher altitudes.

FEA has the potential to simulate all the effects happening on an aerostat. The validation of the model and the accurate feeding of parameters, however, necessitates detailed understanding of the structure-property relationship of the material system used for the aerostat and to implement the complex theoretical analysis J. D. Hunt [44] carried out on FEA by nonlinear optimization of the deviation and vibration encountered in the tethered satellite on balloons at higher elevation using continuum mechanics [45]. It soon became an important technique which makes the analysis much easier to perform and in turn significantly reduces the operating costs and interruptions in data acquisition [46].

The theoretical analysis becomes challenging for the large deformation and time dependency for the polymeric materials due to their viscoelasticity behaviour. The time dependence is critical and cannot be neglected, as an aerostat is subjected to large strain for the extended period of time. D. Wakefield [47] performed the nonlinear viscoelastic analysis on super- pressure balloons of NASA and finite element work of structural analysis on an inflated system. The linear and nonlinear analysis on aerostat by static deformation behaviour were carried out by I. Khan et al. [48]. It has been highlighted that this type of analysis requires the consideration of nonlinearity into account, as linear results could not sufficiently represent the stresses encountered in the real system. The overall FA is represented in Figure 11.4 considering static analysis.

It has been further elaborated that the dynamic analysis to obtain the overall result in static condition is itself not sufficient. A. Rajani et al. [28] performed the analysis of dynamic stability in an aerostat system by incorporating the concepts of the dynamic tether, apparent mass, and allowing six degrees of freedom for motion. The reliability of the evaluation of actual lifetime of a rubbery material by FEA is critical to ensure the safety and reliability of the rubber components [49] and the lifetime prediction become challenging of such material. Le Gac et al. [50] reported the lifetime prediction of the polymeric materials used in the pipe under sea water, where it was exposed to hydrostatic pressure and temperature of the pipe. R. Kunič et al. [51] performed the analysis of degradation dynamics on polyurethane-based coating on aluminium for a thickness-sensitive solar absorber. The lifetime prediction is also influenced by the other materials incorporated into the matrix, which ultimately changes the life of the product. X. Gu et al. [52]

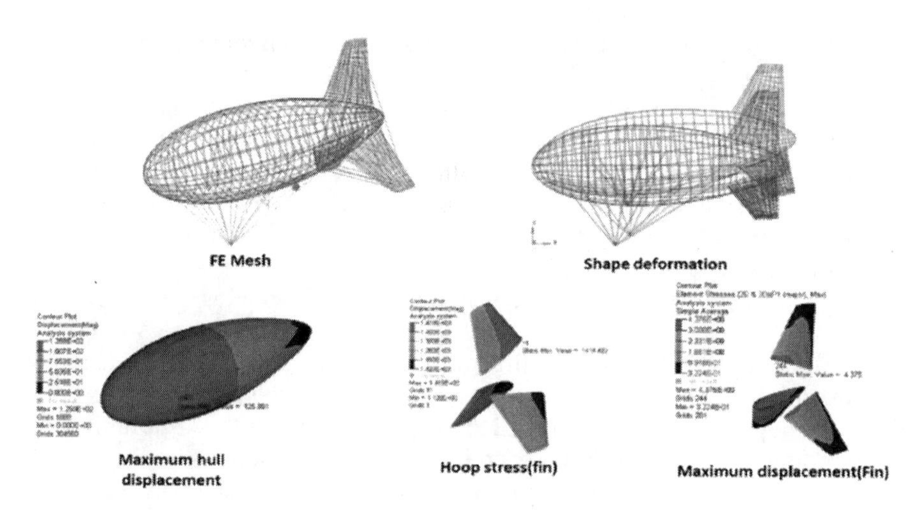

FIGURE 11.4
Finite element analysis for stress and strain distribution [48].

have studied the role of nanoparticles on the lifetime prediction of ZnO/polyurethane nanocomposites, and it has been observed that the material degrades fast when nanoparticles act as a photocatalyst and can shield the matrix when nanoparticles behave as a photo-stabilizer.

The different modelling approaches attempted so far have focused on a particular issue regarding material properties or on a parameter of an aerostat application. The time-dependent viscoelasticity and dynamic analysis of a complex geometry such as an aerostat are insufficient; moreover, the effect of weather conditions on stability and permeability are not well investigated. The life prediction of rubbery materials is carried out using the time-temperature superposition principle and the Arrhenius method. The materials used to manufacture an aerostat are a combination of textile and rubber, thus the prediction of life based on only the rubbery portion without considering viscoelasticity may not serve the purpose. Furthermore, the changes in composition of elastomers makes the analysis even more difficult.

11.4 Material Challenges for LTA Systems

Generally, airborne objects fly at the stratosphere level, at a height 10–50 km above the earth's surface, and the exposure to ozone, UV, and cosmic radiation are very high at this level, which results in loss of material strength and barrier property by degradation of the polymeric system. A tethered aerostat operates at a much lower attitude; however, it encounters clouds, and the moist environment significantly decreases the lifetime of polymeric materials. The atmospheric pressure at higher altitude is very low; sometimes it shows 20–1000 times lower than that we experienced at the surface. The rain and the wind in the depth of the clouds also alter the pressure at the higher altitude and dynamic pressure changes produced enough stress on the flexible materials to generate microcracks at the inflated conditions. The temperature varies very widely with the altitude, and at a sub-zero temperature, polymeric material becomes brittle with the resultant loss of flexibility of the airship. Table 11.1 represents the critical requirements of materials for LTA [7, 53].

The flexible multi-layered coated or laminated textiles are currently used in a non-rigid aerostat/airship envelope; however, it has been realized that a single layer may not be able to meet all the requirements. The typical hull material consists of four main layers (as shown in Figure 11.5): strength layer, protective layer, adhesive layer, and gas barrier layer. In the strength layer of the envelope, generally a woven fabric (either single or multi-layered) is used, and usually the fabric is prepared using high-strength polyester or advanced high-performance fibres like Zylon, Spectra, Kevlar, Vectran, M5, etc. The characteristic feature required for strength layer is high strength, light weight, good bondability, and increased service life at a wide range of temperature [7, 54].

TABLE 11.1

Desired Properties and Features for the Materials Used in Airborne Platforms

Sr no.	Properties	Features
1	Resistance from harsh weather conditions	To protect the system from environmental degradation
2	H_2/He gas retention/barrier property	To minimize lift loss and maximize flying cycle
3	Low temperature flexibility	To reduce the chance of developing cracks
4	Light weight	To minimize the requirement of lifting gas and to maximize the payload
5	High strength	For enhanced lifetime
6	High tear and bursting resistance	To reduce the chance of damage
7	Abrasion resistance	To reduce the chance of damage
8	Good bondability/sealability	For robust operation and to reduce the chance of gas leak
9	Good joint strength	For robust operation and to reduce the chance of gas leak
10	Service life	About 3–5 years in current state of the art

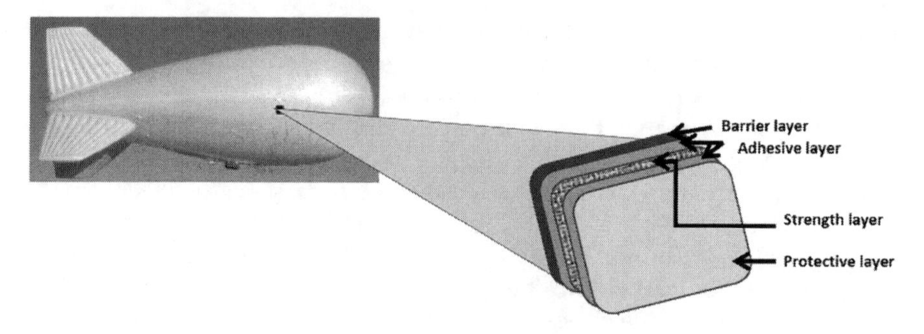

FIGURE 11.5
Schematic showing the different layers present in the hull of the aerostat.

A suitable polymeric material is used for the protective layer at the top of the aerostat/airship envelope and the main properties required for the material suitable as a protective layer are weatherability, thermal reflective/emissive properties, low temperature flexibility, and good bondability. The Tedlar, Tedlar with aluminium coatings, PVDF, and Teflon are currently used in the protective layer [7, 54]. It is generally a very thin layer which is used to join or bound various layers together to make a multi-layered composite structure. The next important composition is the adhesive layer, for which polyurethane and sometimes acrylic or epoxy-based formulations are usually used. The main properties required for this layer are compatibility with other layers, good bondability, and low temperature flexibility. The most important component of the aerostat/airship is the gas barrier layer, which is often the innermost internal layer of the envelop, and universally suitable polymeric materials with required reinforcement are used. The main feature required for the suitable barrier material are very low hydrogen/helium gas permeability, good bondability, and low temperature flexibility. Generally, ethylene vinyl alcohol copolymer (EVOH), polyester (Mylar), polyurethane, etc., are frequently used in the gas barrier layer with graphene/clay as reinforcement agents [7].

Although the polymeric system used in a particular layer is able to perform specific purpose individually, there are plenty of processing and operational issues when they are integrated in a multi-layered composite structure. The frequently encountered problems are the delamination of layers due to poor bondability and that most of the polymers are not heat sealable. Poor performance in terms of low temperature flexibility is one of the significant issues which retards the lifetime of the airborne platforms, and sometimes neat polymer is not able to fulfil the multifunctional requirement even after reinforcing with required fillers. The salient properties of thermoplastic PU which provide an excellent microstructural feature and make it a popular choice for gas permeability applications are enlisted in Table 11.2.

TABLE 11.2

Salient Properties and Microstructural Features of Thermoplastic Polyurethane

Sr No.	Properties	Features
1	Wide varieties of chemical composition and molecular weight	Freedom of choosing the flexibility with desired properties
2	Hard/soft segment ratio	Micro- to nanoscales domains for better control over the alignment of polymeric chains
3	Crystallinity and amorphous segments	Better control over flexibility and mechanical properties
4	Aromatic to nonaromatic and ether to ester versions	Control over the environmental stability and rate of degradation
5	Orientation and packing of polymer chains	Lower void space for lower permeability
6	Low glass transition temperature (Tg) and the segmental motion of PU chains	Better low temperature flexibility to avoid microcracks in the polymeric films
7	Low density of PU	Lower weight and higher payload
8	Degree of cross-linking	Higher stability with increased flexibility
9	Optimum free-volume content	Better control over the diffusion of gas molecule and Tg
10	Surface characteristics	Better control over the structure property relationship

11.5 Future Perspective and Concluding Remarks

The tethered or untethered helium aerostat, which has been known for a decade, is being revitalized by the advent of new materials and technologies, particularly nanotechnology. The aerostats have been used for the varieties of applications like supporting aerial imaging cameras and radar and surveillance devices for remote monitoring of security and patrolling. The aerostats are receiving renewed attention in the scientific and surveillance communities and are contemplated as the most important considerations in tactical planning. The aerostat is of paramount importance for the reconnaissance in armed forces and naval and air forces as well as for homeland security and antiterrorism.

The careful designing and fabrication of an aerostat leads to considerable enhanced operational life and increased surveillance efficiency, but the stable materials at harsh environmental conditions limit the high performance of the aerostat. Functional materials play a very significant role in aerostat technologies and need to be designed and developed in the forms of multi-layer coatings, laminated fabrics, flexible sheets, etc., that could be deployed for very long periods, thus reducing operating costs and interruptions in data acquisition. The airborne time of typical tethered aerostat is generally limited by harsh weather conditions at higher altitude. The intense UV light reduced aerostat life, and also balloons are not reliably able to survive high winds due to 'dimpling'. The loss in envelope shape is also reported, as the inflated fabric is unable to resist high surface pressures. The use of synthetic materials and laminates with high strength-to-weight ratio, such as nylon and polyester, coupled with weather-resistant, heat-sealable, and impermeable coatings, such as polyurethane, has improved the survivability and reliability of modern aerostats. Materials and technology of such a nature are not readily available commercially.

The traditional industrial flexible products like rubber, elastomers, and polymeric materials are looking forward for advanced high-performance spin-off under harsh environment and extreme conditions. The advent of new technological advances encouraged the R&D institutes to actively pursue research on materials science and technology and understanding of the science of raw materials and compounds for durability and excellent performance thresholds. The use of nanomaterials as fillers in a rubber or polymer matrix provided an unprecedented opportunity to modulate the characteristics properties in dynamic conditions on a bespoke formulation. The optimized nanocomposite is also lighter and shows improved strength and elongation with increased thermal conductivity.

The airborne surveillance through the aerostat is a well-established technology, and world-leading companies like Lockheed Martin Corp., TCOM, and Raven Aerostar commercially provide the general aerostat balloons. In the traditional commercially available aerostat, the power for the payload is pumped up through the tether cables from a generator. The generator retards the mobile surveillance operation and hence seriously limits the surveillance capabilities. In view of the non-availability of state-of-the-art aerostat materials technologies, there are pressing requirements for development of an innovative material solution, meeting the applicable processing and manufacturing standard. There are lots of developmental materials programmes to produce a material system and process parameters for making UV/ozone-stable gas barrier layers of the hull structure of an aerostat for enhanced operational hours. Further, it is also learnt to integrate the flexible solar cell or to suitably convert the top layer of the hull structure for light harvesting for enhanced surveillance capacity. Hence the future developmental works should target reducing the total weight of the airborne aerostat in terms of batteries and other energy devices to meet the power requirement of payloads coupled with weather-resistant, heat-sealable, and impermeable coatings for enhanced aerostat life (Figure 11.6).

The use of LTA platforms for reconnaissance and surveillance over long periods can be designed by generation of power on board through

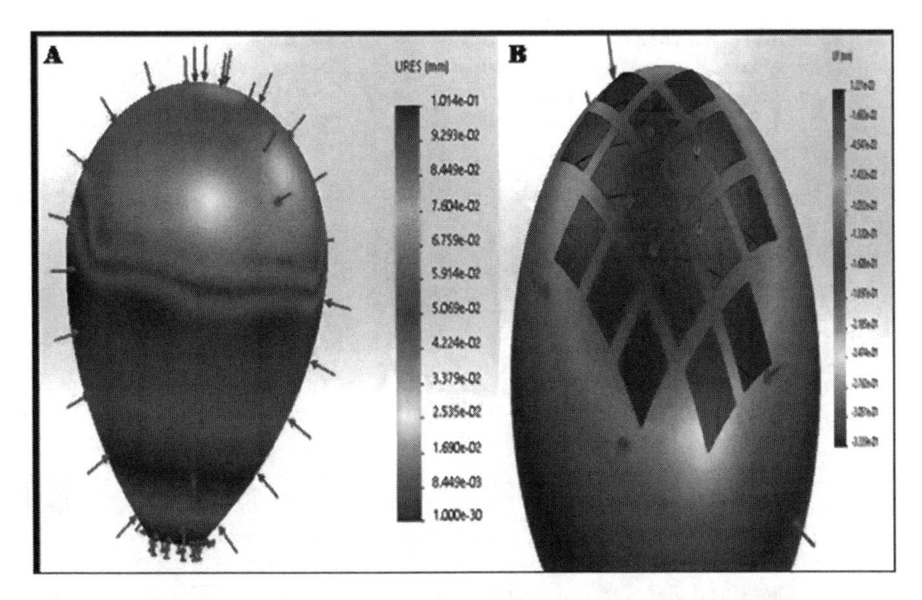

FIGURE 11.6
The FEA analysis (A) to identify the stress accumulation on the balloon surfaces and the integration of flexible solar cells (B) on the top surface of the balloon.

photovoltaic arrays. The flexible photovoltaic materials are able to drastically reduce the weight of airborne platforms without hindering the operational and aerodynamic requirements of the airborne platform. The key advantage of combining an extra photovoltaic layer on an aerostat is the protection from UV degradation of polymeric skin to reduce the gas leaking rate, which in turn will increase the operational hours of robust surveillance capacity with onboard power generation. The solar-powered aerostat can have a range of devices like light sensors, temperature sensor, current and voltage sensors, strain gauge, humidity sensor, camera (VIS, IR), anemometers, etc., for effective reconnaissance, weather forecasts, meteorological conditions, and also on-board health monitoring of the airborne platform.

References

1. Bolkcom, C. 2006. Potential military use of airships and aerostats "FY2000–FY2005 TARS budget", Air Combat Command, USAF.
2. Hsu, E. 2004. JLENS to proceed on accelerated path: JROC OKs requirements for army's future elevated sensor aerostat, *Inside Missile Defense*, 10: 2–3.
3. Ahmad, I., Shah, K., and Ullah, S. 2016. Military applications using wireless sensor networks: A survey. *Int. J. Eng. Sci.* 6: 7039.

4. Đurišić, M. P., Tafa Dimić, Z. G., and Milutinović, V. 2012. *A survey of military applications of wireless sensor networks. Mediterranean Conference on Embedded Computing (MECO)*, Bar, MO, 196–199.
5. Wilson, J. R. 2010. Return of the Military Airship. *Defense Media Network*. https://www.defensemedianetwork.com/stories/return-of-the-military-airship/
6. Joshi, M., Adak, B., and Butola, B. S. 2018. Polyurethane nanocomposite based gas barrier films, membranes and coatings: A review on synthesis, characterization and potential applications. *Progress in Materials Science* 97: 230–282.
7. Zhai, H., and Euler, A. 2005. *Material Challenges for Lighter-Than-Air Systems in High Altitude Applications. AIAA 5th ATIO and 16th Light. Sys Tech. Balloon Syst. Conf.*, 7488.
8. Das, B. R. 2010. UV radiation protective clothing. *Open Text. J.* 3: 14–21.
9. Kathirvelu, S., D'Souza, L., and Dhurai, B. 2009. UV protection finishing of textiles using ZnO nanoparticles. *Indian J. Fibre Text. Res.* 34: 267–273.
10. Lee, Y. C. 2008. Long mission tethered aerostat and method of accomplishing. US Patent-US 2008/0265086 A1, 1–9.
11. Katangur, P., Patra, P. K., and Warner, S. B. 2006. Nanostructured ultraviolet resistant polymer coatings. *Polym. Degrad. Stabil.* 91: 2437–2442.
12. Durney, G. P. 1980. *Concept for Prevention of Catastrophic Failure in Large Aerostats. Proceedings of the AIAA International Meeting and Technical Display on Global Technology 2000*, Baltimore, MD.
13. Kassim, M. E. B. 2008. Designing and analyzing preliminary parts of an aerostat. B. Tech Thesis, Faculty of Mechanical Engineering, Universiti Teknikal Malaysia Melaka.
14. Li, J., Lv, M., Sun, K. and Zhang, Y. 2016. Stratospheric aerostat-A new high altitude scientific platform. *Curr. Sci.* 111: 1296–1297.
15. Liggett, P. E., Sinsabaugh, S. L., and Mascolino, J. I. 2009. Conductive seam cover tape. United States Patent -US 2009/0220726 A1, 1–5.
16. Joshi, M., and Chatterjee, U., 2016. Polymer nanocomposite: An advanced material for aerospace applications. In *Advanced Composite Materials for Aerospace Engineering, Processing, Properties and Applications*, eds. S. Rana and R. Fangueiro, 241–264, Cambridge: Woodhead Publishing, Elsevier.
17. Khudyakov, I. V., Zopf, D. R., and Turro, N. J. 2009. Polyurethane nanocomposites. *Des. Monomers Polym.* 12: 279–290.
18. Adak, B., Joshi M., and Butola, B. S. 2018. Polyurethane/clay nanocomposites with improved helium gas barrier and mechanical properties: Direct versus master-batch melt mixing route. *J. Appl. Polym. Sci.* 135: 46422.
19. Mondal, M., Chattopadhyay, P. K., Chattopadhyay, S., and Setua, D. K. 2010. Thermal and morphological analysis of thermoplastic polyurethane–clay nanocomposites: Comparison of efficacy of dual modified laponite vs. commercial montmorillonites. *Thermochim. Acta.* 510: 185–194.
20. Miller, J. I. 2005. The design of robust helium aerostats. M. Tech Thesis, Department of Mechanical Engineering, McGill University, Montreal.
21. Kim, H., Miura, Y., and MacOsko, C. W. 2010. Graphene/polyurethane nanocomposites for improved gas barrier and electrical conductivity. *Chem. Mater.* 22: 3441–3450.

22. Liggett, P. E., Carter, D. L., Dunne, A. L., et al. 2013. Metallized flexible laminate material for lighter-than-air vehicles. United States Patent -US 8,524,621 B2, 1–10.
23. Masteikaite, V., and Saceviciene, V. 2005. Study on tensile properties of coated fabrics and laminates. *Ind. J. Fibre Text. Res.* 30: 267–272.
24. Pierleoni, D., Xia, Z.Y., Christian, M., et al., 2016. Graphene-based coatings on polymer films for gas barrier applications. *Carbon* 96: 503–512.
25. Miller, T., and Mandel, M. 2000. *Airship envelopes: Requirements, materials and test methods. Proceedings of the 3rd International Airship Convention and Exhibition*, Friedrichshafen, Germany.
26. Cui, Y., Kundalwal, S. I., & Kumar, S. (2016). Gas barrier performance of graphene/polymer nanocomposites. *Carbon*, 98: 313–333.
27. Krausman, J. A., and Petersen, S. T. 2013. *The 28MTM tactical aerostat system: Enhanced surveillance capabilities for a small tethered aerostat. Proceedings of the AIAA Lighter-Than-Air Systems Technology (LTA) Conference, AIAA 2013–1316*, Daytona Beach, FL, 1–11.
28. Rajani, A., Pant, R. S., and Sudhakar, K. 2010. Dynamic stability analysis of a tethered aerostat. *J. Aircr.* 47: 1531–1538.
29. Nuraje, N., Khan, S. I., Misak, H., and Asmatulu, R. 2013. The addition of graphene to polymer coatings for improved weathering. *Polym. Sci.*, 2013: 514617.
30. Ram, C. V. and Pant, R. S. 2010. Multidisciplinary shape optimization of aerostat envelopes. *J. Aircr.* 47: 1073–1076.
31. Raza, W., Singh, G., Kumar, S. B., and Thakare, V. B. 2016. Challenges in design and development of envelope materials for inflatable systems. *Int. J. Text. Fashion Technol.* 6: 27–40.
32. Shamini, G., and Yusoh, K. 2014. Gas permeability properties of thermoplastic polyurethane modified clay nanocomposites. *Int. J. Chem. Eng. Applic.* 5: 64–68.
33. Shim, E. 2010. *Coating and Laminating Processes and Techniques for Textiles: Smart Textile Coatings and Laminates*. CRC Press/Taylor & Francis Group, Boca Raton.
34. Singha, K. 2012. A review on coating and lamination in textiles: Processes and applications. *Am. J. Polym. Sci.* 2: 39–49.
35. Sinha, M. K., Das, B. R., Kumar, K., Kishore, B., and Prasad, N. E., 2017. Development of ultraviolet (UV) radiation protective fabric using combined electrospinning and electrospraying technique. *J. Inst. Eng. India: Ser. E* 98: 17–24.
36. Sivakumar, A., Murugan, R., Sundaresan, K., and Periyasamy, S. 2013. UV protection and self-cleaning finish for cotton fabric using metal oxide nanoparticles. *Indian J. Fibre Text. Res.* 38: 285–292.
37. Wang, Y., Wang, H., Li, X., Liu, D., Jiang, Y., and Sun, Z. 2013. O_3/UV synergistic aging of polyester polyurethane film modified by composite UV absorber. *J. Nanomater.* 2013: 169405.
38. Rabea, A. M., Mirabedini, S. M., and Mohseni, M., 2012. Investigating the surface properties of polyurethane based anti-graffiti coatings against UV exposure. *J. Appl. Polym. Sci.* 124: 3082–3091.
39. Baginski, F., and Collier, W. 1998. Energy minimizing shapes of partially inflated large scientific balloons. *Adv. Sp. Res.* 21: 975–978.

40. Baginski, F., Chen, Q., and Waldman, I. 2001. Designing the shape of a large scientific balloon. *Appl. Math. Model.* 25: 953–966.
41. Baginski, F. E. 2002. A mathematical model for a partially inflated balloon with periodic lobes. *Adv. Sp. Res.* 30: 1167–1171.
42. Anderson, W., and Dungan, M. 1994. *Numerical analysis concepts for balloon analysis. 32nd Aerospace Sciences Meeting and Exhibit, 32nd Aerosp. Sci. Meet. Exhib.*, Univ. of Michigan, Ann Arbor, MI.
43. Deng, X., and Pellegrino, S. 2012. *Computation of Partially Inflated Shapes of Stratospheric Balloon Structures. 49th AIAA/ASME/ASCE/AHS/ASC Structures, Structural Dynamics, and Materials Conference.* 7–10 April, Schaumburg, IL.
44. Hunt, J. D. 2008. Structural analysis of aerostat flexible structure by the finite-element method. *J. Aircr.* 19: 674–678.
45. Zahariev, E., Delchev, K., and Karastanev, S. 2006. Suppression of deviations and vibrations of tethered satellite. *Mech. Based Des. Struct. Mach.* 34: 389–408.
46. Miller, J. I., and Nahon, M. 2008. Analysis and design of robust helium aerostats. *J. Aircr.* 44: 1447–1458.
47. Wakefield, D. S. 2009. *Non-Linear Viscoelastic Analysis and the Design of Super Pressure Balloons: Stress, Strain and Stability. AIAA 20th Aerodynamic Decelerator Systems Technology, 18th Lighter-Than-Air Systems Technology and Balloon Systems Conferences*, Seattle WA.
48. Khan, I., Sharma, R. K., and Sati, S. C. 2015 *Geometric Nonlinear Analysis of Flexible Structures Using Radioss Software. Altair Technology Conference*, India, 1–7.
49. Woo, C. S., and Park, H. S. 2011. Useful lifetime prediction of rubber component. *Eng. Fail. Anal.* 18: 1645–1651.
50. Le Gac, P., Choqueuse, D., Melot, D., Melve, B., and Meniconi, L. 2014. Life time prediction of polymer used as thermal insulation in offshore oil production conditions: Ageing on real structure and reliability of prediction. *Polym. Test.* 34: 168–174.
51. Kunič, R., Mihelčič, M., and Orel, B. et al. 2011. Life expectancy prediction and application properties of novel polyurethane based thickness sensitive and thickness insensitive spectrally selective paint coatings for solar absorbers. *Sol. Energy Mater. Sol. Cells.* 95: 2965–2975.
52. Gu, X., Chen, G., Zhao, M., et al. 2010. Role of nanoparticles in life cycle of ZnO/polyurethane nanocomposites thermo-mechanical properties of ZnO/PU films before UV exposure. *TechConnect* 1: 709–712.
53. Liao, L., and Pasternak, I. 2009. A review of airship structural research and development. *Prog. Aerospace Sci.* 45(4): 83–96.
54. Dever, J., Banks, B., Groh, K., Miller, S. 2005. Degradation of spacecraft materials. In *Handbook of Environmental Degradation of Materials*, ed. Myer Kutz, 465–501. William Andrew, Elsevier, Burlington, MA.

Index

E

F

G

H

I

J

K

L

U

V

W

X

Y

Z